Let the

EASTERN
BASTARDS
FREEZE
in the
DARK

Let the
EASTERN
BASTARDS
FREEZE
in the
DARK

THE WEST VERSUS THE REST
Since Confederation

MARY JANIGAN

ALFRED A. KNOPF CANADA

PUBLISHED BY ALFRED A. KNOPF CANADA

Copyright © 2012 Mary Janigan

www.randomhouse.ca

Knopf Canada and colophon are registered trademarks.

Library and Archives Canada Cataloguing in Publication

Janigan, Mary
Let the eastern bastards freeze in the dark / Mary Janigan.

Issued also in electronic format.

ISBN 978-0-307-40062-8

1. Regionalism—Canada, Western. 2. Federal-provincial relations—Canada, Western—
History. 3. Canada, Western—Politics and government. 4. Natural resources—Canada,
Western. I. Title.

FC3209.A4J36 2012   971.2'02   C2012-902094-X

Text and cover design by Leah Springate

Printed and bound in the United States of America

10 9 8 7 6 5 4 3 2 1

For Tom Kierans. Always.

CONTENTS

—

Preface

—

IT WAS A SULTRY DAY IN LATE JUNE when Stéphane Dion marched into the boardroom of the *Globe and Mail* to explain his proposal for a federal carbon tax. The Liberal leader was idealistic and fervent. He was also impatient and implacable. His tax—which would particularly affect oil and natural gas producers—would raise more money in the West than it would in other regions. Much of the cash would be earmarked for the fight against national child poverty. When asked how Westerners would react to this federal cash grab, the former political science professor was dismissive: it would be "good for them."[1] Good for them? They would be forced to diversify their economy, came the response.

Stephen Harper would make mincemeat of Dion. The prime minister would compare his opponent's scheme to the National Energy Program of the early 1980s, dismiss it as "insane," and warn that it would "screw everybody."[2]

What kind of language was this? How had a theoretical discussion about the best way to curb greenhouse gas emissions degenerated into this squabble between the Montreal-based Opposition leader and the Calgary-based prime minister? How could these politicians recklessly push regional buttons? Why was a provincial carbon tax permissible—Alberta had imposed a small tax on excessive emissions—while a federal tax was an intrusion, even for the many Albertans who deplored the Oil Sands emissions? It was almost as if Harper and Dion were speaking in code, eliciting responses that were bred in the bone. Less than four months later, Dion would lose the 2008 federal election, emerging with one seat in Manitoba, one seat in Saskatchewan and no seats in Alberta.

The issue of resource control had conjured up another regional divide.

This book was born on the day that Dion preached to the unconverted. It is the tale of how the West was colonized, as immigrants streamed onto Prairie land that Ottawa was virtually giving away. It is also the tale of how the West was almost lost when Ottawa would not release its iron grip on the West's lands and resources, largely because the Rest of Canada ferociously objected to any transfer.

Most important, this is the forgotten story of Canada. Fights over resource control are woven into the very fabric of the nation. Maritimers do not remember that their premiers once claimed they had bought the West, fair and square, so they owned the West's lands and resources. Residents of Quebec and Ontario have no idea that their premiers once demanded *much* higher subsidies if the Prairie provinces secured control over their lands and resources. British Columbian residents do not know that their premiers once hotly insisted that their twentieth-century claims to huge chunks of federally controlled land trumped the West's long-standing demands.

The federation was dysfunctional. The West's battle for control over its lands, minerals, oil and natural gas, forests, and waterpower brought out the worst in every region. There are no heroes in this narrative— although the wily politician who brokered temporary peace pulled off a remarkable coup. The cast of characters includes larger-than-life leaders who towered over their times, whose every utterance was once chronicled but who are only remembered now in obscure place names. There was high drama and great anger; there was greed and jealousy. There was political clumsiness. The stress likely contributed to the death of one premier. Another waged an epic door-slamming argument with a prime minister. The provinces fought with each other *and* with Ottawa, as alliances formed and shifted. Politicians said and did both foolish and damaging things.

The battle lines over resource control were drawn early on the Canadian frontier. Disputes first flared in the mid-nineteenth century when Ottawa paid 300,000 British pounds to the Hudson's Bay Company, and acquired the vast terrain of Rupert's Land and the North-Western Territory. When the Métis resisted that transaction, the federal government carved a truncated Manitoba out of its new turf, and set about the business of settling immigrants on its acquisitions. Ottawa wanted a colony, docile and rich; it kept legal control of the resource wealth in Manitoba and the renamed Northwest Territories. The West disputed Ottawa's schemes: virtually from the beginning, Manitoba and Territorial legislators sought more power and more money. Also from the beginning, Ottawa resisted, making only minor concessions to keep the peace. And from the beginning, the Rest of Canada opposed any federal transfer, turning the West into a pawn in their disputes with each other and Ottawa.

It was an unfortunate quarrel that spanned generations. Westerners were embittered because the other provinces controlled their resources. The four original partners—Quebec, Ontario, Nova Scotia and New Brunswick—retained full control over their *own* resources when they agreed to confederate in 1867. British Columbia kept resource control when it joined in 1871; so did Prince Edward Island in 1873.[3] The constitutional inequality rankled.

The quarrels reached a nadir at a now-obscure federal–provincial conference in November of 1918. Only eight days after the end of the Great War, the nine premiers trooped to Ottawa to discuss the daunting challenges of the peace. On Ottawa's official agenda were three issues: the problem of settling the returning soldiers on the land; the difficulty of luring more settlers onto the land; and the Prairie provinces' request for the transfer of their resources. The conference spanned four days, and the ensuing minutes are brief.

But the gathering unexpectedly became a microcosm of everything that was wrong within the federation—and everything that remained wrong. Into those four days, decades of past quarrels were compressed, and decades of embittered claims were foreshadowed. In their fierce letters to each other, tabled at the conference, the premiers reflected generations of grudges. Tellingly, Prime Minister Sir Robert Borden, who had convened the gathering, was not even there: he had just arrived in London, England, for the peace talks. The West was an afterthought. The discussion of resources became a lowlight in a debate that can be traced from Confederation to the modern-day wrangling over the pollution and profits of the Oil Sands.

The West of the twenty-first century is now the regional powerhouse of the federation. Through good times and bad times, the Alberta and Saskatchewan economies have chugged along nicely. Manitoba has been a steady performer as well. The former underdogs are now top dogs— and the fights over pipeline routes across British Columbia's lands to the Pacific coast and emissions controls at the ravaged Oil Sands have inten- sified. Meanwhile, advocates of economic growth dispute the warnings of environmentalists.

Perhaps the best hope of compromise today could come from an understanding of those quarrels of yesterday. Resources have been a great blessing and a great curse. Few Canadians have understood this double-edged heritage: the ongoing drama of Quebec nationalism has often overshadowed the role of resources in the nation's destiny. But the struggle over their ownership and control has run like a fault line through federal–provincial relations, and through the provinces' relations with each other, zigzagging across the decades with astonishing persistence. In 2012 Ontario premier Dalton McGuinty complained that the West's oil and gas exports were pushing up the value of the Canadian dollar and damaging his province's economy. Such "simplistic" talk deeply offended Westerners.[4] Saskatchewan premier Brad Wall aptly described McGuinty's claim as "unnecessarily divisive."[5] Then New Democratic Party leader Thomas Mulcair claimed that the West's booming resource sector was bad for the nation's economic health. The politicians were using the same language of regional discontent about resource wealth, and repeating the same mistakes, as their predecessors.

In retrospect, the November 1918 conference was a seminal event— when angry debates over the nation's dreams bequeathed a legacy of ill will. The conflict haunts us still. And it is not likely to go away.

THE NOVEMBER 1918 CONFERENCE:
WHERE THE WEST WAS ALMOST LOST.

—

A MID-NOVEMBER GALE WAS HOWLING, splattering chill rain, as dignitaries clambered into government motorcars outside the Château Laurier Hotel. Nine provincial premiers, their senior cabinet ministers and staff were proceeding to the Victoria Memorial Museum, which had been the headquarters for Canada's government since a disastrous fire in the Parliament buildings in 1916. For days, first ministers and their high-powered retinues had been bustling into town, settling into the elegance of the railroad hotel, poring over briefing papers, and bracing themselves for brisk discussions. Now, on November 19, 1918, they were shuttling to one of the most critical Dominion conferences since Confederation. The sheer size of the delegations alone guaranteed attention—as the *Ottawa Citizen* boasted, the "Largest on Record."[1]

The politicians could draw little comfort or safety from their numbers. More than four years after the first blaze of the guns of August, there had finally been eight days of peace, but the dignitaries had no time to savour this rare moment of grace. Canada's armies had hunkered in the trenches of Europe during the Great War. Their sacrifices had raised the nation's stature and clout abroad, but the price of victory was steep. Canadians were exhausted, depleted and numbed by the scale of the carnage. During the war's last three months alone, 6,800 Canadians and Newfoundlanders had been killed and 39,000 wounded. Families mourned loved ones. Disabled veterans, some with horrifying injuries, were already on the streets. Spanish influenza was sweeping the country—it would eventually kill 50,000 Canadians—and Red Cross wagons now prowled Ottawa.

Food shortages loomed. Domestic workers were restive, and union militancy was growing. Alberta coal miners had organized; Vancouver steel workers had staged strikes; union militants were reaching out to immigrant labourers at Maritime steel mills. Tens of thousands of soldiers were coming home, very soon, and they needed training, jobs, homes and land. The challenges were daunting, almost overwhelming. The Canadian economy had to shift from wartime to peacetime almost overnight. But economics was a fledgling art, and no one knew how to attain stability and prosperity after a great war.

The country's politicians had to acknowledge that their world was dangerously different. Those imposing premiers in their stiff shirts and austere ties were really supplicants in dignified dress. They had arrived in Ottawa, elegant caps in hand, because they needed more federal money and programs to meet their upcoming obligations. It took little imagination to picture the imminent arrival of veterans and new immigrants needing adequate schools, highways, bridges and hospitals. Newcomers would want work to support their families, housing and affordable food. They would want government aid if, in desperate need, they could not find work. No one knew how veterans and new immigrants would fare as they jostled for room in an adjusting economy: survival of the fittest was not a prospect that any premier could contemplate with equanimity. So the politicians were now gathered in Ottawa, ready to finalize skeletal schemes to stave off disaster.

Unfortunately, they aimed low. Their answer to the dilemma was a halcyon fantasy, right out of a nineteenth-century handbook on useful work. They planned to funnel both new immigrants and returning veterans out of the dangerously volatile cities and onto pastoral farms, particularly in the West. This was a collective self-delusion. It was known that good farmland was scarce and that the economy was industrializing. But, in the final months of the war, elected officials and bureaucrats had hatched schemes to expropriate so-called unused lands, often at the expense of Aboriginal owners, and sell them to prospective settlers for minimal down-payments. The new arrivals, they reasoned, could grow crops, tend animals, breathe the healthy outdoor air and prosper, far away from the discontented denizens of the big cities. Those plans were among the last gasps of the back-to-the-land movement, and the premiers surely wrestled with secret doubts about their efficacy. But as they filed into the museum's first-floor mineral gallery, once the Hall of Invertebrate Fossils, where the Senate usually convened, they were seeking solidarity. Co-operation might be their salvation, if both levels of government could work together to fend off social disruption.

Two of the three items on the official agenda were land-settlement schemes. But there was a third: the Prairie provinces' campaign for control of their natural resources. That would put the worm of discord in their midst. Although few Canadians grasped its importance at the time, in retrospect this Western demand was the most significant issue on the table, and it required deft handling. The other six provinces had always administered their own resources. They regulated their lands, minerals, waterpower and forests with the scrawl of a pen, and they pocketed the royalties. But Manitoba, Saskatchewan and Alberta had never exercised that control; that power was vested in Ottawa. For almost fifty years, Ottawa had handed out Western lands to railways, mining and forestry companies and settlers. Ottawa had attracted those newcomers with extraordinary offers of 160 acres of virtually free land, and *federal* officials had distributed that land to new arrivals at Prairie railway stations. In the beginning, Ottawa was simply doing what fledgling Western governments could not do: lavishly promoting Canada abroad, and efficiently settling the new arrivals. To soothe Western resentment, Ottawa transferred extra money to Prairie governments to compensate for those lost resource revenues. It was an imperfect bargain, between very unequal partners.

Long before November 1918, however, that arrangement had become an enormous irritant to Western governments. In 1913, fed up with their inferior status, those provinces had created an informal "Gang of Three"— although the premiers would have scorned such slang—to press their case for control. Now, before other first ministers could endorse new schemes to put more Western lands into the hands of immigrants and veterans, they wanted to settle the issue. The Prairie provinces wanted the power to allocate their own land on their own terms; and they wanted to decide for themselves which firms got timber leases and permits for oil and gas exploration. (Although the first massive oil discovery in the West did not occur until 1947, far smaller wells had hinted at the region's enormous energy potential.) It was a malign trick of history that had left them powerless, they maintained, and it was time to rectify that injustice. The West's hour had come. Or so they thought.

The other premiers, of course, were in no mood for sensitive negotiations over resources that the federal government had administered for almost five decades. Each needed to leave the conference with good news for his own province after the bleak wartime years. The Western premiers wanted equality, *and* they wanted compensation for the lands and resources that Ottawa had already disbursed. Surely, the other premiers felt, they could not be expected to endorse a giveaway to the West without obtaining something for everyone? True, the Western provinces wanted something but, then again, they *all* wanted something. What made Westerners so different? From the perspective of the six other premiers, when the sky was falling, Westerners wanted to change the name on the deeds to their lands underneath. Should Ottawa deign to bestow any favours on the West, something should be done for the others, who had, in their view, bought the West and picked up much of the tab for its development. Westerners countered that they had given up so much land and resources for the nation's well-being: *they* were Confederation's victims. Equality was their mantra. They were also impolitic and impatient, secure in the rightness of their demands. The issue fit uncomfortably onto the agenda.

Neither the federal representatives nor the other premiers grasped the depths of Western discontent. As the acting prime minister, Finance Minister Sir Thomas White, was extending a warm welcome to all delegations, the other governments were simply getting ready to table lists of their own needs.

Perhaps a good host could have placated his guests. But their official host, Prime Minister Sir Robert Borden, had issued the invitations to the conference and then chosen to go off to Europe for the peace talks. After seven years in office, including the four war years, he was worn out. The introduction of conscription in 1917 had strained the nation's seams, fraying the ties between French and English Canadians. On the home front, many Canadians saw spies everywhere, openly doubting the loyalty of immigrants from enemy nations. Many of those immigrants were themselves grappling with the unease of divided allegiances. Borden's Union government of Liberal and Conservative luminaries, cobbled together in late 1917, was in constant danger of disintegration. His ministers, separated by party loyalties, were suspicious of each other and exhausted by the weight of their responsibilities.

Borden was always putting out fires. His diaries throughout 1918 depict a man hounded by incessant petitioners and problems. Inflation was eroding the buying power of the working classes, fostering discontent and enhancing the appeal of labour unions. Finance Minister White had deftly imposed taxes on incomes and business profits in 1917, as the costs of war inexorably mounted. But, in early 1918, he was ill and anxious to resign; it took all of Borden's persuasive skills to retain him. The prime minister had barely acknowledged the previous First Ministers' Conference of February 1918 on the emergency shortages of food and farm labour. He did, however, note that, after the formal meetings, a delegation of Western premiers had requested future discussions on the issue of natural resources: "They did not press for immediate consideration but asked that the matter be brought up before the session of 1919."[2] Shortly thereafter, Borden hastened to Washington to deal with Canada's balance of trade, the exchange situation and the use of electrical power from Niagara Falls. He was endlessly sidetracked. During the spring, he fended off British efforts to break up the Canadian Corps, dealt with the painful fallout in Quebec from conscription, tackled the operations of the Canada Food Board and watched anxiously as the German offensive broke through Allied lines.

So much was happening. In late May, Borden left for an Imperial War Conference and an Imperial War Cabinet in England. By August 24, he was home, tackling problems that had accumulated in his absence. He grabbed a brief holiday, headed off to meetings in Washington and then wearily returned to his desk. His degree of attention to the Western premiers' continuing laments can perhaps be discerned from a brief notation in his memoirs about that early fall: "The usual mass of correspondence awaited me; numerous discussions as to the labour situation; arrangements for a forthcoming Conference of provincial Premiers; preparation for a speech at Toronto (October 27th) appealing for the Victory Loan . . . ."[3]

Borden was weary of the ferment in domestic matters. Perhaps more crucially, he was rarely, if ever, inspired by a grand vision of how the federation might function, and what different levels of government could do. He was simply not interested in brokering compromises among his adamant colleagues in Confederation, nor was his stand-in, Finance Minister White. Although the November conference was only the second official federal–provincial gathering during his tenure as prime minister, it could not outweigh the attraction of the preliminary peace conclaves in London.

Borden leapt at the opportunity to leave, and he went with an agenda. In 1917, in England, the prime minister had already pushed to expand Canada's role at any upcoming peace negotiations. The Dominion would no longer function as a mere adjunct of Britain. After the sacrifices of war, he argued, Canada should have its own seat at the peace negotiations. The Dominion should be a full-fledged member of any new international organizations. Nationalism was in the air. In early 1918, U.S. president Woodrow Wilson had proposed a League of Nations that could provide "mutual guarantees of political independence and territorial integrity to great and small states alike."[4] Ever loyal to Britain and its empire, Borden nevertheless emerged from the war as a resolute Canadian nationalist with a mission: in future, should Britain declare war, Canada would no longer automatically acquiesce. Borden saw no contradiction in this position. The Dominion had matured. Should anyone overlook Canada's new stature or want to turn back the clock to a more convenient pre-war world, Borden wanted to be there to stop such assumptions. That required his presence abroad.

But there was also the lure of those European lights. As an earnest New Brunswick lawyer from a relatively humble background, the prime minister was dazzled by the limelight, and by those who occupied it. He loved to hobnob. On November 19, 1918, when hostilities broke out at home over the West's request for resource control at the First Ministers' Conference, Borden sat in Parliament's Royal Gallery to hear the King's address, attended a meeting of the British Privy Council and had an audience with King George V. When the premiers concluded their proceedings on November 22, he was contentedly writing to Sir Thomas White in Ottawa, passing along the latest news about demobilization, reconstruction and the now-largely-forgotten Siberian Expedition. Borden did not return home until May of 1919; his memoirs dealing with this time abroad make no reference to the Western premiers, or to their discontent.

That may seem like an astonishing oversight today, but Borden had been hearing about the West's discontent over resource control since he first entered Parliament in 1896. He was inured to the Prairie premiers' complaints. They had first formally discussed the issue with him in October of 1913, and, even though they were deflected as the conflict devastated their communities, throughout the war the issue never really went away. In the meantime, other provincial governments had spelled out their vehement objections to the West's demands. When it became clear that victory in Europe was almost inevitable, the Western trio reinvigorated their alliance, asking Borden to consider their demands before first ministers rubber-stamped any land-settlement schemes.

The preoccupied prime minister must have felt he had complied with that request, in his fashion, because their demands became the first item on the agenda of the November 1918 gathering. Borden had clearly decided that there was nothing to lose. His senior ministers could handle the conference discussions about the settlement of soldiers and new immigrants; his cabinet had already endorsed those schemes. On resource control, the Western premiers would have the chance to put their case in person to their peers. Nothing would or could come of this: Borden already knew Ottawa had no extra money in the kitty to buy peace among the provinces. In fact, there would be a shortfall on the federal books— which could rise as high as $100 million—if Ottawa picked up the costs of soldier settlement, demobilization and reconstruction. A general subsidy increase for all provinces was unthinkable. But assumptions that the status quo could continue indefinitely constituted a great failure of collective imagination.

Few politicians in the rest of Canada in 1918 appreciated how mightily the West had changed in a mere few decades. Time had moved at warp speed on the Prairies. In the early 1870s, when Ottawa had affirmed its control over the West's rich resources, huge buffalo herds still grazed across the land. Throughout that decade, as those massive herds vanished, Aboriginal bands signed seven treaties with Ottawa, staking out their lands in the midst of an ever-growing number of settlers. In the summer of 1877, Ontario MP James Trow, chairman of the House of Commons Immigration and Colonization Committee, set out on an inspection tour of Manitoba and the Northwest Territory. He wanted to ascertain if the land, which Canada had recently acquired from the Hudson's Bay Company, was suitable for settlement.

It was an arduous trip through the "Far West." Trow's reports, first published in his local *Stratford Beacon* newspaper, offered an "unvarnished account."[5] His were discouraging words. Trow floundered over muddy and often-impassable roads, urged his horses across wobbly bridges, and grumpily pointed out the advantages that a railroad would bring. He bemoaned the swarms of mosquitoes and bulldog flies, which "cut like a lance."[6] Trow had big dreams that the West would ultimately provide a "happy home for millions of the surplus population of Europe."[7] But he concluded that this could not occur before railroads snaked across the Prairies.

Less than a decade later, the railway spanned the nation. Canadian Pacific financier Donald Smith drove the last spike in the transcontinental track in late 1885, near the Eagle Pass summit in British Columbia. It was an astonishing feat. The railroad funnelled newcomers across the West, enticing those in search of a better life with the prospect of prosperity in return for hard work. Immigrants from the British Isles, Central Canada and the United States staked out claims, cleared the land and nurtured crops or tended vast herds of cattle. Among them, often clumped together in group settlements, were Mennonites, Germans, Hungarians and Scandinavians.

In 1896, Interior Minister Clifford Sifton redoubled federal efforts to attract immigrants. He freed encumbered settlement lands, simplified the acquisition procedure and allowed Western settlers to buy adjacent land at reduced prices. Most important, the new Liberal minister also encouraged promotional efforts to recruit agriculturalists in Russia, Romania and the Austro-Hungarian Empire. Booking agents scouted for prospective immigrants abroad. As new strains of wheat and ingenious agricultural systems allowed settlers to combat the prairie perils of flood, drought and early frosts, more than three million people immigrated to Canada between 1896 and 1914—the plurality settling in Alberta and Saskatchewan.

The pace of change was astounding. When Alberta premier Arthur Sifton first revived his province's campaign to control its own resources in 1911, his government could scarcely keep up with the clamour for roads, bridges and telephone lines. His province was spending almost three times the per capita amount of Ontario and Quebec. Until the economic bust of 1913, Calgary had its own university. The city hosted a symphony orchestra. In May 1914, Calgary Petroleum Products struck oil in the Turner Valley, roughly seventy kilometres southwest of the city. Across the West, automobiles sputtered along the streets, between imposing sandstone buildings that housed upstart oil companies or professionals such as lawyers and engineers. Prosperous grain marketing co-operatives clustered around railroad branch terminals. Many of the new arrivals, including those who hammered out a living in the dangerous coalfields, never found their pot of gold. But the promise of equality of opportunity usually restrained and isolated their protests. Historian Donald B. Smith perfectly captured the mood: "Before the economic crash of 1913 the prairie working class believed that imagination, sacrifice and hard work would give all who tried a new start."[8] As preached from pulpits across the nation, faith, hope and charity were also "how-to" guides to social advancement.

The West was alive. Its network of institutions, including schools, churches and community organizations, was thriving. Its provincial governments were now competent replicas of their Eastern brethren. But Ottawa still controlled the resources and the resource revenues.

———

The Prairie premiers' campaign to remedy that inequality might have escalated more rapidly during the second decade of the twentieth century if war had not broken out in Europe, with Prairie governments rallying to the British cause. Their grain growers fed the citizenry at home and the troops abroad. Their Anglo-Canadian youth readily enlisted, although most soon lost their early idealism. (Most, too, were mere fodder: of the 630 soldiers from what became the South Saskatchewan regimental region, only 105 were left in the field at armistice.[9]) Western forests and coal mines, along with their workers, were exploited to feed the war machine. It was an arduous effort in stressful times.

Although the majority of Westerners were fervent patriots, at least in the beginning, the war turned neighbours against each other. Foreign names were automatically suspect. In the course of the war, the popular Calgary restaurant Cronn's Rathskeller morphed into Cronn's Restaurant, then Cronn's Café and finally into the Cabaret Garden. The son of German immigrants, Richard J.A. Prechtl, recalled that his parents survived on their northern Saskatchewan homestead during the war only because their closest neighbours, the Garnett family, "stood back from the prejudices of war" and helped them.[10] "[But] the necessary propaganda and the casualty list brought the tragedy of war into every home," he wrote.[11] Work camps housing resident aliens from "enemy" nations, mostly Ukrainians, dotted the Prairies. Soldiers in uniform, en route to the Front, thronged train stations across the West. As the war progressed and grim news came from the trenches, their loved ones feared the worst. Canadians' spirits, so high with the declaration of war in 1914, deflated.

But there was a particular undercurrent of unease and resentment in the West, and it grew stronger during the war. Inflation soared, while Canada's protective tariff wall initially maintained the high prices of farm machinery. In April 1918, to the consternation of grain growers, Ottawa extended conscription to farm workers who had previously been exempt. The harvest was poor, and food and fuel were in short supply. Resource firms were circling Western assets, angling for federal leases. In July of 1917, Shell Transport Company Ltd. boldly applied for exclusive oil and natural gas rights over a large swath of northern Alberta and the

Northwest Territories; the proposed tract spanned 328,000 square miles, an area roughly 75,000 square miles larger than the entire province of Alberta itself.[12] Ottawa ignored the request during the war, but Albertans were aware of the corporate wolves hungrily circling their turf. It was becoming clear that oil and gas were the fuels of the twentieth century, powering ships, planes and motorcars. No one yet grasped the extent of the West's vast resource wealth, but many firms were eager to drill for black gold.

As those companies applied to Ottawa's Ministry of the Interior for exploration rights over Western tracts, the Prairie premiers' distress deepened and their patience frayed. While the Dominion had grown in international stature during the war, the West had also matured within Confederation. It had emerged bloodied, battered and proud from the conflict. The federation had tapped the West's resources for the war effort. Now, the Western premiers were determined to obtain the recognition that their sacrifices merited; they wanted to administer their own resources in this postwar world, and exploit them for the benefit of *their* residents. Like Borden at the European peace talks, they were at the November 1918 conference to assert their rights and to shore up their new status among their peers. Borden did not pay sufficient attention to detect the parallels; he was too preoccupied to even attempt to broker a compromise prior to the conference.

Meanwhile, the Western premiers had convinced themselves that the first ministers' gathering represented their golden opportunity to win resource control. Those premiers were not naïve souls in the big city. They were canny political veterans on their home turf. But they were tone-deaf to the effect of their demands on the other premiers. Nor were there any spin-doctors to polish their leaden rhetoric. The leadership of the Gang of Three had completely changed during the war, but the new trio resolutely espoused their predecessors' demands, and their very words. That position, first spelled out in a terse letter to Sir Robert Borden in 1913, had marked the creation of the Prairie premiers' historic united front. That letter remained the only statement of their claims in November of 1918. But its prose did not sparkle, nor was it easy to catch on first reading what the Premiers were actually demanding. Today, the archaic language sounds even more obscure:

> It has been agreed between [sic] us to make to you, on behalf of said Provinces, the proposal that the financial terms already arranged between the Provinces and the Dominion as compensation for lands should stand as compensation for lands already alienated for the general benefit of Canada, and that all the lands remaining within the boundaries of the respective Provinces, with all natural resources included, be transferred to the said Province, the Provinces accepting respectively the responsibility of administering the same.[13]

In clear English, the premiers wanted their resources, *and* they wanted their additional federal subsidies to continue, in compensation for the resources that Ottawa had already used. This ambitious request had not marked an auspicious start to their campaign in 1913; nor did their fellow first ministers' hearts grow fonder over the ensuing years. The Western premiers had obviously heeded the call of unity, at the expense of clarity and perhaps common sense. But their successors at the November 1918 Conference clung to that letter.

The other premiers were in no mood for generosity, or for grand gestures. They were not visionaries ready to remake their world; the so-called reformers were really moderates. Since Confederation, even the provinces that theoretically controlled their own resources had tussled with Ottawa over the extent of that jurisdiction. But this demand from their former colony was *not* just another episode in the long history of such fights. When they perused the West's demands for the resources that all Canadians had once purchased, the premiers from the Rest of Canada could see only avarice. In particular, they fiercely objected to the request for continued extra subsidies after the resources were transferred. They had little patience for the Western leaders and their narrative of unappreciated contributions and inequality. Unresolved grudges simmered in the souls of all premiers as they journeyed to the November 1918 Conference.

———

If the premiers wanted good publicity for their progress in tackling federation grievances, their timing could not have been worse. Although Canadian newspapers chronicled their four-day gathering, the first ministers were often pushed off the front pages by reports of the German army's retreat from conquered ground, or speculation about the terms to be imposed on the fallen enemy. The premiers were out of the spotlight in their straight-backed chairs at the stately Victoria Memorial Museum, dubbed "The Castle" (now the Canadian Museum of Nature), and many Canadians were scarcely aware of the gathering. Their discussions were held behind closed doors; they issued no official communiqués. The museum's location in a former cow pasture on the outskirts of Ottawa, a mile south of Parliament Hill, reflected their isolation, obscuring the complexity of the issues and the scope of their efforts. There could be little public grandstanding. A mere day before his fellow first ministers walked into the museum, on the other hand, Prime Minister Borden had arrived in London to the cheers of throngs at the railway station. Perhaps prophetically, in an interview there, he warned that the problems of peacetime could be even more daunting than the challenges of war. He was sure, however, that Canadians would handle the burdens "with equal courage, resolution and confidence."[14] It was as much a prayer as a rallying cry.

As Western resources were the first item on the agenda on November 19, it's tempting to picture the three Western premiers arrayed at the heavy wooden conference table, flanked by their retinues, brandishing the demands from 1913 that their predecessors had bequeathed. There was Manitoba's Tobias Crawford Norris, who had led his Liberal party to victory over the scandal-tainted Conservatives in 1915. A bulldog of a man, Norris had migrated from Ontario to Manitoba to homestead, but had gradually shifted to a career as a professional auctioneer, famed across the West for his persuasive skills. First elected in 1896, finally claiming the premier's chair in wartime, he had transformed his government into an unlikely centre of reform activity. In 1916 Manitoba became the first province to allow women to vote. Norris lengthened the school year and made attendance compulsory. He passed minimum wage legislation, established a workmen's compensation board and regulated industrial conditions. He launched a public-nursing system and government-backed farm credit. He even balanced his province's books. He was a whirlwind, but no radical.

Norris was probably the most resolute member of the Gang of Three, but his formidable skill at sales patter had little effect on his fellow premiers. He was certainly the most nettled by their indifference to the unique character of the West's grievances. More than two years after the 1918 conference, when Arthur Meighen had succeeded Borden as prime minister, Norris could still barely contain his rage. In a sizzling memo to Meighen, crafted with adept political spin, he wrote as if he were still at the museum conference table. Some Western delegates, he observed, "were impressed by the lack of accurate knowledge or acquaintance with the historical or constitutional basis of the relationship between the Dominion of Canada and the Prairie Provinces with regard to those Natural Resources."[15] Given the first ministers' tradition of generally decorous behaviour with one another, those were fighting words in December 1920.

The other two Prairie premiers shared Norris's staunch convictions. Saskatchewan's William M. Martin, the province's second premier in its brief history as part of Confederation, another transplanted Ontario resident, had moved west to practise law in Regina in 1903. Elected to the House of Commons five years later, he became a stout defender of Western interests, pushing for more railways, lower freight rates and freer trade with the United States. In October 1916, the provincial Liberals drafted him as a replacement for retiring premier Walter Scott. A mere nine months after he assumed office, his Liberals romped to victory at the polls with a substantive majority. Like Manitoba's Norris, Martin had a reformist streak; he introduced compulsory school attendance and overhauled the provincial courts. He also maintained that Borden's tariff and resource policies unreasonably penalized the West. Still, for the sake of the war effort, he took the politically risky step of throwing his support behind the prime minister's Union government in the federal election of 1917. Martin therefore came to the museum conference table with grateful allies and considerable clout.

The final member of the Gang of Three was Alberta Liberal Charles Stewart, who had replaced Arthur Sifton as premier when Sifton had resigned to join Borden's Union government in 1917. Stewart had been a pioneer farmer, settling in the central Alberta community of Killam after a storm destroyed his family's Ontario farm. His life had been no easier there: his shack was cold; the warmest spot was the kitchen table where his wife kept the baby. To survive, Stewart had supplemented his income by working as a bricklayer and a stonemason, joining work crews that were laying foundations for the Canadian Pacific Railway. He later ventured into real estate and the sale of farm implements. In 1909, he won election to the provincial legislature by acclamation. Three years later, he was the first minister of municipal affairs; in 1913 he was in charge of public works. Stewart backed the creation of a farmer-run co-operative to operate grain elevators. He was a firm advocate of public ownership of the province's utilities, and a long-time supporter of Alberta's campaign to control its resources. In January 1918, he undoubtedly shocked his former employer when he called for a rate freeze, or the nationalization of all railroads. Weeks later, a member of his own caucus denounced his failure to get the issue of natural resources on the agenda of the February First Ministers' Conference. But Stewart was merely waiting for peace to reactivate his long-time campaign for resource control. Alberta politics never lacked drama.

These three Liberal premiers had survived the rough, scrappy politics of the West. None of them was a pushover; nor were they shy about stating their case. They wanted control of their lands and natural resources as soon as possible, and they wanted their current level of subsidies to continue as compensation for past usage. They were certain that Ottawa was taking in far more from resource revenues and land sales than it was spending to administer their resources. Their extra subsidies, they argued, were a mere pittance compared to that federal profit. But such assertions did not translate easily into ringing appeals. As the conference began, *The Globe* tartly summarized the prevailing indignation: the Western premiers

"Want Lands And Money, Too."[16] To the uninitiated, that demand simply sounded greedy.

The premiers of the other provinces were clearly unimpressed. But the topic would not go away. Over the next few days, it reappeared fitfully, disrupting the proceedings and prompting all premiers to write barbed letters to each other. These contemporaries were an extraordinary collection of characters, and they were not reticent about countering Western claims with their own frustrations and grievances. Perhaps the most fascinating participant was new to these meetings, but his folksy ways could not conceal his steely core. British Columbia's John Oliver had inherited the job after his exhausted predecessor H.C. Brewster died from pneumonia on his way home from the gruelling First Ministers' Conference on food shortages in February 1918. One of Brewster's last official acts in Ottawa had been a futile attempt to add his name to the Gang of Three's request for a future meeting on resource control, because *his* province had its own complaints about federal resource management.

In the nineteenth century, the B.C. government had transferred lands to Ottawa so that the federal government could link the Pacific coast with Central Canada by rail. Those railway lands spanned a corridor across the province, spilling out for twenty miles on each side of the proposed route. They also covered 3.5 million acres within the Peace River region. Now British Columbia wanted the return of the unused lands, and full control over any resources within those lands. This was not a new issue. British Columbia had been complaining about Ottawa's disposal of its lands since the CPR reached the coast in 1885—but the campaign for their return only gathered steam in the early twentieth century. Premier Oliver maintained that his claims were "of an even stronger character" than the West's demands.[17] He was determined to fulfill Brewster's aspirations, and prod the issue onto the conference agenda. Such exaggerated assertions likely explained why the Western premiers had apparently dodged his predecessor's efforts to join their campaign after the February conference. After all, to them, the railway lands were simply an anomaly: British Columbia had controlled its resources since it entered Confederation in 1871.

Oliver did not accept adversity, or refusals, meekly. Born in an English farming community, he had left school at eleven to labour in a lead mine. When the mine closed, his family emigrated to Ontario. By 1877 Oliver was at work on a Canadian Pacific Railway survey crew on the British Columbia mainland, tucking aside funds for a farm. He prospered in agriculture, but was drawn to municipal and then provincial public office. Given his background, this was not an easy ride. More sophisticated politicians in the British Columbia Assembly derided him as a yokel when he arrived in 1900; they belittled his rough tweed clothes and rougher grammar. But Oliver painstakingly mastered Assembly procedure and its hard-hitting ways, and he became skilled at marshalling his thoughts. When the Liberals won power in September 1916, he took over the agriculture and railway portfolios, retaining his lifelong skepticism about the conduct of private railway companies. A year later, he ensured that returning veterans had the chance to own and develop rural property by enacting legislation that would become the model for federal settlement plans at the November 1918 conference. But Oliver did not lose his determination to recover his province's railway lands.

The Western premiers were equally wary of the big wheeler-dealers at the November 1918 gathering: Ontario and Quebec always had their own agendas. The two provinces were regarded as Confederation's fat cats, and their respective financial situations were relatively comfortable. But neither government believed that it had adequate resources for the immense challenges ahead. In 1906 Ottawa had raised provincial subsidies from their paltry nineteenth-century levels. The British North America Act of 1907 had inscribed this new deal in constitutional stone, superseding the subsidy arrangements at Confederation. But time and inflation waited for no provincial treasurer. On the eve of the Great War, an anonymous Ontario official pointed out that in 1912 Ottawa had transferred to the provinces the equivalent of only 10 per cent of its customs and excise revenues, down from 35 per cent at Confederation. Federal revenues had soared since 1867, but transfers had not kept pace.[18] Worse, some provinces were receiving less federal money per capita than other provinces, and Ontario and Quebec were particularly disadvantaged.

The unfairness grated. Ontario premier Sir William Hearst was a northern lawyer and a strict Methodist from Sault Ste. Marie. He espoused temperance so ferociously that his government had turned the mere possession of alcohol outside the home into a criminal offence. Until his predecessor, Sir James Whitney, died of a heart attack seven weeks into the war, Hearst had been the minister of lands, forests and mines. He was well aware of the value of natural resources: his government's annual budget statement for 1917, unveiled in February 1918, showed the largest surplus in the province's history, mostly thanks to the enormous proceeds from a new tax on nickel. Such windfalls infuriated the Western premiers, but Hearst was impervious to any comparison.

The Ontario premier needed more money to develop his postwar economy, and he would not accept policies that continued the unfairness in federal subsidies. He wielded clout at the November 1918 conference, and he knew it. As a representative of the nation's economic heartland, he grandly assumed that his fellow Conservatives in Ottawa would do nothing without his blessing. Hearst had blandly told Borden in a terse letter prior to the conference that Ottawa's dealings with the Western provinces were none of Ontario's business. But he added, "If, however, your Government should decide to grant the request of the Governments in question, such action would, I submit, necessarily open the whole question of subsidies payable by the Federal Government to the Provinces."[19] Those delicately oblique words carried a tough warning: if the Western provinces obtained control of their resources and kept their extra subsidies, Ontario wanted more money, too. At the very least, Hearst had set Ontario's price for its consent to any transfer.

Quebec premier Sir Lomer Gouin was a wily survivor who had navigated his province's treacherous politics since 1905. The Liberal lawyer from Montreal had resisted Borden's efforts to draft him into the federal Union government of 1917, largely because he opposed conscription. By the end of the war, after successfully steering a tricky line between Quebec nationalists and his federalist inclinations, Gouin was at the height of his influence. His authority within his own party and his dominance over the opposition were almost absolute.[20] There was even speculation that he might replace the ailing federal Liberal leader, Sir Wilfrid Laurier. Quebec had weight at the conference table beyond its status as one of nine provincial governments. Its major city, Montreal, was the financial capital of Canada. The province represented one of the two peoples—French-speaking and English-speaking Canadians—who had united to form Canada in 1867.

Gouin had never hesitated to flex his powers. During his years in office, he had clamped an embargo on the export of pulpwood from Crown lands, hiked stumpage fees for forestry companies and offered long-term leases on waterfalls.[21] His government had run small surpluses, year after year, partly because of those increased resource revenues. The Western premiers could only fume. Now the Quebec premier was quietly mulling a costly program of postwar reconstruction, based partly on the development of hydropower, and this plan required more revenues. If Gouin was silent on the West's demands prior to the conference, he was, as we shall see, surprisingly decisive soon after the museum's ornate doors swung shut on the private conclave, tabling an explosive resolution about what should be done with the West's extra subsidies.

The Maritime provinces also could not be counted on to support the Western premiers. Nova Scotia, New Brunswick and Prince Edward Island considered themselves particularly aggrieved within Confederation. Like the West, they had long memories for slights. They were still simmering because Ottawa had extended the boundaries of Manitoba, Ontario and Quebec in 1912 by gouging land out of the federally controlled Territories. They brushed aside the harsh geographic reality that there was no land adjacent to their boundaries. They wanted compensation anyway. And there was no reasoning with them. Decades of scarcely concealed resentment lay behind that position: Ottawa had used its tariffs to protect central Canada at the expense of the Maritimes. Their machinery cost more, their resources were at the disposal of Central Canadian plunderers, and their export markets were endangered. Decades after the 1918 conference, Prince Edward Island's premier A.E. Arsenault was still angry, and still carrying grudges:

Millions of our money are [sic] sent each year through the banks, insurance companies, and the trust companies to Montreal and Toronto, which are the headquarters for all our national business organizations. Those millions are used to build up Ontario and Quebec at our expense. . . . Our factories have disappeared and today we purchase everything from Central Canada except our food. We purchase from the manufacturers of Ontario and Quebec everything we wear from the shoes on our feet to the hats or caps on our heads. Every piece of farming machinery comes from those provinces; every appliance, every bit of furniture, stoves, kitchen utensils, crockery, radios, automobiles, tractors. . . . On the other hand, those provinces take but little from us.[22]

The demands of the Western provinces at the Ottawa conference brought out the worst in the Maritimes. Confronted once again with the Prairie premiers' terse one-page letter from 1913, the three Maritime premiers united as an opposing gang of three. By November of 1918 Nova Scotia's George Henry Murray had occupied the premier's chair for an astonishing twenty-two years. (He would eventually set a Canadian record of twenty-seven consecutive years.) As the senior member of the premiers' club at the conference, the Liberal lawyer held fast to his belief that his province should be first in line for any federal goodies. A deeply cautious man, Murray had been obliged to run deficits between 1913 and 1915, partly because coal royalties and federal subsidies were no longer sufficient to meet his province's financial needs. To his dismay, he had been forced to resort to new taxes. Now, he was adamant that Ottawa address Nova Scotia's smouldering resentments before the momentum of wartime industrialization in his province inevitably ebbed. In his view, Maritimes subsidies should be raised *before* Ottawa considered the West's imprudent, even impudent, demands.

Murray's cohorts in this contrary-minded gang shared his militancy. They were united in their pose as Confederation's real victims. New Brunswick premier Walter E. Foster was a fierce advocate of Maritime rights. His Liberals had won power in 1917, and Foster nurtured a seemingly inexhaustible list of regional grievances, including the slow pace of port development and the tariffs that had curbed his province's trade with the United States and the Caribbean. Foster was well aware of his province's slipping fortunes. A former bank clerk, he had vaulted into the position of managing director at a merchandise and dry goods firm after marrying the owner's daughter. He knew commerce, and had been president of the Saint John Board of Trade. At the nub of Foster's complaints was a bitter reality: New Brunswick was losing clout within Confederation, even though it had been one of the original four partners in 1867. Foster was low-key and generally conciliatory, but he was determined to do whatever was needed to right the imbalance.

The third Maritime leader, Aubin-Edmond Arsenault, was Canada's first Acadian premier. A native Prince Edward Islander, and the scion of a prominent Conservative family, Arsenault took over the premier's office in 1917 when his predecessor was appointed to the bench. Even in wartime, Prince Edward Island seemed sleepy. During Arsenault's tenure, the most heated debates were over a proposal to open key roads to motorcars on selected days of the week and the decision to restrict leases to the province's depleting oyster beds. But the war spared no one: farmers had constantly begged Arsenault, to his distress, to exempt their sons from service overseas, especially during the harvest. There was little that the premier could do: Prince Edward Island's young men had fought hard. Now, in peacetime, his priority was to show that the Islanders' sacrifices had been worthwhile, and their economy would strengthen. Prophetically, he saw his island as a tourist destination, and he needed special federal help to fund his ambitious dreams.

These nine premiers were all big men in their own provinces. They treated each other with courtesy, and with respect. They understood that each of them served at the whim of the voters, and that none of them, with the probable exception of Quebec's Gouin, was secure on his perch. They were not looking for controversy. So it would be charitable to ask what the Western premiers were thinking when they presented their case for the return of their resources at this high-stakes conclave. It was a gamble but they likely felt they had little choice.

There was always a catch in their case. The Western premiers did not view this First Ministers' Conference as the best venue for their hearing; nor did they want to discuss their demands with all the participants. Weeks before peace in Europe, on September 27, 1918, Alberta's Premier Stewart had telegraphed Borden; he emphasized that his understanding, when he had seen Borden the previous summer, was that the prime minister "would call the *Western Premiers* together at an early date to discuss natural resources."[23] Stewart was well aware of Ottawa's postwar agenda for Western lands. Indeed, he pointed out that any discussion about the transfer of resources would affect Ottawa's land-settlement schemes, so he suggested "as early a date as would be convenient."[24] The Alberta premier wanted to limit the negotiations to the federal government and the Prairie provinces, but Borden ignored him. Instead of scheduling separate talks between the West and his government, he put the issue on the November 1918 Conference agenda, and he notified all premiers about the upcoming negotiations. Now every first minister had a "say" in the West's demands. This did not bode well.

Worse, the optics of the agenda were now terrible. The three Prairie premiers were advocating the transfer of huge swaths of land and resources to their governments, while all governments were scrambling to sate the intense demand for farmland. On November 21, in mid-conference, there were news reports that more than 105,000 members of the Canadian Expeditionary Force in Europe had "expressed the definite wish to take up farming in Canada after the war."[25] Almost three-quarters of these aspiring farmers did have prior agricultural experience, but fewer than 41,000 had been working on farms when they enlisted. Almost half of the applicants were willing to work for wages to gain experience. But, sooner or later, these veterans would want to buy their own land. The Prairie premiers' demands appeared to be out of kilter with the nation's immediate priorities.

Today, it would be too easy to conclude that Borden set up the Western premiers for failure. After all, in theory, if Ottawa opted to transfer jurisdiction over their resources to the West, the decision would be worked out between Ottawa and those three provinces. So why did

Borden involve the other six premiers? Something was at play here that permitted the other six premiers to grandly assume they had the right to set the conditions for any transfer. The past did have a long shadow. In 1870 Ottawa had transferred 300,000 British pounds to the Hudson's Bay Company when it surrendered its charter rights to Rupert's Land. Nearly fifty years later, many Canadians asserted that Eastern taxpayers had "bought" the West, fair and square, and that they therefore deserved a voice in any decision to give *their* resources to the Prairie governments. Unwilling—and likely unable—to broker a compromise before the November 1918 Conference, Borden simply left the first ministers to present their conflicting claims.

It was a cop-out, a feeble avoidance of responsibility. But then, of course, Borden clearly assumed there would be no consequences. He was wrong. Virtually from the start, Canada's federation had been creaky: the Maritimes nurtured economic and political grievances; British Columbia complained that it could not get Ottawa's attention or support; Quebec scrupulously guarded its language and cultural rights; Ontario craved greater influence. The West's clumsy request for resource control at the November 1918 Conference sharpened these reactions in its Confederation partners, and they became the pawns in a federal–provincial and inter-provincial free-for-all.

True, even at that stage in Canada's development, Ottawa was far more capable of dealing with settlement, immigration and nation building than were the three Western provinces. Granted, Ottawa did not have the money to raise its general subsidies, and buy peace. But, in its perennial disregard for the peripheries versus the centre, the federal government miscalculated badly. The outcome deepened Western hostility. The consequences of such failures reverberate still—despite the West's modern-day economic clout.

Today, few Canadians remember the First Ministers' Conference of that wintry November of 1918. Its apparent successes were short-lived. Its failures were orphans, and no politician wanted to accept responsibility or blame for them. The Western premiers' mission to wrest control of their resources from Ottawa did not succeed. But the reasons behind

their lack of success at that time are far more fascinating than any dissec-
tion of a rubber-stamped outcome. There had always been narratives of
provincial grievance and self-important posturing. But in 1918 the first
ministers put their cards on the conference table, and those grievances
became legends. Decades later, long after Ottawa had ultimately transferred control of those resources to the Western provinces in 1930, Prince
Edward Island's Arsenault recalled the conference, and he wrote only
about the resource talks. The land-settlement schemes, which most first
ministers viewed as their primary agenda item in 1918, were long forgotten. Incredibly, even in the early 1950s, Arsenault was still enraged that
the West had eventually gained control.

What did *not* happen in November 1918 marked a pivotal change in
Confederation. Today's provincial politicians can all sound like those
self-important premiers, parroting the grievances that plagued the
November 1918 Conference. Ignore the convoluted dialogue, and today's
Westerners could be those same Prairie premiers. And although
Newfoundland and Labrador did not join Confederation until 1949, its
modern politicians too sometimes eerily echo the sentiments of the
Western premiers in the Hall of Invertebrate Fossils. Listen to politicians
from Quebec, British Columbia, Ontario and the Maritimes: they occasionally sound like clones of their counterparts in 1918. The legacy of the
November 1918 gathering is everywhere. It can be heard whenever there
is a mere whisper of a federal carbon tax, or whenever any "outsider"
criticizes the Oil Sands. And it all began when Britain decided to divest
itself of its onerous responsibility for the vast lands of its North American
empire in the west.

RIEL VERSUS MACDONALD:
CONTENDERS FOR THE WEST. 1857 TO SUMMER OF 1870

—

THE BATTLE LINES AT THE NOVEMBER 1918 CONFERENCE were drawn up long before the frosty standoff at an Ottawa museum. Dissent over the control of Western lands and resources first erupted in the late 1860s as Canada was negotiating a deal to buy the huge territory on its western and northern flanks. The scheming encompassed three nations, entangling devious American diplomats, haughty British aristocrats, Prairie colonists and Métis hunters, and anxious Ottawa politicians. But, at the centre, circling each other warily, were two determined men, Prime Minister Sir John A. Macdonald and Métis leader Louis Riel. It was their confrontation on the cusp of the West's entry into Confederation that set the pattern for decades of dangerously volatile relations between Ottawa and the West.

Both men were aware of the immense value of the West's resources, and the urgent need to control them. However, their plans could not have been more different. The wily Sir John A. nurtured the vision of a nation that spread across the continent, fencing the Americans within their existing northern boundaries, thwarting their dreams of annexation. If he could secure title to those vast Western lands from Great Britain and the Hudson's Bay Company, he could lure Anglo-British settlers to the frontier, and stake out the railway that would stitch together their worlds; the mere promise of a railroad across prairie lands would convince British Columbia to join Canada. The residents who already lived in Rupert's Land and the North-Western Territory would become members of Canada's fledgling federation of four provinces, whether they liked it or not. It was their destiny: in 1867, the Canadian Constitution had actually made provision for their eventual entry. That "Big Picture" approach, which would carve out a nation from the frontier, was to become tiresomely familiar to generations of Westerners, who would complain that Ottawa always ignored their views.

Far from Macdonald's Ottawa perch, Louis Riel was back in his Red River birthplace in July 1868 after a decade in Eastern Canada and the American Midwest. The Métis politician had not yet heard the voice of God like a thunderbolt, designating him as the prophet of the New World. But this educated bilingual man, who was so quick to bristle at any trace of Eastern condescension, did not like the changes that had taken place in his frontier community during his absence. Both the French-speaking and English-speaking Métis were slipping into second-class status in their homeland, and they were desperate for news of Great Britain's plans. Would the growing numbers of English-speaking immigrants from Britain and Canada and the United States displace them? Did they actually have title to their thin lots stretching back from the riverbanks that they had inhabited for generations? Who would govern them if the Hudson's Bay Company—"the Company"— relinquished its charter from the British Crown? Riel was only one-eighth Aboriginal and his relations with church and state would fluctuate wildly throughout his lifetime, but he always fiercely identified with the French-speaking Métis and their Roman Catholic God. It was clear to him that, if the Hudson's Bay Company surrendered its right to administer the territories, the Métis needed to secure control over their lands and resources to survive.

Later generations have depicted Riel as either traitor or patriot, visionary or madman. That is partly because his own views shape-shifted over the decades, as did his strategies. But there was one cause that he consistently espoused during a lifetime of dealings with Ottawa: the Métis should have title to their homesteads, and local governments should administer the land and the resources. It was a demand for respect and recognition— *and* for local control over any new settlement. In the late 1860s Riel's campaign against Ottawa's high-handed behaviour would *briefly* unite many Red River residents, including the English-speaking Métis.

Riel's Red River coalition could not last: the Métis leader was too mercurial, and he would eventually commit a fatal outrage that would splinter the Red River community and turn many Canadians against the settlement's demands. Yet this unlikely and flawed hero almost single-handedly willed the Province of Manitoba into being. His demands were a direct challenge to Macdonald's vision of Canada, but the prime minister went along with most of them because he wanted to subdue the resistance without slaughter—and because he needed to control the Prairies in order to woo British Columbia into the federation. In return for provincial status for Manitoba, however, the veteran politician forced Riel to accept what would become the template for sixty years of federal domination over Western lands and its resources. Western alienation took its first breaths during those dramatic days of the first Riel uprising.

Diplomacy was a lost art in the tense Western spring of 1869. It was clear that the sovereignty of the Hudson's Bay Company over the huge territory of Rupert's Land was no longer tenable. The firm's monopoly over the fur trade had effectively evaporated in 1849 when Riel's father, Louis, had rallied Métis support for a trapper who had been charged with violating the company's privileges. (The trapper was found guilty but no penalty was imposed.) Now new settlers were streaming onto the lands around Fort Garry, in what is now Southern Manitoba, and they were staking out homesteads on territory that the Indians and the Métis regarded as their own. Other newcomers were spilling across the prairies. The Company's insufficient government, which exercised jurisdiction over the lands draining into Hudson Bay, could not handle such unanticipated challenges.

There were rumours that the Company's new owners who had quietly purchased the firm's shares in 1863 were ready to surrender their almost-two-hundred-year-old Royal charter, which granted control over the lands and the resources. But there were always rumours on the Red River frontier, and the local newspaper reports were usually little more than gossipy speculation. The esteemed Bishop Alexandre-Antonin Taché of St. Boniface was busy preparing for his upcoming trip to the Vatican. Hudson's Bay Company governor William Mactavish, who administered Rupert's Land and Assiniboia, was gravely ill. Anyone who might know what plans were afoot for the people and the control of their rich lands was not talking.

In the absence of facts, the inhabitants fretted. Across the Red River from the company's Fort Garry trading post, Louis Riel observed their distress. An intelligent and deeply religious man, born in Red River in 1844, Riel had a deep attachment to his community. Bishop Taché had recognized Riel's potential during his school days. In 1858, anxious to find Métis priests for the territory, the prelate had dispatched the promising young student to a Montreal seminary for a classical education. Seven years later, Riel withdrew from the college without graduating, apparently on account of depression following the death of his father. He worked as a clerk at a Montreal law firm, and then stopped in Chicago and St. Paul in Minnesota as he headed west. By July 1868, he came home to a community where Protestant English-speaking settlers dominated the French-speaking Catholic Métis.

Those newcomers looked down on his people as mere hunters and half-breeds. And they made little secret of their disdain. Poet Charles Mair had barely unpacked in November 1868 when, in a sensational letter to his brother in Lanark, Ontario, which was reprinted in Toronto's *Globe* and the *Montreal Gazette*, he disparaged the "racket of a motley crowd of half-breeds, playing billiards and drinking" in the village of Winnipeg.[1] He ridiculed the Métis women as having only a totem pole for their family tree. And he castigated the five thousand Métis men who were dependent on food donations during that hard winter of 1868–69: "It is their own fault, they won't farm."[2] Although Mair would later become a champion of the Aboriginal cause, he limited his praise during those early months to the "inconceivably rich" lands with their loamy soil and abundant woods.[3]

Ironically, Mair, a former Queen's University medical student, was not in the West to farm. He was the paymaster on the Fort Garry section of the Dawson Road, which Canada was rashly clearing across Rupert's Land, from Fort Garry to Lake of the Woods, despite the heated objections of the Hudson's Bay Company. Without plowing an inch of arable soil, the Anglo-Canadian bureaucrat had effectively penned a recruit-ment letter for the folks he had left behind. His message was clear: the lands were ripe for the claiming from the existing Métis residents, who didn't even bother to exploit them. It was Canada's destiny to expand into this paradise.

Riel saw the condescension and he seethed. He was almost certainly the author of a biting letter signed "L.R." to Quebec newspaper *Le Nouveau Monde* in early 1869. The letter was a response to the disdainful Mair, and L.R. challenged the accuracy of his assertions about the relatively mild weather—"very often the ice sets by All Souls' Day"—and Mair's pie-in-the-sky estimates of the annual incomes of the English-speaking farmers. The Métis were not the only people going hungry in the hard winter of 1868–69. Many people "of all colours" had received food aid, including Englishmen, Germans and Scots. L.R. resented Mair's "barely civilized" comments about Métis women. But he was particularly offended by his contempt for all Métis. "If we had only you [Mair] as a specimen of civi-lized men, we should not have a very high idea of them."[4] Riel privately decided to play a public role "when the time should come."[5] He would not have long to wait.

The Rupert's Land settlers were anxiously discussing their many pos-sible futures. Did the Americans intend to annex Rupert's Land in response to the pleas of a willing faction within the Red River settle-ment? Would Great Britain pay a lump sum to extinguish the Company's charter rights, oust the Company governor, and extend its colonial rule over Western lands? Would the British government be capable of resist-ing any American takeover when London was so far away from the west-ern frontier? Or would Her Majesty's Government find a way to end the Company's centuries-old charter, and then shuffle the people and their lands with a flourish of the royal pen into the new Dominion of Canada, as another faction within the territory vehemently advocated? This last scenario appeared to be the most likely: in the fall of 1868, settlers had heard that the Hudson's Bay Company was discussing the fate of its charter in London with representatives of the Canadian and British gov-ernments. But nobody in the region knew the facts.

They remained in the dark for months. And when the Hudson's Bay Company finally agreed in March 1869 to cede its royal charter over Rupert's Land, no one in authority bothered to notify the Westerners. On the surface, the agreement seemed to be an arcane legal rite-of-passage in which the recipient would return its centuries-old gift to the giver in exchange for concessions and 300,000 British pounds. Britain, in turn, decreed that, as soon as the Company surrendered its charter, it would transfer Rupert's Land along with the adjoining North-Western Territory to Canada, which was the real behind-the-scenes purchaser. Britain merely acted as an intermediary.

That legal sleight-of-hand would bedevil Canada for decades. At the November 1918 Conference, the premiers from the other six provinces would maintain that *their* taxpayers had picked up the tab of 300,000 British pounds for the West, while Western taxpayers had scarcely contributed. (Taxpayers in all provinces, including newcomers such as British Columbia, paid off that debt over the decades.) Westerners would seize on the legal intricacies of the transaction. They would claim that the 300,000 pounds had not bought the land: the exchange of money had merely extinguished the company's charter rights over the land. Meanwhile, as part of the deal, Ottawa had also given away 45,000 acres of *their* land to the Hudson's Bay Company as well as the right to claim one-twentieth of the territory's arable lots. This combination of grandiose claims and legal nitpicking would be poisonous.

The deal that so offended Riel and his Métis supporters had actually been in the works for more than a decade. Since the late 1850s, without consultation with the Rupert's Land residents, Britain, the Hudson's Bay Company and the Province of Canada had explored their legal options. In 1857, before Riel had even left for his Montreal studies, a special committee of the British House of Commons had examined the Company's handling of its vast lands in British North America. In addition to its Royal Charter to Rupert's Land, the Company also held exclusive trading rights for the North-Western Territory, which sprawled to the north and west of Rupert's Land, and those rights were up for renewal in 1859.

The committee concluded that Parliament should only renew the firm's monopoly trading rights in areas where there was no immediate prospect of permanent European settlement. But the Company's rule over large tracts of Rupert's Land, including the Red River district, could no longer suffice.

That left the tricky question of who would take over. Perhaps unsurprisingly, the committee members happily accepted the Province of Canada's pitch to assume control over part of the territory. Canada should have whatever lands it wanted for settlement "in her neighbourhood," the British MPs grandly declared, as long as it could open up communications and provide local government.[6] The British government did not act on this recommendation. But it did *not* renew the company's monopoly over trade in the North-Western Territory. Those rights lapsed in 1859. Now the only obstacle to the West's eventual union with Canada was the Company's Royal Charter to Rupert's Land.

Governments initially dawdled because the West was so rugged and so far away from the Province of Canada. By the mid-1860s, however, the bloody American Civil War was over, and the Republic was pushing west into the frontier. Britain had to do *something*: the grim reality of nineteenth-century colonialism was "Use It or Lose It." When the provinces of Canada, Nova Scotia and New Brunswick formed a federation in 1867, the British North America Act included a clause that anticipated the Dominion's expansion. If Canada wanted to admit the colonies of Newfoundland, Prince Edward Island and British Columbia, the Parliament of Canada *and* the respective colonial Legislatures first had to endorse the union. A British cabinet order would complete the transaction because Canada could not amend its own constitution. For Prairie turf the procedure was even easier. If the Parliament of Canada asked Westminster for the admission of Rupert's Land and the North-Western Territory to the Dominion, a British cabinet order alone would effect the deed. Remarkably, unlike the rules for the colonies, the BNA Act did not require the prior consent of anyone in Rupert's Land or the North-Western Territory.

Canada was now quick to act. In mid-December 1867, when the Dominion was less than six months old, Parliament asked Queen Victoria to unite Rupert's Land and the North-Western Territory with Canada. Parliament also asked for the authority to pass legislation for the "future welfare and good Government" of the territorial residents.[7] Once again, there was no mention of any need to consult the inhabitants. The obvious implication was that the Hudson's Bay Company would do the talking on their behalf, or at least on its own behalf, during the upcoming negotiations to cede the charter.

In the beginning, when the talks commenced in London in the fall of 1868, the British Colonial Office acted as an intermediary between the Company and Canada's delegates, Defence Minister Sir George-Étienne Cartier and Public Works Minister William McDougall. But there was never any doubt about the identity of the real purchaser. As the British under-secretary of state for the colonies, Sir Frederic Rogers, wrote to Company governor Sir Stafford Northcote in late February 1869, the negotiations were "really between the seller and the buyer, the Company and the Colony." The British government was merely the channel "between these two real parties to the transaction."[8] The terms were finalized in March 1869.

Fortunately for Canada, there was an escape clause in Britain's original law to implement that pact: the surrender of the Hudson's Bay Company charter to the Crown would be null and void if Britain did not transfer the lands to Canada within a month after the Company surrendered them. That astute lawyer Sir John A. would soon need that "out." As the signatures were scratched on parchment, and as the foundation for two fiercely competing narratives was established in the West and the Rest of Canada, Canadian politicians in London scrambled to secure a loan of 300,000 pounds.

In June 1869, Canada passed legislation for the temporary government of Rupert's Land and the North-Western Territory. The Act, which anticipated the transfer, was terse. There was no mention of control over the West's public lands or resources: it was assumed that Ottawa would administer them. The future territory's lieutenant-governor would need Parliamentary approval for any territorial laws, institutions and ordinances. The residents of Rupert's Land and the North-Western Territory simply came with the package. As historian Donald Creighton observed, Prime Minister Sir John A. Macdonald "was trying to keep as closely as possible to the idea of a Crown colony."[9]

The West was the land bridge to British Columbia. With those vast prairie expanses included, Macdonald could convince British Columbia to join Canada. With possession of the West, the prime minister could promise a railway, linking Central Canada with the coast. Any objections from the territorial settlers were not just unwelcome: they were dangerously obstructive. They could hamper the success of the Confederation League in British Columbia, which vehemently advocated union with the Dominion of Canada along with the construction of that vital rail link.

Throughout that summer, thousands of miles away from that apparently done deal, Westerners were still speculating about the terms of any possible pact. The uncertainty was debilitating and dangerous. Anxiety naturally focused on who would administer the lands and the resources, and who would have the right to use them. For the Métis, the descendants of French fur traders and Indian mothers, a traditional way of life was at stake. In a territory where land titles were often held by "the ancient custom of the country," any change might be threatening.[10] Even in Assiniboia, that small parcel of Rupert's Land that would become the core of today's Manitoba, most lots were held "by squatter's right," which the Assiniboia courts had recognized. New purchasers only paid for the sellers' improvements, not for the land itself.[11] Without the reassurance of deeds from land registry offices, it was not even clear that the only official land titles, which the Hudson's Bay Company had granted in the early nineteenth century, could withstand legal challenge.

The Métis were deeply attached to their ancestral homes on thin river lots that fell back from the shores of the Assiniboine and Red rivers. Those lots were close together for mutual protection, and for help in times of sickness or need. Many families had held their lands for generations, and they often engaged in part-time farming. Some were even full-time farmers. But the new settlers who were heaving their belongings off wagons and steamboats often did not respect those ancestral rights.

The prospect that a new government would enforce the claims of condescending Ontario colonists over *their* ancestral lands and resources was chilling. The uncertainty dragged into the summer of 1869, although the Red River residents had now seen newspaper reports that the Dominion of Canada would assume control of their world.

The first firm indication of their fate, and Canada's intentions, was not reassuring. In August 1869, at the behest of the Canadian government, Colonel John Stoughton Dennis and his party of surveyors marched into the Red River settlements to carve out lot lines. His mission was presumptuous and clumsy: the Dominion had not taken formal possession of the Company's lands. Worse, Dennis was staking out 160-acre lots within tidy thirty-six-square-mile townships that were subdivided into parcels of one square mile. Future settlers would be able to slip into new homesteads with precise measurements and land titles. There was only one hitch: those thin Métis river lots did not fit within Ottawa's cookie-cutter shapes. Dennis was imposing his grid on Métis land, and their claims appeared to be non-existent.

Such provocation even disconcerted some English-speaking settlers from Ontario, who had been accustomed to having a voice in their provincial government. Taken aback by their protests, Dennis made futile attempts to reassure the residents. He even spoke at length with the charismatic Louis Riel, whom local Métis respected for his education and experience. The surveyor privately insisted that Métis rights would be safeguarded. But Dennis also accepted the hospitality of Dr. John Christian Schultz, an Anglo-Canadian businessman and land speculator, who aggressively advocated union with Canada and who had made little secret of his disdain for the Métis lifestyle. Dennis's close relationship with Schultz was politically inept, and it raised Métis hackles. Company Governor Mactavish had already privately advised Ottawa that the presence of the surveyors would arouse Métis and Aboriginal fears, but Ottawa had ignored him. Now, no one in authority, not even Mactavish, would publicly guarantee that Métis land titles would be respected, or that treaties would be forged with the Aboriginals.

This was a remarkable oversight. While the authorities were silent, the surveyors' stakes told a vivid story. Riel grasped his mission, and in late August of 1869, he mounted the steps of St. Boniface Cathedral after Sunday mass to urge resistance. The Métis could not tolerate this provocation: the Canadian surveyors were contemptuous of the original settlers and the authority of the Company, which still constituted the legitimate government. "They sought to seize the best lands of the *métis*, especially at Oak Point."[12]

Eyewitnesses supported his tale. Within days of the surveyors' arrival, the director of St. Boniface College, Abbé Georges Dugas, penned a nervous letter about Dennis's activities to the absent Bishop Taché, who was en route to Rome. "His project has set the country on fire . . . the people of Oak Point all came to warn him not to set foot in the neighbourhood if he wished to keep his head on his shoulders."[13] The priest lamented the Dominion's lack of concern for the sensibilities of the settlement. His rhetorical question was barbed: "Since one sees every day in the papers that the matter of Confederation between Red River and Canada is not yet concluded, why does the government permit these gangs of adventurers to come and spread disorder among our people?"[14] The priest's complaints were among the first in what would become a familiar litany: Central Canada was oblivious to Westerners' opinions.

Dennis and his surveyors kept working, although they retreated to the southern boundary of Rupert's Land. The Métis and their supporters remained incredulous and angry. How could they be sold so casually? Had Canada actually bought them and their lands and their rich resources? Would they have a voice in their fate? At the very least, Canada's dominion should not extend over the West until the deal with Great Britain and the Hudson's Bay Company was finalized in London *and* Métis rights over their lands were officially recognized.

Their fears, however, were soon confirmed. On September 21, 1869, the *Nor'-Wester* newspaper in Red River passed on a report from an unnamed English newspaper: Canada's public works minister, William McDougall, would become lieutenant-governor of the territory. The transfer of jurisdiction would soon be complete, and Ottawa would establish a "local government for the territory."[15] This was the final straw. Métis hunters and boatmen pulled together a military force based on their formation for the buffalo hunt. On October 11, Riel and a small party of Métis halted the surveyors as they worked their way northward from the American border across lands that the Métis viewed as their own. On October 19, with the support of a local priest, Joseph-Noël Ritchot, the French-speaking Métis created an elected National Committee, and those Métis hunters and boatmen put their force under its direction.

Canada remained largely oblivious to the unrest. But events were moving fast. On October 26, 1869, in a one-paragraph item, the *Nor'-Wester* cited a Minnesota newspaper report that William McDougall had arrived in St. Paul on October 12. The new lieutenant-governor would likely arrive in Winnipeg "towards the end of the present week."[16] It was almost a social note. But it did not escape Riel's attention. As the elected secretary of the Metis National Committee, he eclipsed the nominal president, John Bruce. And he was now alert to every threat.

Riel was the natural leader during those first tense months of confrontation with Ottawa. Hurt pride and righteous grievances were a dangerous combination. Today Riel is hailed as the Father of Manitoba, the young firebrand who fought for the dignity of his people and the control of their lands. But at the time of the 1869–70 Red River Resistance, he was a lightning rod for dissent and controversy. Contemporary accounts portrayed him as mercurial, charming, moody, diligent and easily irritated. The Company's senior Canadian official, Donald Smith, who would act as a federal emissary during the Resistance, saw him as trigger-happy, given to threatening death in the face of any provocation, and impervious to reason. But Smith would also concede that new Canadian arrivals had in fact seized lands from their Métis and Aboriginal owners. Such behaviour had convinced the long-time inhabitants that "avarice and selfishness" were "the new order of things."[17] Riel had a lot to be angry about.

Certainly, British-born newcomers did not treat him or his fellow Métis as equals. In his vivid account of his reconnaissance mission throughout the West during the Resistance, British Army officer William Francis Butler would describe Riel as a "short stout man with a large head, a sallow, puffy face, a sharp, restless, intelligent eye, a square-cut massive forehead overhung by a mass of long and thickly clustering hair, and marked with well-cut eyebrows—altogether a remarkable-looking face . . . *in a land where such things are rare sights.*"[18] Riel, who was only in his mid-twenties, spoke earnestly to Butler about his devotion to the people, his wish for peace and his benign intentions toward the soldier. Butler saw only an unstable, so-called half-breed in moccasins.

Throughout that chill autumn of 1869, Westerners still subsisted on rumours, which the precious mail or the local newspaper could not dispel. Life was tough for everyone, French- and English-speaking, despite Mair's glowing reports. The settlers raised their own livestock, and grew their own vegetables. Buffalo was a staple, particularly for the Métis. Too many people, of all backgrounds, drank too much. Violence was all too familiar. Merchant Alexander Begg kept a diary during the 1869–70 Red River Resistance, but the snippets from daily life that he described could have happened anytime during the mid-nineteenth century in the West. There were dancing sprees with wild fiddlers. Begg noted that settler Thomas Johnson's body had been found frozen stiff near his own door; he had been shot dead after a drinking binge. The climate was a constant concern—Begg noted the weather with every entry—and it hampered communications with anyone outside the territory. Rupert's Land was more than half a world away from the posh salons of London. In many respects, it was centuries away. It was small wonder that, in London *and* Ottawa, its residents were afterthoughts.

For Ottawa, the grubby business of payment remained. Canada had to find 300,000 British pounds for the Hudson's Bay Company, and it had to transfer the money before Her Majesty's Government would shuffle Rupert's Land and the North-Western Territory into the Dominion. It was a huge amount of money for a new nation, and it was no easy task to raise those funds, as the Maritime premiers at the November 1918 Conference would grimly recall. Canada's customary lenders balked, and Britain had to guarantee the loan on condition that the principal would be repaid by April 1, 1904. This financial wheeling and dealing took time. The transfer, which was originally slated for October 1, 1869, was delayed until December 1. Sir John A. and his government hoped for a low-key handover: somehow, after years of manoeuvring, Canada would acquire the West.

Macdonald did not reckon on Riel and his Métis militia, however. When news of Lieutenant-Governor McDougall's imminent arrival reached Red River, the resistance proceeded with calculated haste. In late October, an armed group of Métis expelled McDougall soon after he had crossed the border, forcing him to retrace his steps into North Dakota. Then on November 2, the Métis seized the pivotal trading post of Upper Fort Garry, where downtown Winnipeg now stands. That same day, Riel invited English-speaking residents to elect twelve representatives from their parishes for a convention. But ten days later, Governor Mactavish, who was still in charge of Rupert's Land until the transfer on December 1, demanded that the Métis disarm. Riel ignored him. A week later, when the convention assembled, the Métis leader asked the English- and French-speaking representatives to form a provisional government. Although Riel repeatedly reassured the delegates about Métis loyalty to the queen, he could not obtain the consent of the English-speaking residents for that government or for his first proposed Bill of Rights, which included protection of Métis culture and lands.

But the greatest blunder belonged to the assertive McDougall. Macdonald's biographer Richard Gwyn has described the prime minister's selection of McDougall as lieutenant-governor as "one of the worst choices of his career . . . . [McDougall was] utterly unfit for so demanding and delicate a task."[19] Macdonald had dispatched his former cabinet minister with instructions to tread softly, and to discern Western sentiments. Stewing at the border, however, McDougall ignored those directives. On December 1, without permission from Ottawa, he crept across the border, solemnly proclaimed Canada's jurisdiction over the West, and scuttled back. Remarkably, he also ordered former surveyor Colonel Dennis to muster an armed force, which included a small group of Aboriginals and fervent pro-Canada settlers, to restore order. The Métis captured them on December 7, and a day later, thoroughly provoked, Riel and his Métis established a Provisional Government, and Riel soon became president. The Métis leader declared that the Hudson's Bay Company had forfeited its legitimacy as a government when it transferred the inhabitants of Rupert's Land without their consent to Canada, through an agreement in London, "which it did not even deign to communicate to its people."[20] The sale had violated the law of nations, and a people becomes free "when the sovereign to which it was subject abandons it, or subjects it against its will, to a foreign sovereign."[21] Canada's Dominion was not welcome.

Nonetheless, the prime minister could not remain passive when almost 2.9 million square miles of Rupert's Land and the North-Western Territory were at stake. When sketchy news of the uprising first filtered into Ottawa in late November, Macdonald had done what any shrewd lawyer would do: he checked on the whereabouts of those 300,000 pounds. Mercifully, they were still under Canada's control. This was a huge advantage. Macdonald promptly refused to deliver the cash to the Hudson's Bay Company, and he declared that Canada would not accept the transfer of the lands on December 1. Flustered, British Colonial Secretary Lord Granville protested that the Company could no longer govern the unruly Territories. He reminded Macdonald that, under the law, the British government had to transfer those Territorial lands to Canada within a month after the Company surrendered its Charter rights to the Crown. But the Company would not surrender its rights until it had pocketed the 300,000 pounds.

Macdonald stood his ground. As long as Canada did not convey the funds to the Hudson's Bay Company or accept the transfer of the Territorial lands and resources from Britain, the Company remained the official government. Effectively, the Prime Minister was putting pressure on the Company *and* on the British government to fix the mess. Canada could not flounce into the Territories now when so much had gone wrong: the surveyors' intrusions, the pro-Canada party's strident agitation and Lieutenant-Governor McDougall's mistakes. Lord Granville fumed, but Sir John A. would not budge. Indignant, the prime minister wrote to his unofficial agent in London, John Rose: "I cannot understand the desire of the Colonial Office or of the Company, to throw the responsibility of the government on Canada just now." Lord Granville did not realize, he said, that the hasty imposition of Canadian authority would only provoke more resistance, and strengthen the pro-American faction within the Red River settlement. "It would so completely throw the game into the hands of the Insurgents and the Yankee wire pullers."[22]

The British simply did not get it. Eleven days after that letter of December 5, 1869, Macdonald penned a sizzling memorandum to explain his decision and to castigate the bunglers. The diplomatic formalities were observed: with cabinet approval, the prime minister wrote to the governor general, who relayed his sentiments to Lord Granville. But there was no flowery language in Macdonald's missive, and no attempt to cloak his exasperation. Canada would not pay 300,000 British pounds as long as the inhabitants of Rupert's Land were staging an armed rebellion. No one in the British government or the Company had mentioned *that* possibility when two Fathers of Confederation, Sir George-Étienne Cartier and William McDougall, had hammered out their agreement in London. "It [Rupert's Land] *was* in a state of tranquillity," Sir John noted archly, carefully italicizing his verb, "and no suggestion was made of the possibility of such tranquillity being disturbed."[23]

Obviously the feisty inhabitants did not relish the notion of being sold without explanation and without their consent. Surely Company officials could have said something, Sir John wrote, because it was their duty to explain "the precautions taken to protect the interests of the inhabitants, and to have removed any misapprehensions." Instead, the Company had done nothing. "The people have been led to suppose that they have been sold to Canada, with an utter disregard of their rights and position."[24] He added, perhaps a trifle disingenuously because he had put the equivalent of a twenty-first-century stop-payment on those 300,000 British pounds: "This is not a question of money—it may be one of peace or war."[25]

Macdonald proceeded cautiously. He ordered the impetuous McDougall back to Ottawa, and he dispatched three representatives on a peace mission to the Territories. Two of them were acquainted with the people and the countryside: Rev. Jean-Baptiste Thibault, who had spent almost four decades among the Métis, and Colonel Charles de Salaberry, who had explored the Red River area in the late 1850s. They were to "disabuse the minds of the people of the misrepresentations made by designing foreigners."[26] Macdonald also dispatched the diplomatic Donald Smith, who could obtain easy access to Fort Garry because he was the Company's senior Canadian executive. He asked Smith, as Ottawa's Special Commissioner, to bolster the resolve of the ailing Governor Mactavish, and to work with "the loyal and well-affected portion of the people for a restoration of order."[27] Smith was no fool: he carried cash to shore up that loyalty. He reached Fort Garry on December 27, 1869.

Smith took his time before revealing his real mission. The Provisional Government was largely Métis, and Smith was convinced that he could not negotiate with Riel or its members—partly because Riel's forces were holding Smith as a virtual prisoner in Fort Garry. Instead, Smith asked Riel to call a mass meeting of the settlement, at which he would divulge Macdonald's instructions. Riel complied, and Smith delivered Ottawa's assurances of generous treatment to the settlers. Mollified, the residents appointed forty English- and French-speaking representatives to a convention, which met in Fort Garry on January 25, 1870.

Riel was on edge, and it showed. But he did remain focused on the need for *local* control of the lands and resources. He knew that Canada's four provinces, in accordance with British tradition, did in fact administer their own lands and resources. He wanted the same status for the Territories. Smith reported that Riel visited him on February 3, 1870, while the convention was meeting. The Métis leader asked if Canada would admit Rupert's Land as a province. Smith replied that he could not answer with certainty. When Smith had left for the West in December 1869, Ottawa had intended to admit the lands as a territory, but "no doubt it would become a Province within two or three years."[28] In response, as Smith told Secretary of State for the Provinces Joseph Howe, Riel threatened him.

There were other rejections. On February 4, 1870, when Riel urged the convention of English- and French-speaking delegates to demand provincial status, he actually read aloud the list of provincial powers in the Canadian Constitution. That may not have made for sparkling debate. But, as the *New Nation* newspaper noted, Riel singled out the provincial right to manage and sell public lands, as well as timber and wood. "This, he [Riel] alluded to, as one of the most important, as far as we are concerned," the paper reported.[29] But the convention voted against demanding immediate provincial status.

A day later, on February 5, 1870, Riel asked the delegates to nullify any existing deal between the Hudson's Bay Company and Canada. When he lost that vote as well, he accused several delegates of being traitors. "Words very nearly came to worse," Winnipeg trader Begg observed gloomily. "Riel said the convention had beaten him this time but he would beat them yet."[30] The diarist recorded the local gossip that Riel was drinking, and that he had threatened the life of Governor Mactavish. Begg charitably decided that Riel was probably excited over the loss of his motion to nullify Canada's deal, and had "given way to expressions that he otherwise would not have done."[31] In fact, Riel rarely touched alcohol.

The convention did agree, however, to put together the community's demands in a second Bill of Rights and to send delegates to Ottawa to negotiate the terms of the transfer. When the convention concluded, on February 10, 1870, Riel pushed Governor Mactavish to recognize his Provisional Government as the temporary government. When Mactavish reluctantly gave his blessing, English-speaking residents agreed to support it. Riel then cobbled together a new version of his Provisional Government with English- and French-speaking members on a twenty-four-person elected council. Riel assumed the role of president, and he selected three councillors as members of his executive.

The time seemed propitious for peace talks. Macdonald now understood that the Métis were not rebelling against the rule of the Hudson's Bay Company, which theoretically remained in force; nor did they object to Britain's control. They simply would not accept Canada's unilateral terms for their inclusion: they wanted to negotiate. In the spring of 1870, at Smith's prompting, Riel dispatched his three handpicked delegates to

Ottawa for official talks on the Territory's fate. They brought another list of conditions about control of the land.

It was that list, drafted and redrafted, that constitutes one of Riel's lasting legacies to the West. Four times over four months, under the Métis leader's careful eye, Red River groups with varying memberships drafted different versions of a Bill of Rights. And, each time, the demand for respect for local land ownership and control stood out within the texts. In December 1869, in the first Bill of Rights, the Métis had insisted that the laws of the federal Parliament would not be valid unless their local government agreed. They had also asked for "all privileges, customs and usages existing at the time of the transfer to be respected."[32] Two months later, when the convention of elected French- and English-speaking delegates had rejected the idea of immediate provincial status, they had nonetheless insisted that "the Local Legislature of this Territory have full control of all the public lands" within a specified area around Upper Fort Garry.[33] Any future *local* government would call the shots on the lands and the resources.

Riel ignored the convention's opposition to immediate provincial status. In a third Bill of Rights that he and his key councillors in the Provisional Government drafted, he insisted that the Territory could only enter the Dominion as a province, and that the province should have full control of its public lands. Finally there was a fourth Bill of Rights, which Bishop Taché, who had hastily returned from Rome, apparently inspired. That final version repeated the demands for provincial status and resource control; it added requests for a provincial senate and for guarantees for Roman Catholic schools.

Taken together, the import of these four Bills of Rights was clear. Before the Territorial inhabitants would join the Dominion of Canada, they wanted assurance that they would control their resources, including their lands, and they insisted upon a voice in their fate. No one could sell them without their "say." Riel was the first Western Father of Confederation to insist upon local control of the land and the resources. Long before the Western premiers took their stand at the November 1918 Conference, Louis Riel was on the barricades.

———

Their demands could not have come at a worse time for Sir John A. Macdonald, who simply wanted to claim his hugely convenient colony. In retrospect, it was the deal of the century. Canada was paying roughly $1.5 million, or 300,000 British pounds, for Rupert's Land, which was far less than the one million pounds that Company investors had originally demanded. That territory spanned the area that drained into Hudson Bay; its boundaries stretched to what is now the Alberta–British Columbia boundary. The North-Western Territory, which Canada would freely receive from Great Britain, spilled across the northern reaches of the Prairies and across today's Nunavut, the Yukon and the Northwest Territories. Canada was incorporating a land mass that was nearly six times its existing size. It was an act of breathtaking colonial presumption. But, at the time, most Canadians simply assumed that they were buying a property that could politely be termed to have potential.

It was not land, however, for the timid or the impatient. Transportation and communication facilities were primitive, and it was exceedingly difficult to travel west from Toronto across the rapids and bogs that curled north of Lake Superior. Methodist minister George Young instead trekked into Fort Garry during the summer of 1868 from the south, across the Canada–United States border. It was a circuitous route by steamship, railway and then hard slogging with packhorses. Young scrambled up riverbanks, negotiated rutted roads, and fended off mosquitoes and bulldog flies. His observations were hardly the stuff of tourist brochures as he crossed by ferry into the "sorry scene" of Fort Garry and the adjoining town of Winnipeg. Sticky Red River mud was everywhere. There were no sidewalks or crossings, so travellers floundered through the mire. There were "a few small stores with poor goods and high prices." There was one tiny tavern. There were no rooms to rent in the few acceptable dwellings, and he could see no churches or schools. "Population about one hundred instead of one thousand as we expected—such was Winnipeg on July 4, 1868," Young recorded gloomily.[34]

The countryside was no better. A locust plague of biblical force had devastated the crops. Flour prices were astronomical; oats for horse feed went for two dollars per bushel. The buffalo hunters "were despairing of success."[35] During that first 1868–69. winter in Winnipeg, Young hunkered down amid intense cold, blizzards, deep snowdrifts and famine. A few months later, that self-important newcomer Charles Mair would blame the famine on the Métis' refusal to farm. But Young, who was actually there during the summer of 1868 and throughout the ensuing winter, recounted that everyone was hungry because the crops had vanished beneath "air-filling [locust] clouds . . . of detestable devourers" and the buffalo hunt had failed.[36] The experience tested Young's faith itself.

Other accounts of those early days on the land were equally daunting. A few years after the Riel Resistance, a hardy clump of Mennonite scouts from Russia would reach Winnipeg on a mission to explore potential settlement sites within a few dozen miles of the town. Their forays would be difficult. It rained, and the roads were terrible. Clerks at the Oak Point government post refused to shelter the foreigners. John Funk, an Old Mennonite minister from Indiana who had joined the party, wrote gloomily to his wife about the brackish waters humming with mosquitoes. "I forget all my compunctions against drinking a little wine when I taste the water."[37]

MP James Trow, on his inspection tour, barely got out of Winnipeg because of "the unusual rainy season and the shocking state of the roads."[38] He fended off wild dogs, shivered sleepless in his tent while the drums of Saskatchewan Aboriginals thumped through the night, patted his trusty revolver for reassurance, skirted dwindling clumps of buffalo and mingled with isolated clusters of American and British settlers. The doughty MP detected potential in Manitoba, and advised Ottawa to get cracking on the construction of a railroad.

The key word was always *potential*, because few doubted that the land was bountiful and rich in resources. From their vastly different perspectives, both Riel and Macdonald saw the West as a source of future wealth. Riel's insistence on control of the resources was designed to protect Métis claims to their rich lands against the inevitable waves of immigrants. He was volatile, and there were already signs of the mental instability that would eventually afflict him. But he and his Provisional Government were financially prescient. Both the third and fourth versions of their Bill of Rights demanded that Canada appoint a Commission of Engineers to explore the territory. That commission would report within five years to the *local* Legislature on the extent of mineral wealth. The last three versions insisted that Ottawa respect *all* existing properties, rights and privileges.

The odds were stacked against Riel, of course, because he was up against Macdonald, a fervent proponent of federal control. The prime minister was intent on protecting his prime real-estate deal against all contenders. Other threats were present, too. American expansionists were dangling invitations to join their republic, and some strong advocates of annexation, such as Irish-American William Bernard O'Donoghue, were members of the Provisional Government. During the crucial winter months of early 1870, American-born Henry Robinson, who favoured annexation, edited the *New Nation* newspaper—which was essentially an organ of the Provisional Government. The American consul in Winnipeg, Oscar Malmros, and the Treasury agent at the U.S. customs house in Pembina, Colonel Enos Stutsman, were confidential advisers to the Provisional Government. Those ties were strong: the fiery O'Donoghue, who was government treasurer, would eventually break with Riel when he declined to ask the United States government to intervene in the resistance on behalf of the Métis. Riel retained his faith that Queen Victoria would support the Métis cause.

The U.S. State Department even appointed lawyer James Wickes Taylor, an expert on the Territories, as a secret agent to watch developments. Taylor was of the view that America could fulfill its great destiny if it swallowed the lands of the British Northwest. He belonged to a syndicate that had extended a branch line to the border. Now he had a scheme: if Red River residents were to become dependent on rail links and trade with Minnesota, which could happen if an American branch line were extended into the settlement, political links would surely follow. On January 5, 1870, Taylor wrote to the general manager of the Grand Trunk Railway, Charles John Brydges, and reviewed Canada's blunders. Ottawa had not promptly guaranteed popular rights and infrastructure improvements when news of the sale became public; McDougall's foolish attempt to send a spur-of-the-moment army into the field in December 1869 had united the population under Riel's leadership. The success of the improbable rebellion had "lifted it beyond the domain of burlesque." The situation, Brydges wrote, "suggests to almost every one I meet the possibility of a treaty with England and Canada" that would cede the Territories to the United States.[39]

Macdonald was well aware of the threat, however—and he had an ace up his dapper sleeve. The dispatch of his emissaries did not preclude military intervention. As he had warned the British government in late 1869, Ottawa was constructing boats to ferry a military force to the territory if his overtures failed. He called for the "hearty co-operation" of the Hudson's Bay Company and Her Majesty's Government in that endeavour, and he eventually secured the promise of British assistance.[40] But it was not until the late winter of 1870, when the conflicting allegiances in the Territories finally erupted in violence, that Macdonald formally opted to send troops, if only to keep the peace.

The violence was pointless. In February, soon after the Provisional Government had agreed to send negotiators to Ottawa, the reckless pro-Canada party staged a rebellion, and Métis troops promptly captured those pro-Canada advocates. The operation strengthened the Provisional Government and Riel's apparent legitimacy. At Smith's behest, Riel even pardoned a leader of the pro-Canada forces after a Métis court sentenced him to death.

Now was the time for caution and patience. Instead, the edgy Riel, perched atop a powder keg, soon lost his grip on himself and his Métis forces. His first loyalty was always to the Métis cause, and he was hypersensitive to any perceived insult to his people. He also resented any challenge to his personal authority, and Rupert's Land was now crawling with emissaries with competing agendas. The residents themselves were divided.

The spark came all too soon. Among the pro-Canada prisoners, there was an Irish-born Orangeman, Thomas Scott, who had arrived from

Ontario in the summer of 1869 and found work as a labourer on the Dawson Road. Scott did not bother to conceal his contempt for his Métis guards, and he demanded a trial. On February 28, the incensed guards beat him. Then they conducted a trial on the charge of insubordination. The verdict was "guilty," and the sentence was death. Riel rejected Smith's plea for clemency: "We must make Canada respect us."[41] As president of the Provisional Government, he approved the execution. It was a fatal and needlessly brutal misjudgment. Methodist clergyman George Young, who had grumbled about Red River's rustic life, would never forget his harrowing hours with the disbelieving prisoner. (Fifteen years later, his military officer son would escort Riel to his trial in Regina as a traitor.)

On the morning of March 4, 1870, a Métis firing squad clumsily shot Scott, leaving him to die on the ground. The execution splintered the fragile factions in the Territory. Tales of Scott's suffering and Riel's cruelty spread across Canada like a wild prairie fire. Ontario embraced Scott as a martyr and disdained Riel as a murderer. Sentiment was high. Riel lost public sympathy within Canada, and the Métis cause was besmirched. Scott's execution deepened the derision of British army officer Butler when he encountered Riel in Fort Garry that summer. The reconnaissance officer dismissed his captor's "theatrical attitudes and declamation." He saw only a poseur: "This picture of a black-coated Métis playing the part of Europe's great soldier in the garb of a priest and the shoes of a savage looked simply absurd."[42] What he did not recognize was that Riel had reached his limit.

Macdonald was now caught between the sympathies of French Quebecers for the Métis, whom they viewed as an oppressed francophone minority, and English Canadians' fury at Scott's execution. He opted to negotiate for his prize. At stake were the future homesteads that Canada could dole out to new settlers, the timber licences that it could sell, and the leases for coal or other minerals that it could offer to the highest bidder—or the best political friend. Macdonald had itemized those plans in December 1867 when he asked the British Parliament for the Territories: Canada would colonize the fertile lands around the Saskatchewan, Assiniboine and Red

River districts; it would develop the North-Western Territory's abundant mineral wealth; and it would open the lands to commercial traffic from the Atlantic to the Pacific. Now the wealth was there for Ottawa's taking, and the prime minister would not back down on the crucial issue of land and resource control.

Macdonald clearly understood the stakes. In mid-April of 1870, a select committee of the Senate examined the conditions in Red River, including the climate, soil, population and resources. The witnesses were not dispassionate observers: all of them, including the caustic Charles Mair, wanted to join Canada, and they severely criticized Riel. The Senators peppered them with detailed questions. What was the length of the grasshoppers? About an inch and a half, said farmer John James Setter from Portage la Prairie. Was the wheat of good quality? Yes, said farmer Joseph Monkman from St. Peter's Parish, and the barley was even better. When did the snow generally commence? About the tenth of November, asserted that rash leader of the pro-Canada party in Red River, Dr. John Schultz. Did the soil require frequent renewal? "Wheat has been grown in the same soil for 40 years, and succeeds," affirmed Mair. "The farmers never use manure."[43] Were the Indians hostile? There were merely a few cases of slaughtered cattle, said James Lynch, who had settled on the shores of Lake Manitoba. The Senators were dazzled.

In late April, on the same day that Riel's delegates finally sat down to negotiate in Ottawa, the committee issued a rave review. Vast Western tracts were available for cultivation. The region stretching from the American border to the northern banks of the Saskatchewan River was grand for growing wheat and vegetables. The drawbacks were the distance from rivers and railroads, the absence of agricultural markets, the occasional grasshopper infestations and the cold. But, the report added, the cold was not *very* cold. If Canada could carve viable passages around the northern shores of Lake Superior to Red River, the area would provide a "very desirable home" for immigrants, and it would "enhance the prosperity and promote the best interests" of Canada.[44]

Such federal aspirations likely unsettled Riel's three negotiators— Father Joseph-Noël Ritchot from the Métis community, Alfred H. Scott from the pro-American party, and Judge John Black from the English-speaking community. Their meetings in Ottawa were initially delayed because authorities had arrested Ritchot and Alfred Scott for the murder of Thomas Scott after the three men arrived on April 11. Those charges were eventually dismissed. On April 25 they finally met with Macdonald and his principal lieutenant, George-Étienne Cartier, the politician who had almost single-handedly tugged Quebec into Confederation, and who had negotiated the agreement to purchase Rupert's Land over six long months in London. The talks were not official. Macdonald and Cartier would not recognize the existence of Riel's Provisional Government. As a compromise, the three men were deemed to be representatives of the North-West population.

Both sides were under intense pressure. British colonial secretary Lord Granville had sternly reminded Ottawa on April 23, 1870 that the nearly four hundred British soldiers and more than seven hundred Canadian militia under the command of British colonel Garnet Wolseley would not march from Ontario to Fort Garry until John Rose, Macdonald's unofficial agent in London, was "authorized to pay 300,000 pounds at once!" Granville wanted to see "Her Majesty's Govt. at liberty to make transfer [of the lands to Canada] before end of June."[45] Macdonald was also anticipating a visit from British Columbia delegates in May to consider the prospect of entering Confederation. Without the promise of a railroad to British Columbia, which required control of the Territorial lands, Canada could not woo the colony. (On July 20, 1871, mollified by the promise of a railway, British Columbia would officially join Canada.) Riel's Provisional Government representatives, in turn, were uneasily aware that Dominion forces were heading west for the putative purpose of keeping the peace. They did not know that Britain would not permit the troops to march until the deal was done.

Genuine peace would be welcome. In Red River, the settlers gleaned news of their fate from hearsay, the sporadic mail service and local summaries of U.S. reports. Trader Alexander Begg dutifully recorded each day's unreliable news tidbits. The Sioux were threatening the settlers at Portage la Prairie. Maybe. A man named Burr had tried to flee without paying his debts, but a guard had seized his property. Probably. Riel was putting on airs. Perhaps. But it was more likely that his behaviour appeared erratic because he needed to control the competing factions.

Every event unsettled Begg. Riel was unpacking the hapless former lieutenant-governor McDougall's furniture for use in the Provisional Government House. The mail could not reach Fort Garry because the roads were inaccessible. Riel was squabbling with his Irish-American supporters, who objected to the raising of the Union Jack over Fort Garry. American troops were en route to the Canada–U.S. border to guard it. Some Provisional Government soldiers had been paid, refused to re-enlist, and were spending their salary on drink. A key member of the pro-Canada party "with greedy eyes" had checked out land around Lake of the Woods "on which to pounce and speculate when the Canadian Government should come in."[46]

Tensions were high, and on April 27, 1870, as Macdonald and Cartier and the three North-West representatives grappled with the issues in Ottawa, Begg was frank: "A good deal of anxiety is felt amongst all classes to hear news from our delegates."[47] Riel was mostly holed up in the Red River fort, as uneasy as every other resident.

Finally, in Ottawa, the two sides reached a compromise. Macdonald had rejected the delegates' demand that Rupert's Land and the North-Western Territory should enter Canada as a single province with all of the rights and privileges of the other provinces. But the prime minister accepted that he *would* have to create a province. His solution was to limit the land mass. The new province would span only a minuscule portion of the Territories: roughly 180 miles from east to west, and 90 miles from north to south. It would be called Manitoba, a name likely derived from a Cree word meaning "strait of the spirit." Sir John told the House of Commons that it meant "The God who speaks—the speaking God."[48]

The delegates, in turn, agreed that Ottawa could retain control of the land and the resources "for the purposes of the Dominion."[49] As Sir John A. later told the House of Commons, Canada had bought the land for "a large sum" from the Hudson's Bay Company, and federal ownership of the resources was "of the greatest importance to the Dominion."[50]

Ottawa required Western land for Aboriginal reserves: it also needed a clear pathway for the railway to the Pacific. Britain's special envoy at those talks, Sir Clinton Murdoch, agreed with Ottawa's decision, and the Provisional Government's request for control over its lands and resources was deemed "clearly inadmissible."[51] If the inhabitants controlled the land, they might "prevent the construction of a Pacific Railway, and . . . impede the ingress of immigrants."[52] Ottawa could not permit those Westerners to thwart its National Dream. The Provisional Government delegates accepted this inequality in provincial status.

The deal was done.

Macdonald wasted no time. He introduced the Manitoba Act in the Canadian House of Commons on Monday, May 2, 1870. The act stipulated that the lieutenant-governor of Manitoba would also act as the lieutenant-governor of the North-Western Territory, with the assistance of an appointed council. A day later, Colonel Wolseley commenced his journey to the Red River settlement from the Ontario harbour of Collingwood. His troops would make forty-seven portages and run fifty-one miles of rapids before they arrived in Winnipeg on August 24.

The legal dominoes fell neatly. The Hudson's Bay Company submitted its deed of surrender to the British Colonial Office on May 9. Sir John Rose arranged to transfer 300,000 British pounds to the Bank of England for the Company's credit. The Manitoba Act received Royal Assent on May 12. On June 23, 1870, Her Majesty's Government in London officially accepted the Company's surrender of its charter rights, and set a date of July 15 for the transfer of those lands to Canada. On June 24, the Provisional Government in Red River endorsed the Manitoba Act, after the leader of the North-West delegation to Ottawa, Father Ritchot, triumphantly reported on the delegates' success to the Legislature. On July 15, 1870, the Manitoba Act took effect.

The relief across the nation, especially in Ottawa and Winnipeg, was palpable. Even Riel had no objections to the Act. He was magnanimous. "Let us still pursue the work in which we have been lately engaged—the cultivation of peace and friendship, and doing what can be done to convince these people that we never designed to wrong them but that what has been done was as much in their interest as our own."[53] He was, in fact, deluding himself.

So why did the three North-West delegates accept Ottawa's conditions, and why did Riel endorse that decision? Ottawa had adroitly grasped the deepest concern of the Manitoba residents, especially the Métis: What would happen to their property rights when Canada controlled the Territories? Accordingly, the Manitoba Act of 1870 confirmed everyone's existing land titles, recognizing the claims of "all persons in peaceable possession of tracts of land at the time of the transfer to Canada."[54] It also set aside 1.4 million acres "for the benefit of the families of the half-breed residents."[55] It was a trade-off, as Bishop Taché would explain decades later: to keep control of the lands and the resources, Ottawa dangled the offer of land for the Métis.

Riel would not have been so jubilant if he had been able to peer into his future. In August 1870, as Wolseley approached Fort Garry, Riel learned that Canadian militia members wanted to lynch him because of his treatment of Thomas Scott. He fled, drawing comfort from the knowledge that he had won rights for the Métis. Even that joy was short-lived, however. Mere months after Manitoba's creation, Anglo-Saxon newcomers were selecting more land for themselves—before the 1.4 million acres for Métis settlement had been earmarked. When Bishop Taché protested, Lieutenant-Governor Adams Archibald declared that the Métis should choose the land that stretched immediately behind their existing river lots.

That decision did not last: Sir John A. appointed a zealous Dominion Lands Agent in Winnipeg who did not respect Archibald's decision, and who allotted land randomly in all provincial townships to new settlers. Frustrated and discouraged, many Métis sold their land and moved farther west into what was now called the North-West Territories. In 1872, Ottawa officially decided that land would not be allotted to the Métis near their original river lots, but on areas of the Prairies where they had not traditionally lived. The Métis could not appeal this decision to the provincial government: Ottawa controlled the lands and the resources and would allow nothing to get in the way of national expansion and its ambitious dreams.

Riel's dreams were in tatters. Thirteen years later, in 1885, he would stage an ill-fated rebellion to win lands for the Métis in Saskatchewan. He would be put on trial for high treason, which carried a mandatory death penalty. And he would be hanged.

Today, there is a statue of Louis Riel on the grounds of the Manitoba Legislature. It depicts the Métis leader in a black frock coat, bow tie, Métis sash and moccasins, holding aloft a copy of the Manitoba Act, and it has the approval of the Métis political establishment after two previous sculptures provoked outrage. The first statue, which was set on the grounds of the Saskatchewan Legislature in 1968, showed Riel as a humiliated figure in a makeshift garment. Twenty-three years later, after unremitting Métis protests, the Legislature removed it. The second statue, which was erected on the grounds of the Manitoba Legislature in 1971, showed a naked Riel hemmed in by concrete towers, helpless in the face of oppression. In 1994, that statue was shuffled to the grounds of the Collège universitaire de Saint-Boniface. In 1996, sculptor Miguel Joyal devised the staid depiction of Riel in his statesman-like garb clutching a parchment in his hand. As scholar Albert Braz recounts, some critics view this Riel as boring: the Father of Manitoba's Confederation could be just another politician, albeit the "Great Red Father."[56] But Riel would likely have celebrated the statue. It captures the dignity that he fought so hard to obtain in his lifetime for himself and his Métis people.

NORQUAY AND HAULTAIN: TWO WESTERN CHAMPIONS AND A FUNERAL. LATE 1870 TO 1897

—

BY THE FALL OF 1870, Ottawa had secured its fractious empire. On paper, federal power over Manitoba and the North-West Territories was now supreme. On the ground, however, Ottawa was an absentee landlord. Its marbled halls and wood-panelled committee rooms were thousands of kilometres from its mosquito-infested frontier. Ottawa's instructions to Lieutenant-Governor Adams Archibald, who had replaced the inept Sir William McDougall, reveal an astonishing lack of knowledge. Archibald was required to investigate the "numbers, wants and claims" of the Aboriginals.[1] And he should provide a list of the laws, ordinances and regulations that the Hudson's Bay Company had enforced. What type of currency was circulating in the Territories, and how much was there of it?[2] How had the Company regulated liquor or maintained its roads? Did municipal governments even exist? "You will have the goodness to report also," the instructions added, "on the system of Taxation (if any) now in force."[3] The West was patently a mystery to its new masters. They could not know that, in Red River, Archibald was in fact preoccupied with a "very malignant" smallpox epidemic.[4]

Empire-building on the frontier was hard work. Ottawa wanted to keep the peace, oust the Aboriginals from their homes, slice those indigenous lands into homesteads, entice Anglo-British immigrants to farm, and improve the treacherous transportation. To implement that hard-nosed agenda, it needed to control the West's lands and resources because it was using them as a form of currency. To secure Rupert's Land, the federal government had been obliged to promise Crown land to the Hudson's Bay Company in addition to 300,000 British pounds. In the ensuing decades, it would also bestow land on the Canadian Pacific Railway as partial payment for spanning the West. It would auction off timber leases, and lease or sell mineral rights to cover its administrative costs. It would set aside lands to pay for school construction, and it would offer virtually free land to newcomers. Prime Minister Sir John A. Macdonald had an unshakeable vision of the West's role in Canada, and he needed the resources and lands of the West to realize it. In essence, Manitoba and the North-West Territories were colonies.

From the start of Ottawa's difficult reign in the West, there was resistance. The first rumblings of discontent in the 1870s were relatively mild demands for higher subsidies to compensate for those withheld resource rights. Manitoba premier Robert Davis was able to wring a few paltry concessions from Ottawa after he muttered that he might "have to look to" the United States if he could not get better terms.[5] But for the most part, the early leaders could barely get Ottawa's attention. By the end of that decade, however, the West's discussions with Ottawa would become more intense. As the November 1918 Conference approached, Western premiers could draw inspiration from two strong predecessors, John Norquay and Frederick Haultain.

Those men came from vastly different backgrounds. Manitoba premier Norquay was an English-speaking Métis, born in the Red River colony, the descendant of Scottish fur traders and Aboriginal and Métis residents. More than six feet tall and weighing in at three hundred pounds, he towered over most contemporaries. His career straddled the cusp of time when new immigrants were displacing the old Red River residents. By contrast, the elegant Haultain, who would become the first and only premier of the North-West Territories, was born in Britain as the son of an artillery officer; both of his grandfathers had served under the Duke of Wellington at the Battle of Waterloo. A University of Toronto graduate, he headed west, in 1884 and put out his shingle as the second attorney in the one-horse Alberta town of Fort Macleod.

These two political pioneers were not in power at the same time, which was surely to Ottawa's intense relief: together, they might have represented a serious political threat to the Rest of Canada's schemes for the West in their fierce defence of Western interests against federal intrusions. In the beginning, they did not push hard for control over their resources because their fledgling governments could not properly administer them. Prudence prevailed until Western populations had begun to grow much faster than Ottawa's subsidies. Then, in separate battles, both leaders cited Ottawa's control over *their* resources to support their claims for more cash. If they could not tap their resources for revenues, and if they could not sell their lands or lease their forests, they wanted more funds to handle their expanding populations. Ottawa might have the legal right to Western resources, as a constitutional provision in 1871 made doubly clear, but the two men still bristled when Ottawa handed out what they viewed as *their* governments' rightful wealth. Norquay finally demanded resource control in the mid-1880s, while Haultain waited until the end of the century, *after* he had secured responsible government. Both men chafed at the dependence of their jurisdictions.

Norquay and Haultain were not alone in their protests. In those first decades after Confederation, *all* provinces were asserting their rights against the dominance of Ottawa's strong central government. But the West had no natural allies — as the Western premiers would learn to their chagrin at the November 1918 Conference—because the other premiers maintained that their taxpayers had bought the West. Why would the other premiers empower the colonies that were buying needed goods from their manufacturers? Why would they support Western demands for higher subsidies when their taxpayers were still paying off the loan to buy the West? Norquay and Haultain had little political clout and no legal recourse. As the Manitoba premier would despondently confess near the end of his career, Ottawa was running the Western show, and he was almost a bit actor. Both men would eventually pay a high price for their defiance toward tough-minded prime ministers.

Federal politicians may have been in the dark about the realities of the West in the early 1870s, but they were also in a hurry. Ottawa's supposedly ironclad deed was useless if it could not exert its authority on the ground. American expansionists were still eyeing Prairie lands, and many dismissed the border as a mere inconvenience. Ottawa could not delay. The first priority was settlement, because homesteaders would be tangible proof of Ottawa's control. So in April 1871 the federal cabinet formally divided the West into tidy bundles. The basic unit of land became the section, which was one square mile or 640 acres. One quarter-section was 160 acres. Every thirty-six sections constituted a township, and those townships were given numbers that ran north from the Canada–U.S. border, and westward with the settlers. "This one sweeping decision" from Parliament Hill established all rural addresses in Western Canada with surgical precision.[6]

The surveyors' stakes spelled out Ottawa's supremacy over the lands and the resources. To fend off any challenges from Manitoba, the federal government even secured British consent to a constitutional act that confirmed Ottawa's right to carve provinces out of the Territories *and* to determine their powers. That act had an extraordinarily dismissive preamble. "Whereas doubts have been entertained" about Ottawa's powers, "it is expedient to remove such doubts."[7] That law confirmed the validity of the Manitoba Act, which put resource control in federal hands; it also reinforced Ottawa's right to administer the Territories. From then on, neither Norquay nor Haultain would have any hope of winning resource control in the courts.

Ottawa then set out to attract settlers, preferably hardy souls of Anglo-British heritage who would turn the sod and sow their grain. In 1872 Parliament offered clear title to 160 acres to any adult male or the sole head of any family who lived on the property for three years, cultivated a portion of it, and paid a fee of ten dollars. This offer seemed like a bargain too good to refuse. But Canada faced strong competition from the United States, which had offered free homesteads outside the original thirteen colonies since 1862. Communication was simpler south of the border as well, and transportation was much easier.

Canada's West was a harder sell, partly because it was so difficult to get there. Throughout the 1870s, some settlers took the wagon road that

Ottawa had hacked out of the muskeg from Port Arthur to Winnipeg. During those early years, that road was almost impassable, and repairs were ineffectual. One outraged group of travellers actually wrote to the minister of public works when they reached Red River, itemizing their troubles. Some passengers had taken twenty-three days to make the journey because of delays in setting out from Port Arthur. The accommodations en route were filthy. Food was often unavailable, and what was available often unspeakable. Passengers had to cart their luggage on and off wagons and boats, and stagger over the portages. They sputtered their indignation: "Is this the way to treat people most of them who have left comfortable Canadian homes and are now seeking to better their condition in the new country towards which they are traveling."[8] It was no surprise that many settlers straggled into Manitoba from the south instead, over the Canada–United States border, even though the last leg of that trip entailed a transfer from the comfort of an American rail carriage to a steamer or a lurching stagecoach.

The voyage was only the beginning of a settler's trials. The pioneer trails into the countryside were even worse than that road from Lake Superior. In 1873 missionary Lachlan Taylor struggled to get from Fort Garry to Norway House, where he tended to his Cree flock, or, as he patronizingly described them, "these simple-minded followers of the Lamb."[9] The Methodist pastor regarded his trip in a trader's cramped open boat amid swarms of flies and mosquitoes as the most uncomfortable voyage of his life. His journey into northern Manitoba covered roughly 360 miles, but it took thirteen-and-a-half days. "Time enough to cross the Atlantic," Taylor observed dourly.[10]

Given the poor trails and tricky portages, it was difficult to enforce the law in Western lands. But few immigrants would undertake to settle in disorderly communities, so in May 1873, Parliament established the fabled North-West Mounted Police. Within months, the Mounties had their first mission, when white wolf-hunters massacred Assiniboine Aboriginals near the Cypress Hills in what is now southwestern Saskatchewan. NWMP troops trekked westward over that treacherous route from Lake Superior, reaching Red River in late October of 1873. During the summer of 1874, almost three hundred officers rode into the southwestern Prairies where American traders seeking buffalo hides had been peddling rotgut whiskey—contaminated with everything from tobacco to red ink—to indigenous hunters. Those hides were being carved up into drive belts for Eastern industrial machinery, and the trade was decimating the herds. Before the Mounties could get their murderers, the wolf-hunters fled across the border, and two attempts to prosecute them failed. But soon afterward a network of NWMP posts dotted the West; Fort Walsh, which was built in 1875, was only a short distance upstream from the massacre site. Ottawa now had a visible police presence.

Despite such inducements, however, Ottawa still could not secure enough Western settlers. Canada needed an advertising blitz. In 1874 Ottawa and the four original provinces agreed that the federal Department of Agriculture would manage their scattershot promotions. Immigration agents fanned out across the British Isles and the United States, but their efforts on behalf of the West were largely wasted. Only a few hardy souls ventured onto the isolated Prairies in the 1870s, usually in groups—such as Mennonites escaping Russian persecution or Icelandic herders and fishermen fleeing economic adversity and volcanic eruptions. (Those Icelandic settlers were so isolated on the shores of Lake Winnipeg that they would establish a local council for the Republic of New Iceland.)

The German-speaking Mennonites first settled in blocks in southern Manitoba in 1874. They could not resist Ottawa's offer of virtually free land on which they could practise their religion without interference. Sixty years after Peter Barkman arrived on a slow ox-drawn cart, he would still remember his family's first crude home in a "soddy," a sod hut two feet below ground level. The walls and roof were made of thin tree trunks; clay filled the gaps. His family shared this palatial manse with its livestock. Barkman would recall a grasshopper plague and the intense cold of his first winter, and the time in the spring of 1875 when the community leader drowned in the Red River while heading to Winnipeg to restock supplies. Community elders earned cash by cutting trees into logs. Barkman could be brisk in his recollections: "Mr. Henry

Wiebe got lost in the blizzard and froze to death."[11] Another neighbour's hands and feet were badly frostbitten. Life on the land was marked by hardship but his community would eventually prosper.

In later decades, the Mennonites would regret their casual dispossession of their Aboriginal neighbours from their lands south of Winnipeg. But in the mid-1870s, they assumed Western land was theirs for the taking, largely because the Aboriginals and the Métis were not full-time farmers. Ottawa abetted that view. In 1872, before the establishment of the first settlements, a prominent Mennonite merchant from Ontario had written a pamphlet about Manitoba, which Ottawa distributed to the newcomers. Merchant Jacob Y. Schantz had praised the Métis as obliging and hospitable. The Aboriginals were "quiet and inoffensive. . . . The Indians who once enter into a treaty will keep it to the letter."[12] The message was implicit: the natives were harmless relics. The Mennonites accepted their plots, built homes, and plowed the fertile soil, oblivious to their struggling neighbours.

As settlers pushed westward, federal officials swept ahead of them, securing title to native lands. During the 1870s, Ottawa signed seven treaties with Aboriginal peoples, treaties that confirmed the federal government's title to lands across modern-day Northwestern Ontario and the central and southern portions of the Prairies. The Aboriginals signed reluctantly, but they had no choice: inter-tribal warfare had hindered the creation of strong alliances and smallpox had devastated their societies. Immigrants were already squatting on their lands, staking out boundaries, and sowing crops.

Worse, the buffalo were scarce because of the American market for hides. In the autumn of 1870, British army officer William Francis Butler had stumbled upon thousands of bleached buffalo skeletons on the ground between the Saskatchewan and Assiniboine Rivers. "There is something unspeakably melancholy . . . the Indian and the buffalo gone, the settler not yet come," he remarked.[13] It was the end of an era. By the mid-1870s, Aboriginal buffalo hunters could not find enough meat to fill their carts. Hide-hunters had decimated the herds, leaving thousands of carcasses to rot under the Prairie sun. "There was not the choice that we once had," Métis trader Peter Erasmus mused. "We had to be satisfied with anything that we could get."[14]

The changes meant that Western tribes like the Cree could offer little resistance to Ottawa's treaty-makers. The remarkable Erasmus, the child of a Danish father and a "mixed blood" mother, was at the Fort Carlton trading post on the North Saskatchewan River in 1876 when the Cree wrestled with their fate. As their interpreter, Erasmus heard Lieutenant-Governor Alexander Morris outline Ottawa's offer for the proposed Treaty Six. The legendary Poundmaker, who was then an ordinary band member, objected fiercely. How dare Ottawa offer 640 acres to each family when the Cree owned vast territories? "This is our land!" Poundmaker shouted. "It isn't a piece of pemmican to be cut off and given in little pieces back to us. It is ours and we will take what we want."[15]

The two Cree chiefs, Mista-Wa-Sis (Big Child) and Ah-tuk-a-kup (Star Blanket), and their bitterly divided peoples retired for private talks. The meeting stretched throughout the day. At the end, in a haunting elegy, Mista-Wa-Sis recalled the years when travellers could not move because great herds of buffalo blocked their passage, and he had to choose his campgrounds carefully to avoid the trample of sharp hooves. "I think and feel intensely the sorrow my brothers express." But, he asked Poundmaker and his supporters, "Have you anything better to offer our people?" He had seen the American traders with their long rifles and strong forts who offered rotten whiskey to the Western Blackfoot. Until the NWMP had arrived, anyone who had protested was shot. Meanwhile, the days of the U.S. Aboriginals were "numbered like those of the buffalo."[16]

Then Ah-tuk-a-kup rose to his feet and stood for one long minute with his head bowed. If only the Cree and the Blackfoot had been friends, he mused, they might have combined to secure a better deal. But it was too late, and the Cree signed. Throughout the 1870s and early 1880s, Ottawa officials would move most Western Aboriginals onto reserves. They would even withhold food from those who refused to live on their assigned plots. Ottawa was implacable in its determination.

While Ottawa colonized its new lands, the Red River government, which handled Manitoba and the Territories into the mid-1870s, could barely cope with the near-constant crises. The federal government rarely paid attention because the West was so far away from Ottawa's overworked and uninformed bureaucrats. Perhaps federal officials could focus only on the West's lands and resources (and not its people) in their single-minded dream of empire. Or perhaps Ottawa was complacent because it had quashed the Riel Resistance. Whatever the cause, the attitude became ingrained.

It seriously irked Western legislators. After a few years of Canada's Dominion, Westerners were more suspicious of Ottawa's intentions than they had been during the Red River Resistance. When Canada swallowed the West in 1870, the federal government had insisted on the right to approve *every* decision of the North-West Territories Council. But afterwards, federal officials could not deal with the backlog. In December 1874, irate councillors formally protested that the federal government had failed to accept or to reject any Council laws or resolutions dating back to September 1873.

The inertia created unimaginable difficulties. For one, that backlogged sheaf of ordinances included a Council request to establish postal communications between Fort Garry in Manitoba and Fort Edmonton in the North-West Territories. The service would entail eight 39-day trips per year each way at an annual cost of $18,000. "In Summer the service, each way, would be performed by two men and four horses," the councillors explained. "In Winter two men with Dog Sleds would be required."[17] The service would not pay for itself initially, they added, but it would encourage immigration and settlement. Their proposal was almost plaintive. Mail from home was far more than a luxury—for the lonely homesteader it was a lifeline. Still, federal bureaucrats dawdled.

The Manitoba and Territorial governments were struggling to bring roads, bridges and schools to isolated communities, but they could not get sufficient cash because Ottawa saw them as too inexperienced to spend the money properly. The federal government did not even establish an official Territorial Government until 1875. Thereafter, the Council of the North-West Territories would meet in the small Saskatchewan community of Battleford where a legislative building was under construction.

Two years later, on his Western inspection tour, Ontario MP James Trow harrumphed that Manitoba was mishandling its small federal subsidy. "Many little bridges were swept away by the spring freshets, and were not replaced during the whole season," he sniffed.[18] He would have been appalled to learn that the Manitoba Legislature was using a mace that a soldier had carved from the hub of a Red River cartwheel. Trow, meanwhile, dreamed of empire.

And then along came Norquay. The Manitoba premier was no rebel, but he was the first Western politician since Louis Riel to seriously challenge federal hegemony. Until recently, historians have rarely mentioned his heritage as an English-speaking Métis, resorting to euphemisms such as the "first native-born premier." Even Grant MacEwan, the former lieutenant-governor of Alberta and a proud recorder of the West's past, asserted that the Norquays came from Scotland's Orkney Islands, adding that the depth of Norquay's roots in the West was "not entirely clear."[19] In fact, Norquay was a child of the Red River fur trade, who grew up in a community based on farming and hunting. After a good education in Church of England schools, he became a teacher. Although he played no role in the armed resistance, he represented his parish in Louis Riel's Provisional Government in 1870, and on December 27, 1870, he won election by acclamation to the new Manitoba Legislature.

Norquay was only twenty-nine years old, but the world of his youth in the Red River area was already slipping away. Territorial society was changing fast. By the late 1870s, as new arrivals from Ontario pushed out the original Métis settlers, the West became a hotbed of racial and religious prejudice. The newcomers "changed Manitoba so drastically that, by 1880, the little province was unrecognizable—and unacceptable—to many of the pre-1870 residents, especially those whose first language was French."[20] Norquay's contemporaries, however, rarely taunted him about his ancestry, perhaps because of his huge size and biting eloquence, or because many viewed him as one of the original English-speaking settlers.

The engaging politician soon shook off such labels to become an advocate of compromise among the factions in the Manitoba Legislature.

In 1878 Premier Davis, exasperated by his fights with Ottawa, retired and moved to the United States. Norquay then led his coalition of French Métis and early settlers to election victory. The new premier loved to party. He could wear out a pair of moccasins when he danced until dawn, so he always tucked a spare set in his pockets. He was also a devout Anglican, and a resolute political lobbyist. After 1878 his federal counterparts rarely had a peaceful moment whenever Manitoba's annual subsidies came up for consideration. Norquay badgered Prime Minister Macdonald and his ministers with detailed breakdowns of his province's books. His principal demand was always that Ottawa boost his subsidies, and loosen its chokehold on his resources.

Norquay would get almost nowhere, however, because those resources were at the heart of an intricate set of federal policies to manage the empire. In 1878 Macdonald outlined his National Policy, which included steep customs duties on manufactured goods. Western homesteaders squawked at the high prices for their farm equipment: they were now a captive market because American manufacturers could not compete. But Central Canadians, whose wealth depended partly on manufacturing, were delighted with Ottawa's espousal of Western settlement, cross-Canada rail links and protective tariffs. Macdonald usually equated the manufacturers' interests with those of the Dominion, so he ignored Western complaints. Ottawa was tugging the West into the national economy through cold-hearted calculation *and* distracted neglect.

The rail line too would aggravate the West. When British Columbia had joined Canada in 1871, Ottawa had formally stipulated that construction would start within two years, and finish within ten. It had also agreed to pay $100,000 each year to the provincial government in return for the use of lands along the railway route. But Sir John A. Macdonald had lost power in November 1873, and his Liberal successor had dallied. The delay in building a railroad to bring immigrants and economic growth to the Pacific coast would mark the start of British Columbia's long-running grievances with Ottawa. In 1874, B.C. premier George Walkem actually travelled to England to ask the British government to put pressure on Ottawa. His journey took six weeks, and he hit a brick wall. Four years later, disgruntled members of the B.C. Legislature asked Queen Victoria for the right to withdraw from Canada.

Finally, Sir John A. got the message. When he slid back into power in September 1878, he went looking for builders, and in 1880, a syndicate that included Hudson's Bay Company powerhouse Donald Smith agreed to build the railroad in exchange for $25 million and twenty-five million Western acres. Ottawa could promise Prairie lands to the syndicate because, as designed, it controlled the resources in Manitoba and the North-West Territories. The situation was trickier in British Columbia, however, because that province managed its own resources. British Columbia earmarked a belt of land that extended twenty miles on each side of the planned tracks. Also, because portions of that land were unsuitable for settlement or were already settled, the province eventually transferred to Ottawa another 3.5 million acres in the Peace River District.

In February 1881, Parliament ratified the deal to build the Canadian Pacific Railway. Construction crews drove in the first spike near Bonfield in northeastern Ontario in the spring of 1882. Financier Smith would hammer in the last spike on November 7, 1885 in Craigellachie, British Columbia. At the time, British Columbia was pleased. Decades later, at the November 1918 Conference, the province would refuse to approve the Prairie premiers' demands until Ottawa returned any unused railway lands. In effect, in this endlessly complicated federation, British Columbia premier Oliver would treat the Prairie premiers as hostages.

It was the West's resources that would pay for the West's expansion. In 1881 Ottawa refined its settlement plans into an official six-page scheme that became the colonization template. The even-numbered sections on each township checkerboard were reserved for homesteaders; the odd-numbered sections within twenty-four miles of its main or branch lines were reserved for the Canadian Pacific Railway. (Another seven million acres were set aside for the construction of other lines.) There were exceptions: two sections in each township were earmarked as school lands, and when those lands were sold at auction, the proceeds would go into a trust fund that would pay for school construction. Two more sections were set aside for the Hudson's Bay Company, to help pay for the purchase of Rupert's Land. Ottawa cannily reserved town plots and any resource-rich lands for itself.[21] As always, the federal government guarded its own bottom line.

With every federal decision, Norquay fumed. He understood that Ottawa was tapping the West's resources to foster Western development. But as more settlers finally arrived in his underfunded province— the population had grown from 12,000 in 1870 to 60,000 in 1881—he could no longer meet their needs with federal subsidies. The premier resented Manitoba's position as the only province that could not sell its lands or lease its resources: British Columbia and Prince Edward Island controlled their resources and they had joined Confederation *after* Manitoba. By the early 1880s, as the railway advanced relentlessly across his lands, Norquay became more and more persistent. Ottawa's contract with the CPR had included a guarantee that no competitors could build branch lines running south of its main line for twenty years, but Manitoba voters wanted easier access to a railroad and competitive prices. When the Legislature defiantly enacted legislation enabling it to charter a railroad to build a branch line, Ottawa disallowed the law. Norquay's trips to Ottawa "increased in regularity—one in 1879, two in 1880, another in 1881 and one more in 1882."[22]

Premier Norquay did achieve one significant victory: in 1881, after repeated pleas on his part and from former premier Davis, Ottawa significantly enlarged the province. The Manitoba boundary now extended westward to the modern-day Saskatchewan border; the northern boundary curled around the huge lakes of the Winnipeg basin; and the eastern edge now ran along the still-undefined Ontario border. Manitoba's land mass was roughly ten times larger than it had been in 1870. Norquay was grateful and Manitobans hailed his achievement. But the settlers in those new areas contributed almost nothing to provincial revenues: they were simply an additional burden. Federal subsidies were calculated to meet the needs of established communities, and they could not come close to bringing roads and bridges, schools and agricultural assistance to a few dozen impatient settlers in the middle of nowhere.

The situation called for drastic measures. In 1882 the Manitoba premier wrote an exhaustive letter to the federal cabinet, stating that he wanted an increase in his federal subsidies. If Ottawa did not comply, he would demand control of his province's lands and resources. His annoyance sizzled on the page. Ottawa was liberally handing out his province's free land to immigrants, which was all very well for Ottawa because those immigrants were boosting federal customs and excise revenues when they purchased goods. But it was doing nothing for *his* books. No federal politician, however obtuse, could miss his message: Ottawa was giving away the land that rightly belonged to the province, and then leaving those settlers in the province's care. How could Manitoba help itself when Ottawa was helping itself to provincial resources?

A federal cabinet committee report in 1882 unexpectedly agreed with Norquay's position. Prime Minister Macdonald decided to pay a new subsidy of $45,000 each year in compensation for Manitoba's land and resources. That special subsidy, in lieu of resources, would set a controversial precedent. At the November 1918 Conference, the Western premiers would demand the return of their resources *and* the continuation of those subsidies. The other premiers would fiercely object, partly because their taxpayers had contributed to such subsidies for decades. A victory at the time, those federal payments to Manitoba would bring trouble down the road.

That same cabinet report, however, thwarted one of Norquay's key demands. The federal ministers insisted that Manitoba had not been cheated out of its patrimony. Provinces such as British Columbia had controlled their lands before Confederation, so those lands came with them *as their property* into Confederation. But "the whole of Manitoba was acquired by the Dominion by purchase from the Hudson's Bay Company, and thus became the property of the Dominion."[23] At the November 1918 Conference, when the Eastern premiers would maintain that they had bought the West, they would be parroting the argument that Macdonald's government had perfected.

———

Settlers were gradually trekking onto the rugged Western plains. James Clinkskill had heard glowing reports about Western lands in his struggling Glasgow grocery store, and had resolved to seek his fortune abroad. The twenty-eight-year-old Scot sold his business, wound up his old life, and in February 1882, sailed to Halifax. He reached Chicago by train, and finally arrived in Winnipeg in mid-March where "every hotel was filled to overflowing."[24] But his troubles had hardly begun. He floundered across the Prairies to claim a homestead in what is now southeastern Saskatchewan, where mosquito bites almost drove him mad, and severe thunderstorms flooded his primitive sod hut. His provisions rotted. He had to be constantly on guard because of the "terribly hard" characters that he encountered. The creeks overflowed, and his wagon became mired in the mud. By July 1882, only four months after his arrival, he could take no more. "Being city bred, I was ignorant of the rudiments of agriculture and I decided it was no use to start learning," he would recount thirty-five years later. "It would take too long before I could make any success of it."[25]

Clinkskill decided to become a merchant, and set off on another odyssey. He straggled back to Winnipeg, ordered goods, and then headed to Prince Albert to establish his store. But the steamer that was carrying his goods could not reach Prince Albert because the water level on the Saskatchewan River was too low, so the frustrated merchant left the non-perishables for the winter in Cumberland House, shipped the perishables back to Winnipeg, bought more goods, and then loaded his cargo into a rail car. Although the CPR tracks were steadily advancing, he had to arrange wagon transportation from the last station in Qu'Appelle to Prince Albert. A few months later, he moved to Battleford, which was then the capital of the North-West Territories, and set up shop. The logistics were utterly daunting.

Clinkskill's tales of those first years capture the perils of Western pioneering, and also the charms. One cold winter night, he was welcomed into an Aboriginal teepee, where he gratefully spent the night huddled with native trappers around a smoky fire. Another night, heading home from Prince Albert, he was so afraid of falling asleep in the intense cold that he resolved to ride all night. When his pony halted, refusing to move, Clinkskill realized that he was perched on the edge of a steep riverbank, teetering forty or fifty feet above the ice. He and his fellow settlers seized joy where they could throughout those long winters. They held so-called bachelors' balls in a log storehouse, dancing reels and jigs until morning to the music of Métis fiddlers. Coal oil lamps hung in sconces on the walls. "Many of the men were members of the Mounted Police in their red coats," Clinkskill recalled. "The girls were dressed in the greatest variety of colors: bright green, pink, brilliant red, and canary. . . . The matrons, of course, had their babies along."[26] To Clinkskill, who would become a member of the North-West Territories Legislative Assembly, Ottawa was then a distant non-entity.

But Manitoba's Premier Norquay literally could not afford to give up his campaign for federal attention. In March 1883, two months after he had won re-election, he pleaded poverty again, and he again pointed out that other provinces could use their resource revenues to cover their costs. It was a dramatic, almost biblical appeal. "Indeed, a large addition to the population of the Province would be nothing short of an evil in disguise . . . our revenue being out of all proportion to our necessary expenditure."[27] Norquay noted enviously that Ontario had pocketed almost $1.1 million in revenue from its resources in 1882. Ottawa, he wrote, should grant similar privileges to Manitoba. The premier added that the issue had been thoroughly discussed during the January provincial election, and "the unanimous opinion " of the voters had backed his demands.[28] He was aggrieved, so he was upping the stakes to demand control, but he was making little progress.

The fight continued throughout the first half of the decade. In May 1884, in a nine-page report, a federal cabinet committee outlined the reasons that Ottawa should keep control of the lands and the resources. Manitoba was a huge strain on the federal treasury, and Ottawa had incurred a "very large expenditure" for security because of the Riel Resistance. It had also picked up the tab for extinguishing Aboriginal title to the lands, and it was now supporting those Aboriginals. Ottawa needed the land for the railway and for other public works. The report added a pert observation that surely aggravated Norquay: the federal government would not have built more than a wagon road across the West without "the declaration that the lands of the North-West would bear a considerable proportion of the cost" of the railroad.[29] If Westerners wanted a railroad, Western lands and resources *had* to pay for the CPR.

The ministers complained further about all that Ottawa had to do. The West needed settlers, and the offer of free homesteads attracted them. Their children would need an education, and the sale of Western lands provided money for school construction. To sustain its huge immigration operation, Ottawa was heavily advertising in other countries, and prospective immigrants trusted it as a "well known and recognized Government" that had the right to give away the land.[30] If Manitoba really wanted to control its resources, it would have to maintain that offer of free land, and pick up Ottawa's tab for immigration operations. If Manitoba met those federal conditions, of course, it would be no further ahead financially. Then the cabinet committee turned the knife. When Manitoba accepted the extension of its boundaries in 1881, it had also endorsed the CPR's existing monopoly rights. As a sop to Norquay, the cabinet committee did call for an increase in Manitoba's subsidies, and Macdonald agreed—on condition that Manitoba recognize the deal as a *final* settlement. That stipulation was a stumbling block; the Manitoba Legislature would not approve the condition and the province lost out on the increased subsidies.

Two other incidents prolonged the premier's bad luck. In 1884 Ontario won a long-running boundary dispute with Manitoba in Canada's highest court, the Judicial Committee of the British Privy Council in London. Ontario premier Oliver Mowat had personally argued his province's claim to lands that had traditionally been part of the Red River colony. As a result, Manitoba lost its rich lumbering lands around Rat Portage (which in 1905 would change its name to Kenora). Ottawa also refused to stop disallowing provincial railroad charters. Sir John A. Macdonald wanted to ensure that the CPR did not go bust: without a transcontinental rail service, Central Canadian manufacturers could not reach Western markets. Ottawa had also guaranteed the

CPR's monopoly, and the prime minister wanted to protect Canada's international reputation for financial integrity. Norquay's persistence in chartering railroads would eventually be his undoing.

The premier resorted to flattery. He had officially aligned himself with the Conservatives in 1883, so he headed to Toronto in mid-December 1884 for a convention that celebrated the Tory prime minister's forty years in politics. Macdonald told the gathering about his Western dream. The CPR would be finished in late 1885, and surveyors had already reached Calgary, where "a town is growing. . . . It is a favorite spot for immigrants."[31] Settlers could get fuel from "one of the most magnificent coal countries in the world" at the foot of the Rocky Mountains.[32] They could get lumber from Ontario forests or the Rockies. Then Sir John spelled out why his West was won. "That is the country of the future [and] in that country there will be a sufficient market for our eastern manufacturers for years and years. The Northwest must be an agricultural country."[33] In the decades ahead, generations of Westerners would denounce that mercantilist bias.

At the convention, the Manitoba premier was unctuous. As the *Winnipeg Daily Sun* reported, "Hon. John Norquay Lays it on Very Thick."[34] He did not get far. The premier went from Toronto to Montreal for New Year's Eve, and then hunkered down in the capital, demanding more money and more power. The *Winnipeg Daily Sun* reported, "Norquay says he will not leave until he knows what the answer is."[35] The fun-loving premier filled his time with a busy social calendar: he dined at Government House and scurried back to Montreal for a banquet. But the delays were discouraging. He had apparently not listened when Sir John A. had outlined his dream.

In mid-January 1885, Norquay suggested a compromise: he would settle for an extra $55,000 each year in lieu of control of the resources; Ottawa could lower other subsidies in partial compensation. Ottawa agreed, but nevertheless it *did* lower other subsidies. Macdonald also offered to transfer control over the province's swamplands, along with 150,000 acres as an endowment for the proposed University of Manitoba. Ottawa insisted, however, that the Legislature had to accept the deal as a final settlement during its next session, or it would be null and void. This was exactly what Norquay wanted to avoid. The premier was in a fix. He had aroused voter hopes for resource control that he could not possibly have fulfilled.

On Monday, February 23, the agreement was tabled in the House of Commons. Ottawa politicians were unfocused: on the previous Friday, one MP had horsewhipped another in the Commons lobby after accusing his victim of living in open adultery with his sister-in-law. Much as the *Winnipeg Daily Sun* relished that tidbit, however, it was more scandalized by Norquay's deal. It summarized the agreement, and then added scathingly, "Only That, and Not One Thing More."[36] The newspaper's Ottawa correspondent noted that the subsidy payments would probably be worth more than the resource revenues but "on general principles, however, dissatisfaction is expressed."[37] On April 1, the secretary of the Western farmers' union even asked the governor general to refuse to ratify the deal until after an election because it would be "ruinous to the future prosperity of the Province."[38] To no avail. The Manitoba Legislature endorsed the deal after a stormy all-night session.

Norquay was stung. During the summer of 1885, in a pivotal speech, the subdued premier struggled to explain resource control to Portage la Prairie settlers. He was deeply hurt by Opposition charges that he had betrayed his province. This was not true: Ottawa had him in a legal choke-hold. "The province of Manitoba has a peculiar position," he argued.[39] The British Parliament had ratified legislation that admitted Manitoba to Confederation without resource control, so when Norquay hounded Ottawa about Manitoba's subsidies, he could only appeal to Sir John A. Macdonald's sense of fairness. "Ours was not a legal claim [but] if we expected what was fair, we had a right to claim it."[40]

That appeal had failed. Ottawa would transfer the resources only if the province agreed to offer free homesteads and to respect the lands earmarked for schools and the CPR, the premier explained. Meanwhile, Manitoba would have lost its extra subsidies in lieu of those resources. It would have been saddled with higher costs for public improvements, while Ottawa would have happily pocketed customs and excise revenues. Resource control would have been a lose-lose proposition. Instead, Norquay told his audience, he had settled for richer federal subsidies. The *Daily Manitoban* reported that the premier's audience in Portage la Prairie cheered. Perhaps. But they were surely uneasy at Ottawa's obvious triumph.

Norquay had used the lack of resource control to excuse his government's minimal services to its voters, and he had threatened Ottawa with his demand for resource control to secure higher federal subsidies. Bad move. His threats were empty because he had no political clout: Manitoba occupied only five seats in the 210-seat House of Commons. Worse, Manitoba voters had bought into Norquay's negotiating bluff: they *expected* that he would obtain resource control. Anything less— including higher subsidies — was a defeat. The premier chose a tough time to explain the legal facts of Manitoba's life.

Sir John A. Macdonald had his own problems, however. The West was still not a huge draw for immigrants. In March 1885, Agriculture Minister John Henry Pope presented grim numbers: more than 166,000 people had entered Canada in 1884, but only 30,000 newcomers had proceeded to Manitoba and the Territories. Compounding the problem, fully 6,600 Westerners had *left*, often for the United States. Pope eked encouragement from wherever he could, though. Eighteen artisans and their families from East London had settled in the North-West Territories, he boasted, even though "not one of the party had the slightest experience of agricultural life."[41] His optimism about their progress was foolhardy in the end, as many resilient Scottish crofters would fail in that same area of present-day Saskatchewan.

The fate of a plucky group of Jewish refugees from Russian pogroms was equally dismal. In 1884 the Canadian agent for a London-based Jewish charitable committee, Alexander Galt, had secured plots over the objections of prejudiced Dominion officials and settled twenty-seven Jewish settlers in southeastern Saskatchewan. The settlers' curious neighbours dubbed the colony "New Jerusalem," but the name did not fulfill its biblical promise of prosperity. The homesteaders were twenty-five miles from the nearest railroad. Their crops failed in the first year, and hail damaged them in the second year. The community rabbi was caught in a blizzard; his feet were so severely frozen that they had to be amputated. In September 1889, a fire destroyed their hay. Suspecting arson, the colonists abandoned their land and moved to Winnipeg. But their lack of farming experience was the real culprit.

Undeterred, Ottawa maintained its promotional campaign. In the southwest of the Territories, newcomers were building a cattle industry. Those ranchers included retired NWMP officers who had decided to settle in the West and American cattlemen who had crossed the border in search of bigger spreads. Americans would put their stamp on Alberta's character: U.S.-born farmers and ranchers, who would make up a quarter of the province's population in 1905, brought a "unique ideology of frontier individualism mixed with democratic populism [and] agrarian protest."[42] There was also an improbable group of Englishmen who usually wrote "gentleman" when they listed their previous occupations on their homestead applications. Many were remittance men. Some survived and even prospered. They staged nostalgic foxhunts, using coyotes as prey. But they could be difficult neighbours, and the Canadian wife of one rancher complained about their condescension. "There are so many Englishmen here . . . they nearly all have no tact in the way they speak of Canadians and Canada."[43]

As railway construction crawled westward, transportation became easier. In many regions, Western communities could now help each other, and the North-West Territories government could extend its administrative reach. New settlers followed the railroad, displacing Métis residents who had fled Manitoba during the 1870s. The situation became dangerous: once again, the surveyors overlooked Métis claims when they parcelled out homesteads, and the Métis could not seek a remedy from North-West Territories lieutenant-governor Edgar Dewdney, who was also the Indian Commissioner. The former British civil engineer was swamped with paperwork, and cantankerous. He did not like the living conditions in the Territories and he had scant interest in policy.

He preferred to dole out federal patronage and to allocate federal money. He also had a keen eye for his own welfare. In the early 1880s, he had bought land around Regina from the Hudson's Bay Company, and had then selected that town as the new Territorial capital. Despite fierce Opposition criticism in Ottawa, his selection of Regina was confirmed in 1883.

History has not treated him kindly. Dewdney followed Ottawa's instructions and denied food to Aboriginals who would not grow crops on their reserves. He also paid too little attention to Métis land claims. In mid-1884, Saskatchewan Métis asked Louis Riel to return from the United States to ease their plight. It was an unfortunate strategy, if only because Riel now considered himself a divine prophet. But the times were ripe for protest. Many Westerners, not just the Métis, were offended by the federal government's absentee decrees and resented federal officials, who knew almost nothing about local conditions.

By the time Dewdney finally acknowledged native starvation and Métis unrest in early 1885, it was too late. Riel set up a new Provisional Government in mid-March 1885 at Batoche in what is now central Saskatchewan. That government also drew up a Bill of Rights. The Métis wanted recognition of their land titles, and they wanted residents, not "disreputable outsiders," to fill local federal posts, Settlers should be regarded "as having rights in this country," and Ottawa should administer the land "for the benefit of the actual settler, and not for the advantage of the alien speculator." They also sought more liberal timber regulations. Most important, they demanded the immediate creation of the provinces of Alberta and Saskatchewan "so that the people may be no longer subjected to the despotism of Mr. Dewdney."[44]

It was a none-too-subtle declaration of war on Ottawa and its bureaucrats. It was also a shrewd ploy: as Riel understood perfectly, all provinces, with the exception of Manitoba, controlled their own lands and resources. Ottawa quickly dispatched troops from Winnipeg and Eastern Canada, but those forces arrived too late. On March 26, Métis forces clashed with the North-West Mounted Police and white volunteers near the community of Duck Lake. Twelve policemen and volunteers died, along with five Métis. On April 2, the Cree band at Frog Lake killed nine people to protest their lack of provisions. On April 24, Métis sharpshooters trounced the Canadian militia at Fish Creek. But they could not hope to win the conventional war that Riel insisted upon fighting. On May 12, the Métis in Batoche surrendered to the commander of the Canadian militia, British General Sir Frederick Middleton, and three days later, Riel himself surrendered. On May 25, Cree leader Poundmaker in turn surrendered to Middleton at Battleford. Settlers turned on their Métis and First Nations neighbours. Troops looted and burned Métis settlements, and on November 16 Riel was hanged for high treason. Western lands and resources remained firmly in Ottawa's hands.

The lurid news reports temporarily deterred immigrants, and the rebellion was a public relations fiasco for Ottawa's promotional campaign in the United Kingdom and the United States. The federal government would also discover that the rebellion had fostered another unexpected and unwelcome phenomenon. Hereafter, in their souls, many white Westerners, who had deplored Riel's behaviour and looked down their noses at their Aboriginal neighbours, would come to regard the North-West Rebellion as a reaction to Eastern neglect. Whenever Western farmers tried to buy the lots adjoining their farms, they would fume about Ottawa's remote land registry. They would also resent the CPR's tax-free status and its right to huge tracts of Western land, which often remained unclaimed and unsold. Ottawa's long-distance management of the West's lands and resources would foster the growth of a fierce regionalism.

Norquay had strongly supported the federal government during the North-West Rebellion. But he could not wring more money from the increasingly impatient Sir John A. Macdonald; nor could he secure more concessions. Worse, his Liberal opponents in the Legislature strongly espoused provincial rights. As Norquay's political life grew more uncomfortable, he took chances, and defiantly issued more railway charters. In early 1887, on the strength of informal assurances of Ottawa's financial backing, Norquay transferred $256,000 to the Hudson Bay Railway to link Winnipeg with a northern port. That summer, he also turned the sod for the Red River Valley Railway, which would run from Winnipeg to the Canada–United States border. Macdonald was incensed. "Your bankrupt population at Winnipeg must be taught a lesson even if some of them are brought down to trial to Toronto for sedition," he told Manitoba lieutenant-governor James Cox Aikins.[45] The legislation to charter the Red River Valley Railway was promptly disallowed. Norquay resolved to defy Ottawa, but the CPR secured injunctions to stop Red River Valley work. Ottawa also made sure that the financial markets did not lend money to the defiant province. Ottawa had finally checkmated the Manitoba premier.

Across the nation, however, the drive for provincial rights was gathering steam. Such sentiments were inevitable, particularly among the four provinces that had joined together to create Canada in 1867. Sir John A. Macdonald was an unabashed centralist, and he was rarely hesitant to disallow provincial legislation that conflicted with his vision. Restive premiers were asking themselves: Had they not created the federal government? Was it not their creature? Nova Scotia had tried to leave within months of its entry, but Macdonald had appeased the moderate anti-Confederates with debt relief and a subsidy increase. By 1887, twenty years after Confederation, the provinces had accumulated lists of major grievances and minor requests, such as a call for uniform laws to recover commercial debts.

But the *big* issue—the one that nibbled at the soul of every premier—was the size of the federal subsidy. The Fathers of Confederation had made a trade-off: Ottawa would pay a fixed amount in subsidies and collect all customs and excise duties, which would be the main source of federal funds. At the time, it seemed like a fair deal. Unfortunately, the exact amounts per person were spelled out in the Constitution, and Ottawa used the 1861 census for many provinces to determine the number of residents. The more recent provinces had their per-capita subsidies limited by population caps. That was a mistake because both their economies and their populations were growing. The premiers were now literally paying for their predecessors' lack of foresight. In 1867

provincial subsidies had amounted to twenty per cent of federal revenues, but by 1887, those per capita provincial subsidies were only thirteen per cent of provincial revenues. Quebec's Liberal premier Honoré
Mercier tackled the issue: he invited the first ministers to a conference
on provincial rights in Quebec City.[46] No fool, Macdonald declined to
attend. But, the first interprovincial conference, from October 20 to
October 28, 1887, met in Quebec City.

Norquay unwisely joined this bandwagon, even though he was in a
perilous political position. He was the only Conservative premier at the
gala event—Prince Edward Island's Tory premier had declined, as had
British Columbia's gravely ill premier. Norquay's presence added legitimacy to the proceedings. In the official conference photo, Norquay
sprawls across his ornate chair, his watch chain looping across his straining
vest, his collar crisp, his fingers tapping impatiently. He looks preoccupied.
Perhaps he was worried about personal financial losses that had left him
with a debt of $1,500 to his own government. Perhaps he sensed that those
fierce assertions of provincial rights and powers would eventually become
a double-edged sword for the West. The premiers unanimously adopted
twenty-two detailed resolutions, including demands for higher subsidies.
They also resolved that Manitoba had the right to charter a railroad within
its original boundaries, and that Ottawa had encroached on that right
when it disallowed the charter for the Red River Valley Railway. In an
ominous sign of the fights ahead at the November 1918 Conference, New
Brunswick did *not* support this resolution on Manitoba's rights. Nonetheless, the premiers sent their demands to Ottawa.

The federal government was not amused. Macdonald had finally lost
patience with his fellow Conservative, and he sprang an elaborate trap.
Norquay had transferred $256,000 to the Hudson Bay Railway without
cabinet approval because his treasurer had telegraphed from Ottawa
that he had secured Macdonald's verbal assurance of financial support
during a visit to Ottawa. The prime minister had even acknowledged
the value of the proposed line to Hudson Bay. But Treasurer Alphonse
LaRivière had *not* secured a written promise that Ottawa would transfer
Crown lands to the province so that Norquay could sell those lands to

cover this $256,000 deal. Shortly before the 1887 Interprovincial Conference, Ottawa refused to transfer the lands. When reproached, Macdonald claimed that LaRivière must have been dreaming.

The Provincial Treasury was under water, and the scandal was ruinous. The Provincial Liberals charged that Norquay should have secured the title to the federal lands before he transferred the cash to the company. Macdonald would not budge, despite the premier's desperate appeals. Norquay resigned in late December 1887, and in late January 1888, his Conservatives lost power. Five months later, Norquay died, probably because of appendicitis. The broken man left a small estate. In December 1888, the Judicial Committee of the British Privy Council ruled that Ontario, *not Ottawa*, had the right to control the resources within the lands that Ontario had won from Manitoba in their boundary dispute. Norquay would have despaired. Perhaps historian W.L. Morton best captured the mood of those dark days: "The tradition of grievance which was forming was reinforced."[47]

Ten months after Norquay resigned, the man who would assume his role as the bane of prime ministers became the leader of the Advisory Council in the North-West Territories Assembly. Frederick Haultain had been in the West for only four years, but he had already established a flourishing law practice in Fort Macleod, roughly 165 kilometres south of Calgary. Haultain had left Ontario to shake off his sorrow at his father's death. He had selected southern Alberta as his new home because he had read tales of NWMP exploits around Fort Macleod, and because a former classmate had settled there. The economy-class train trip from Toronto to Calgary had lasted four long days. Then Haultain had taken a jolting stagecoach through lands that the buffalo had once grazed.

The West was a tonic. While Manitoba premier Norquay lobbied Ottawa for increased subsidies, Haultain learned about Western grievances around the dining room table in Henry "Kamoose" Taylor's hotel. It was a remarkable classroom. Each bullet hole in the walls and tables of the barroom had its story. The guests paid extra for candles and soap. By hotel rules, those guests had to be up at 6 a.m. because their sheets were needed as tablecloths. But the folks in Fort Macleod were feisty, and they had headed west from regions where they were accustomed to having a voice in their government. Throughout 1885, as the Northwest Rebellion raged around them, those guests armed for battle against the Aboriginals. But they also discussed their impatience with far-off Ottawa. After Riel's defeat, such talk inspired the elected NWT council members to deliver an angry denunciation of Ottawa's despotism to the territorial capital of Regina.

The elected council members then drafted a fierce twenty-one-point Bill of Rights, which was based on the lists formulated at Haultain's table and which demanded that the Territorial Council assume authority over federal money earmarked for the West. That key request would preoccupy Haultain for years. Inspired by the diners at this dubious establishment, the members also insisted that settlers should have the right to cut trees on Crown land for lumber and fuel. Ottawa should also respect the claims of the existing settlers and squatters who were on the land before Ottawa's surveyors arrived. There was very little difference between the demands of the Riel Rebellion and those of the gathering at the Fort Macleod hotel. Three council members actually ferried their list to Ottawa. To no noticeable effect.

Their demands would eventually escalate far beyond that modest list, however, and Haultain would become their champion. In early 1887, before his thirtieth birthday, he was elected to the Territorial Council. In 1888 he won a seat by acclamation in the Legislative Assembly that replaced the Council. That fall, the new lieutenant-governor, Joseph Royal, selected Haultain as a member of his Advisory Committee on finance; Haultain immediately became chairman.

Thereafter, Haultain was the insistent voice for responsible government in the North-West Territories. In his first decade in office, he did not demand provincial status or resource control. While other Westerners were vexed with Ottawa's disposal of their assets, Haultain shrewdly demanded control of the money that Ottawa spent on Territorial services. He wanted responsible government with individual departments; ministers would have to account to the Assembly for every dime of departmental spending. And he used every procedural trick over the ensuing nine years to win it.

The result was an endlessly inventive drama. Haultain had no political clout: in the late 1880s, the North-West Territories had only four seats in the 215-seat House of Commons. But he never relented. If Ottawa insisted on controlling the resources, he argued, the elected Territorial representatives, not the lieutenant-governor, should control federal grants. Year after year, under Haultain's leadership, the majority of the Assembly asserted its ascendancy over far-removed Ottawa, which Royal, the long-suffering lieutenant-governor, came to personify.

Royal was a Quebecer, born in 1837 to struggling, illiterate parents. After Roman Catholic officials put him through a Jesuit-run college, he became a lawyer and then a journalist who wrote for conservative Catholic publications. In 1870, at the invitation of Bishop Alexandre-Antonin Taché, he moved to Manitoba, where he eventually entered Norquay's cabinet as an unofficial representative of francophone Catholics. Then he headed to Ottawa as the MP for a Manitoba riding. In 1888, in recognition of his loyalty during the Riel Rebellion, Macdonald sent him to Regina. The North-West Territories would be no reward: they would test his resolve to live a prudent and patient life.

Westerners may have forgotten the tales of Haultain's skirmishes, including those with Royal, but they were celebrated at the time—and his strategy of resistance entered Western lore. In the beginning, there was peace: Royal started his five-year term with a generous gesture by consulting the Advisory Council on his estimates. The councillors pored over the spending plans for the small amount of local revenue *and* for the money that the Dominion transferred each year to the lieutenant-governor. Royal would never recover from this gracious act, however. Haultain hailed the arrival of a new right, which the Assembly asked Ottawa to formally recognize. Teacups rattled on Parliament Hill. But Interior Minister Edgar Dewdney, the same feckless politician who had contributed to the 1885 rebellion, ignored the request.

The crisis escalated. In October 1889, the councillors, including Haultain, abruptly resigned because Ottawa did not recognize their right to oversee federal spending. Royal selected another four councillors. Sir John A. Macdonald ordered Royal to roll back the clock and reassert his right to allocate Ottawa's cash unilaterally. But the prime minister's letter was private, so the lieutenant-governor could not cite it. Assembly members therefore blamed him, and Royal was trapped. The Assembly passed a vote of non-confidence in the new councillors. They resigned, but Royal refused to accept their resignations. Haultain charged that the replacement councillors were "tottering pillars . . . who should not be foisted on the House."[48] The Assembly majority then refused to consider the estimates because they only accounted for local revenues. The new councillors again resigned, and this time Royal accepted. The two sides were at a stalemate.

One problem remained: who would authorize the spending of local revenues? The Assembly unilaterally created its own advisory council of two members who could approve local spending. Royal wearily assented. The Assembly then tartly reminded Ottawa that since Territorial resources put cash in federal coffers, Ottawa had "a direct financial interest in good government and public improvements" in the Territories.[49] It was a threat: Ottawa's resource revenues would falter if the Assembly could not meet the settlers' needs. Ottawa should therefore put more cash under the Assembly's control. It should also replace its annual grant with a fixed subsidy, so the Assembly could make long-term capital commitments. Ottawa largely ignored the upstarts, although the Justice Department did call for the disallowance of the two-member Advisory Council.

The beleaguered Royal was left to improvise. The lieutenant-governor was engaging, diplomatic and clearly in sympathy with the push for greater local control. But, by the late 1880s, he was also exasperated, caught between his oblivious masters in Ottawa and an Assembly majority under the leadership of the implacable Haultain. In January 1890, Royal brashly selected *another* advisory council from the Assembly minority. Now Haultain had an easy target. When the Assembly convened in the fall, the deadlock continued. In late October, the majority, under Haultain, refused to allow the new councillors to introduce any motions or legislation. Mere days later, Haultain's forces claimed the right to control the Advisory Council and oversee the spending of Ottawa's grants—and the "Haultain Resistance" was underway. On several occasions Royal wrote to Prime Minister Macdonald, outlining the impasse and damning Haultain. But Sir John A. Macdonald remained unflappable and unhelpful: the events in distant Regina were like the buzzing of a gnat. The session eventually adjourned, and in March 1891, confirming the Old Chieftain's enduring popularity, the federal Conservatives swept all four Territorial seats.

Then everything changed. On Saturday, June 6, three months after that election, Sir John A. Macdonald died after a devastating series of strokes. In Winnipeg, as across the nation, church bells rang dolefully. The *Calgary Daily Herald* rushed a special "very large" late-Saturday edition into print, which the hundreds of people gathered outside its doors snapped up. By Sunday, "scarcely two persons met on the streets without referring to the death of the beloved statesman."[50] An estimated fifty thousand people massed in the streets of Ottawa for the funeral. A day later, the procession to a Kingston cemetery was more than two miles long. And even in the remote North-West Territories, almost every small community held a memorial service to mourn the passing of the man who had tugged them into Canada and his National Dream.

Macdonald's successor, Montreal lawyer Sir John Abbott, assumed his duties with reluctance. Three months later, as his government plowed through a backlog of legislation, it added an incredibly vague clause to the North-West Territories Act: Assembly members could oversee funds that "the Lieutenant-Governor is authorized to expend by and with the advice of the Legislative Assembly or of *any committee* thereof."[51] That could have meant anything, or nothing. A month later, on Halloween, Haultain's majority triumphed in the Territorial elections, and when the Assembly reconvened, Haultain exploited that vague wording.

He secured Assembly consent to create an Executive Committee with powers that went far beyond that reference to "any committee." Fed up with the deadlocks, Royal agreed. In early December 1891, Ottawa allowed the Assembly to allocate the money that it had not already earmarked for that year. Two months later, however, in a futile bid to

reassert his primacy, Royal sent his estimates to Ottawa without consulting the Assembly. Haultain, in turn, asked Ottawa to put its grants
under the Assembly's control. Ottawa agreed. Royal had rolled the
dice, and lost. He was edged aside, and lapsed into passivity. When his
term ended on October 31, 1893, he was exhausted. He returned to
Montreal as the editor of a short-lived newspaper. His last years were
spent "in rented apartments, moving frequently, and [he] ended up in
a boarding-house."[52]

His successor was an Ottawa journalist and former MP, Charles
Herbert Mackintosh, who regarded himself as a gift to Territorial society. Within weeks of his arrival, he airily informed Prime Minister Sir
John Thompson, who had replaced Abbott in December 1892, that the
Executive Committee was "very reasonable," although he could not
understand why the legislators endlessly parsed the Constitution. But he
took a dim view of Haultain—who would eventually and unexpectedly
become his son-in-law in 1906—as "rather inclined to pig-headedness."
He was further unimpressed: "I fear power was yielded and privileges
granted before the proper time."[53] He even compared Territorial politicians to formerly meek maids who defied their husbands when married.
Such patronizing attitudes would blight Ottawa's relationship with the
West for decades.

But Mackintosh did capture one side of Haultain's character—his
impatient conviction about the rightness of his actions—that would
eventually contribute to his downfall. Merchant Clinkskill, who scruti-
nized Haultain during his decade in the Assembly, detected abilities that
set him "head and shoulders over any man that has taken part in political
life in the West." But Clinkskill had one reservation: "Able, clever and
resourceful as Haultain was, he had a failing of being too apt to consider
and decide matters himself."[54]

Haultain certainly gave as good as he got amid many provocations.
He wanted responsible government, in which councillors were in
charge of individual departments and had to account for their ministry's spending. Remarkably, in a significant scuffle with Haultain,
Interior Minister Thomas Daly could not grasp the difference between

responsible government and provincial status. In June 1894, Daly wrote peevishly, "I notice, however, that in one voice you people shout for Provincial autonomy, in another you say you don't want it."[55] The mere use of that undignified phrase "you people" must have raised Haultain's hackles. Haultain replied, "We have never shouted for provincial autonomy, but have consistently said we do *not* want it. . . . We want the autonomy of any self-governing organization."[56] Daly snapped: "It seems to me that you have all the control of your affairs that is necessary under the circumstances."[57] Haultain replied that Daly was confusing form with substance. "We *do* want responsible Government, not provincial Government."[58] Those italics were his. It was an astonishing exchange. Ottawa had not moved very far from Sir John A. Macdonald's original conception of the Territories as a colony. And the Rest of Canada had certainly not challenged that view.

But the West was changing with confounding rapidity. In the federal election of June 1896, some of the Mennonites in Rosthern, Saskatchewan, hesitantly went to the polls. As a Mennonite would later recount, Sir Wilfrid Laurier was the Liberal candidate in his riding, and he was proud that "the votes of the Mennonites contributed to his success."[59] (Laurier later won a by-election in Quebec East, and he would choose to represent that riding.) The Liberals swept into power with a majority, and Laurier became Canada's seventh prime minister. (Remarkably, there had been four Tory prime ministers since Sir John A.'s death.) Laurier's interior minister, Clifford Sifton, resolved to inject new life into the West after decades of sluggish population growth. He peered beyond the United States and Western Europe, reaching out to Eastern and Central European peasants, who were mired in poverty in feudal economies. The promise of free land in a new world was irresistible to many, and emigrants packed their bags.

This new wave of settlers would in turn need more services than the Territorial Assembly could afford. Assembly members fretted about their small grants and uncertain legal powers. In October 1896 Haultain's ally, James Ross, complained that Ottawa treated the Territorial government like a branch of the Interior Department. If Ottawa did not increase its subsidy soon, "the government would come to a stand-still."[60] Ross wanted complete control of the lands and resources, but Haultain was more cautious: he still viewed the demand for resource control as a ploy to extract more federal money.

The Territories' financial woes could not wait. In the spring of 1897, Haultain and Ross trekked to Ottawa to press their demands. Ross had Liberal connections, so the duo received a cordial welcome. Sifton agreed to add $40,000 to the Territories' future annual grants. More remarkably, in June, Laurier granted responsible government. As the first premier of the North-West Territories, Haultain organized his cabinet on a departmental basis. But his government continued to be highly dependent on Ottawa for revenues. Laurier had retained the system of grants—so there was no financial certainty from year to year.

Across the West, the Old Chieftain's dream of agricultural prosperity was becoming a reality. Once-isolated pioneers could now ship their crops to nearby towns. Neighbours were reaching across ethnic lines. In June 1897, the Mennonites in Rosthern celebrated Queen Victoria's Diamond Jubilee, marking her sixty years on the British Throne. The residents decorated the town, and organized a sports day followed by a dance. Farmer Tobias Unruh ferried an organ on his wagon for nearly twenty miles to the festivities. English-born farmer Seager Wheeler played the concertina, while his brother Percy strummed the foot-stomping banjo.[61] The joy was infectious. New settlers wrote home about the promise of Canada, and more immigrants arrived with every train.

But still Haultain worried. His government could barely cope with its existing population. How could it possibly provide services to thousands of newcomers who were forging into the wilderness to establish remote communities? The uncertainty and the unfairness were getting to the premier. He was edging closer to a war with Ottawa that he could not win.

HAULTAIN LOSES SELF-CONTROL—AND LAURIER KEEPS RESOURCE CONTROL. 1898 TO 1905

—

ALONG THE CIRCULAR CORRIDOR OUTSIDE THE OFFICE of the premier in the Alberta Legislature hangs a portrait of an elegant Frederick Haultain. The first and only premier of the North-West Territories faces the viewer. He is in formal dress, his black bow tie looped around the upright collar of his white shirt, his silken lapels glimmering. A gold watch chain crosses his vest. A cigarette smoulders in his right hand.[1] His gaze is affable but coolly appraising. He seems very sure of himself. There is even a whisper of vanity in the pose. The painting is dated 1912, seven years after Ottawa carved the provinces of Alberta and Saskatchewan out of the North-West Territories. There is no hint in Haultain's eyes of regret or thwarted hopes. Perhaps enough time has passed since Sir Wilfrid Laurier's Liberal government exacted its fierce revenge on this provincial-rights advocate. Perhaps the debonair premier is content with his role as an unofficial Father of Confederation, who fought Ottawa to a draw in his campaign for provincial status, until he fumbled into the home stretch.

It was an extraordinary contest. Haultain was made of stern stuff, raised in a family with a strong military tradition. His father was a British artillery officer who later represented Peterborough in the Province of Canada Legislature. Young Frederick was brought up a Presbyterian but at an early age switched to the Anglican Church, which had once aspired to the unofficial status of state religion. His family motto—"He who commands himself, commands enough"—surely inspired his lifelong reserve, although he would lose that self-control during his last years in power.[2] He was the cultured epitome of a White Anglo-Saxon Protestant, who expected to be treated with respect. During the years of legal cat-and-mouse with federal Conservative governments of the 1880s and 1890s, Haultain had coolly played a long game. In those early years, his demand for control of the land and the resources had been a threat designed to pry more money out of the Territories' federal overseers, and he had made steady progress.

Then, on the cusp of the new century, his world turned upside-down. In 1897 the Liberal government in Ottawa had agreed that Territorial ministers could oversee the spending of federal grants; those ministers were accountable for their decisions to the Legislature. But, as Haultain soon discovered, he was now in the worst of all possible worlds, with growing responsibilities and not enough cash to meet them. His demands for provincial status and resource control grew more insistent as his responsibilities escalated and Ottawa penny-pinched. The financial squeeze fuelled Haultain's exasperation.

But it was Ottawa's casual responses to his demands that eventually sent the premier over the edge. Haultain's ongoing battles with Interior Minister Sir Clifford Sifton anticipated the West's battles at the November 1918 Conference. If Haultain took too little account of the challenges that Ottawa faced in Western settlement and national development, Sifton paid far too little attention to the elected representatives of the West's struggling peoples. The minister was often distracted and almost disrespectful. His encounters with Haultain were cycles of raised expectations and dashed hopes, as Ottawa radically altered the world beyond the Legislature's doorstep.

During those early Laurier years, Ottawa effectively espoused Sir John A. Macdonald's approach to the West: the federal government needed to control Western lands and resources because the territorial government might allow parochial concerns to interfere with the national interest. Sifton reasoned that Ottawa had bought the West—just as the six premiers from the Rest of Canada would argue at the November 1918 Conference on behalf of their taxpayers. Accordingly, the federal government had the right to use its property to promote its Big Picture. Even that great con-ciliator, Sir Wilfrid Laurier, could not reconcile Sifton and Haultain, largely because he and his cabinet ministers agreed with Sifton's development strategy. In these early years, the West's battles were with Ottawa—not with the other six provinces—but federal ministers reflected the views of their constituents in the Rest of Canada. The West was Canada's colony.

Laurier assumed that Western estrangement would dissolve as Ottawa brought development to its new territory. Instead, Ottawa's control of the lands and the resources would become a Western symbol of its second-class status—although Laurier would retain the support of Western voters. The implacable Haultain could not deter the reso-lute Sifton, and he could not stop himself from alienating the generally tolerant Laurier. His fate would ensure that generations of Westerners would regard him as a martyred hero in the struggle for Western resource control.

When Sifton breezed into office in 1896, a tornado struck his stodgy ministry. The minister of the interior concluded that Ottawa would never be able to populate the West if it kept focusing its promotion on the United States, the British Isles and Western Europe. Too few people in those nations were sufficiently dissatisfied with their lives to uproot themselves. Furthermore, some Western European nations were actively discouraging emigration. Sifton broadened Ottawa's advertising efforts to include Central and Eastern European farmers, who grabbed at the chance to escape the remnants of serfdom and to own 160 acres of land. The effect was electric. In 1896, roughly seventeen thousand immigrants set foot in Canada. Between 1896 and 1914, more than three million people arrived, twice the number who had come during the previous thirty years. Roughly thirty per cent of those new arrivals took up home-steads in the West, while another twenty per cent found work on the railroads or in the resource industries.[3] The influx was almost over-whelming. Both the North-West Territories government and, to a lesser extent, the Manitoba government were swamped with settlers.

Governments could do little to ease the transition of those exotic

newcomers. Most could not speak English. To the dismay of many established settlers, they even looked different, with their shabby clothes and
frightened eyes. Sifton's agents met the immigrants at the ship, ferried
them to the West, provided free lodging while they selected their 160-acre
plots, and often simply abandoned them to their fate. It was a tough ini-
tiation for the newcomers, but it was also a visionary development
scheme. And, in the beginning at least, it required federal resource control. The elaborate operation depended on Ottawa's ability to allocate
plots from its land registry grid, while covering a portion of its administrative costs with revenues from the West's resources. The federal Liberals
trumpeted the benefits that centralized leadership brought to their promised land. They disregarded warnings about the perils of remote rule, and
brushed aside squawks from Manitoba and the Territorial government.
In their narrative, federal resource control was for the West's own good.

On the ground, as ragtag settlements popped up across the West,
that story did not sound so sweet. Although Haultain had obtained
responsible government, his government did not receive the fixed
annual subsidies that the Rest of Canada pocketed. He depended on
Ottawa's whims for the annual grants that funded the bulk of his budget,
and those grants did not keep pace with the skyrocketing population.
This fresh influx of newcomers in their turn needed roads, schools,
bridges and hospitals. They wanted links with the world beyond their
villages so that they could sell their agricultural produce. In the early
years of responsible government, Haultain was circumspect. He muttered about provincial status and resource control, but he did not
seriously pursue these goals until Sifton shrugged off his pleas for more
cash once too often. Even then, although his demands would grow pro-
gressively more emphatic, he would never be as strident about resource
control as many Westerners. Not even at the end, when he would go
way too far.

In the beginning of the Laurier years, Haultain was preoccupied with
the scramble to educate those alien newcomers and, although he would
not have used the term, to "Canadianize" them. He viewed the federal

government as an irresponsible recruitment agency. But his sarcasm about Ottawa's disregard for his government's challenges was lost on Sifton. Every day, trains chugged into the West, disgorging nearly destitute immigrants headed to their promised lands in Manitoba or the Territories.

The sight even unsettled those who should have been kinder. One Ukrainian Catholic priest, Father Nestor Dmytriw, could scarcely conceal his horror and disdain during a journey to visit his flock in the Lake Dauphin district of Manitoba in 1897. The train from Winnipeg was packed with passengers from different lands, including a Ukrainian woman in filthy clothes and her husband, "a tall, thin man with a pale, depressed, melancholy countenance."[4] From the Dauphin station, the priest struggled across the countryside in a wagon and then on foot, consoling himself with the thought that he was emulating the Apostles' travels. He became lost in forests, terrified of the dark night. He sheltered with Ukrainian peasants who could only afford to feed gruel to their children. He concluded that free land, even though it was undoubtedly fertile, was worth little to a penniless newcomer. "Let no one set out for Canada without a cent . . . here no one gives anything away free," he would write to a Ukrainian daily newspaper in the United States. "For a start in homesteading and for living expenses until such time as one is able to live off his land, he needs at least $300. Without that small capital, one either perishes or is doomed to a life of penury."[5] Sifton's Canadian Dream could be a tough sell.

New arrivals often had to find jobs before they could afford to farm their 160-acre plots. Ukrainian Maria Adamowska settled as a child with her family in Canora, Saskatchewan, in 1899. Her father had brought some money with him, so he bought a cow and a horse. But their resources were meagre, and she spent her first winter in a cave that her father dug into a riverbank and covered with turf. The family nearly starved to death. The following summer, her father found work on a farm near Brandon, leaving Adamowska and her mother to clear and plow almost four acres of land. Her mother planted wheat seeds from their homeland, and waited for the harvest. "We lived on milk," she would later recount. "One meal would consist of sweet milk followed by sour milk; the next meal would consist of sour milk followed by sweet milk. We looked like living corpses."[6] The newcomers were like abandoned children on the Territories' doorstep.

Far to the northeast of the premier's Fort Macleod constituency, Polish peasants, including some who had never left their native village before their odyssey, were settling in remote clusters. Their memories of those first days retained their bitter edge. Jan Plachner had left Poland to work on Rocky Mountain construction sites, and then used that cash to purchase a homestead near the modern-day Alberta community of Starko. Once settled, he sent for his wife and son. The family's home was a sod house in the woods. Plachner and his son spent hours chopping down trees, prying out roots, and clearing the ground, acre by painful acre. The summers could be scorching, and the winters were seeringly cold. As the son would later recall, his mother, who hated snakes, once reached for an old teapot "only to find a [garter] snake winding its way out of the spout."[7]

Haultain wondered if Sifton knew what he was doing to those peasants and to the Territorial government. Although the number of American and British immigrants also increased, many Territorial residents fixated on the newcomers from Eastern and Central Europe, and the sight brought out their worst instincts. The newcomers stood out in a crowd. They dressed differently, they could scarcely communicate, and they clung together for support. Many Anglo-Saxon Canadians could rarely see beyond their prejudices. The editor of the *Saskatchewan Herald*, Patrick Gammie Laurie, denounced the Ukrainians as carriers of infectious diseases: they were "not even fit for Manitoba which takes anything."[8]

The prevailing opinion of the established settlers was captured in a front-page story in the *Daily Nor'Wester* in 1898 under the headline, "More Galicians: Over Seven Hundred Arrived last Night from the East—Bound for Saskatoon and Yorkton." The writer parsed the newcomers for suitability. They were carrying about $18,000 in total; roughly two hundred of them could read and write in their native languages; and their group included Roman Catholics, Greek Catholics and Lutherans. There were hints of a deluge: "These all have relatives in these places, who have per-suaded them to go there." There were insinuations of uncouth lifestyles: "Stockings appear to be a luxury unknown to any of them, and the long boots do not seem to injure their bare feet." And there was condescen-sion: "As a whole, they are somewhat above the average Galicians, both in physique and general intelligence."[9] On that same day, another story was headlined, "Galician Riff-Raff."[10]

Group settlements compounded the prejudice. Members of a Christian religious sect, the Doukhobors, first arrived in Canada in 1899 with Sifton's encouragement, and they settled in modern-day Saskatchewan. In the beginning, they were an uneasy fit, because their culture was *so* different. Former NWT legislator James Clinkskill complained about their pen-chant for bartering in his new dry goods store in Saskatoon. He also grumbled that the Doukhobors refused to conform to Canadian laws on schools and marriage. "These people proved unsatisfactory settlers. . . . Many of them also were religious fanatics," he asserted.[11] It was probably no accident that the sympathetic Doukhobors would later offer food to hungry Japanese-Canadians in internment camps during the Second World War.

Such disdain did nothing to dissuade Sifton, but it added to the pressures on Haultain, who had to fend off the complaints of his Anglo-Saxon residents. The Territorial government could not sell its land; nor could it sell or lease its resources. But it had to find the money to provide services to those alien newcomers. Anger at Sifton's high-handed approach became palpable in Territorial government circles.

Haultain would not let Ottawa off the hook. Even the preparation of the Territories' annual estimates became an excuse to take sly potshots. From the start of the immigrant deluge, the premier was emphatic; he needed funds for roads and bridges. In many districts, settlers were trav-elling long distances to reach markets, which "would not be necessary were intervening streams bridged."[12] He wanted $9,000 to deal with out-breaks of infectious illnesses, which, he said, "in most instances are directly traceable to immigrants who bring such diseases with them."

He needed funds to maintain hospitals, and to establish new ones, and that price tag was escalating along with the number of patients who had no money for their treatment. The implication was that Ottawa was funnelling impoverished, sickly souls into the Territories' care.

The urgency was clear. The Territorial government needed $427,100 in 1897–98, and it could raise only $27,100 from local licences and fees. It depended on Ottawa for the remainder, and Ottawa was only sending $242,879. Haultain was bitter. The Assembly was "expected to provide for educating the young, caring for the sick and destitute, rendering the country habitable by improving roads, bridging rivers, protecting against prairie fires, increasing the water supply."[13] It was a litany that Haultain and his Assembly would chant endlessly over the next eight years. Sifton added his previously promised $40,000 to the annual grant.

But the 1897–98 estimates also showed that the Territorial government was developing administrative skills. Haultain wanted money to hire a law clerk, for instance. To keep order on the frontier, largely because more residents had violated Prairie fire ordinances, he needed more money for the upkeep of jails. He was dividing the settled portions of the Territories into regions with local councils that could handle small road and bridge repairs. His agents were destroying wolves and noxious weeds, and he wanted to collect agricultural statistics. His government needed $100,000 for schools, and an additional $18,000 for six school inspectors. It was on a steep learning curve, and by the time of the November 1918 Conference, when the Western premiers would insist that they could administer their lands and resources, they were likely right.

The Haultain government did not ask for provincial status with these estimates. But cabinet members did add a crucial new demand: they wanted annual *subsidies in lieu of resources*, just like those that Premier John Norquay had secured for Manitoba in the 1880s. After all, they reasoned, Ottawa had given away more lands in the Territories than in Manitoba as payment to the Canadian Pacific Railway and other smaller railroads. This was because, when the CPR finally selected its parcels, there were not enough good lands left in Manitoba. That Territorial request had no effect, but the Territorial government had adopted an unfortunate ploy: invidious comparisons with other provinces would become a no-win game for the West at the November 1918 Conference.

Immigrants were pushing out the boundaries of settlement in their hunt for good lands. Dmytro Romanchych arrived in 1897 on a ship that disgorged enough immigrants to fill twelve trains, each twelve coaches long. Most of his party eventually made their way to Dauphin, roughly 250 kilometres northwest of Winnipeg, to the end of the railway branch line, where they camped in tents beside the station. "The best lands in the immediate vicinity of Dauphin had already been taken up by the English and Scottish settlers who had got there before us," he would recall.[14] He and his fellow Ukrainians had to venture over rough terrain with their Swiss-born federal immigration agent, who communicated with them in German. The party struggled into the northwest through thick brush and forests in search of suitable lots. They tramped past the projected terminus of that branch line, which was a village called Sifton. The formidable minister of the interior had even left his mark on this remote spot.

Sifton's traces were everywhere across the West. As a young university graduate, he had moved from Ontario to Winnipeg in 1880, where he articled in law. Two years later, he and his brother Arthur set up a law practice in Brandon, roughly 200 kilometres to the west. But Clifford Sifton was a born politician. He first embraced Western agrarian protest movements, which became a basis for the fledgling Manitoba Liberal Party. It was those Liberals, including Sifton's father, John, who almost defeated Premier John Norquay in 1886 with their vehement defence of provincial rights. Two years later, Clifford Sifton would win that Brandon seat and join the provincial Liberal cabinet. It was an ironic debut for the man who would aggressively defend federal resource control against Western claims. But, by the time Sifton entered the Laurier cabinet, he had adopted a pan-Canadian vision that would thwart Haultain's bid for mastery of the West.

Sifton was a bulldozer. The Department of the Interior had assumed the Agriculture Department's responsibility for immigration in 1892.

This role made sense because the Ministry of the Interior also controlled the West's public lands, and in theory would lure immigrants from their homes, shuttle them west, and assign them to their allotted plots. In practice, before Sifton took over, the ministry's promotional campaigns had been aimed at former Canadians in the United States, who usually did not want to return, and at residents of the British Isles, who often scorned the notion of eking out a living from the soil.

Sifton wanted farmers, not manual labourers. On the cusp of boom economic times, he stressed the agricultural opportunities for Central and Eastern European peasants, opened more immigration offices, sponsored guided tours, and offered bonuses to steamship companies, railroads and colonization groups for every farmer or domestic worker that they brought to Canada. After the 1896 election the Liberals had fired the former Interior Ministry bureaucrats. Immigration agents in the West now earned commissions—not salaries—based on the number of newcomers that they settled. When that disdainful priest Nestor Dmytriw breezed through Winnipeg on his way to Dauphin in 1897, he had nothing but praise for Sifton's bureaucracy: "The new administrative staff, with its very energetic chief [Clifford Sifton] deals with immigration matters promptly, efficiently, and with bureaucratic formality."[15]

That was as close to a rave review as Sifton could have wished for as he force-fed thousands of people into the West. But there was a darker side to Sifton's politicking. The interior minister doled out jobs to Western Liberals, and expected them to work for the party during elections. Worse, although no charges were ever proven, Sifton's wealth was obviously increasing, through his interests in lands and other securities and through his alleged "use [of] of his position and talents as a broker or investment agent for others."[16] His biographer D.J. Hall tracked Sifton's drive to overhaul the Joint Stock Companies Act. First, Sifton proposed legislation that would eliminate the need for detailed annual corporate reports—a move that would protect his private business operations from scrutiny. When Laurier shied away from the bill in response to Conservative complaints, Sifton personally pushed it through the House of Commons. It was not a commendable

performance. His opulent Ottawa home had a music room, satin fur-
nishings, a billiard room, valuable paintings and a library. Such tales
about the former small-town lawyer did not play well among his
Brandon electors, and allegations would haunt him after he left cabinet. The contrast with the honest Haultain, who never amassed a
personal fortune, was striking.

Sifton made no apologies for his lifestyle, or for his dream. "Here,
then, we have the situation in a nutshell," he proclaimed, "a vast and
productive territory becoming quickly occupied by a throng of people
who will be called upon to take up the duties of citizenship almost at
once, whose successful pursuit of agriculture will make them financially
independent, and who in a short time will constitute a most potent
factor in the national life of Canada."[17] Sir John A. Macdonald's centralist
vision was reincarnated in a politician who would never see eye-to-eye
with Haultain: Sifton could never willingly surrender control over the
land and resources that furthered those plans.

In the final years of the nineteenth century, Sifton even tackled the
Canadian Pacific Railroad, which had already become a source of lasting
Western grievance. Under the terms of its contract with Ottawa in 1881,
the CPR was exempt from most taxes, permitted to build branch lines
wherever it wished, and granted a twenty-year monopoly on the construction of any line that ran south to the American border. (Sir John A.
Macdonald had cancelled that monopoly in 1888.) The CPR had also
received $25 million and 25 million acres of Western land that had to be
"fairly fit for settlement." Crucially, it had also secured an exemption
from taxes on those lands, and that exemption would extend for twenty
years after the land was selected or until it was sold. CPR officials were
no fools, and did not select their lands until settlers arrived. When the
settlers improved their lands, they inadvertently pushed up the value of
the adjoining CPR turf.

The homesteaders were infuriated: if they wanted to buy more good
land, they often had to pay the CPR's inflated prices. Furthermore, their
homesteads were needlessly separated from those of their fellow settlers
because of lands that the CPR had claimed. Other smaller railroad

companies had made similar, if less lucrative, deals with Ottawa. As the protests mounted. Sifton pushed the railroads, particularly the CPR, to select their remaining claims. By 1905, most of the remaining railroad lands would be earmarked. These privileged lands would pay for a large portion of the West's railroads.

So here was Haultain, the irresistible force in Regina, going up against Clifford Sifton, the immovable object in Ottawa. Something had to give. On January 14, 1899, Haultain sent his annual estimates to Sifton, along with a plea. He needed $535,000 in total, and he could not count on more than $30,000 from local sources. He also upped the ante. His government, he wrote, had now "at its command all the official machinery necessary to carry on the most intricate matters of government."[18] In short, Ottawa should allow his Territorial government to deliver every service that provincial governments performed. The premier was not angling for full provincial status; nor would his government interfere with Dominion efforts to settle the Territories. But, with enough money, his government could take on "the successful maintenance of these burthens."[19] It was a plausible letter. It was also apparent that Sifton could make the premier go away for another year if he simply sent more cash. Haultain waited. And waited.

And then he got his answer, buried in the middle of page 23 of the Saturday, July 15 edition of *The Globe* newspaper. In reply to a question in the House of Commons, Sifton explained that the Liberals had already raised the Territorial grant by $40,000 when they took power. "Although he [Sifton] recognized the needs of the Territories," he could not increase the grant again "at present."[20] It was a huge blow to Haultain, and he promptly telegraphed Sifton: "Globe of Saturday last reports you to effect that there will be no increase to Territorial grant this year, sincerely hope this is not the case." How could his government provide new settlements with schools, roads, bridges and a competent administration? Then he summarized the problem that would eventually lead him to betray his family motto, and lose command of himself. "We are confronted with impossible conditions unless Federal Government recognizes that its duty does not end when large and dependent settlements are planted in the West."[21]

Sifton took two full weeks to respond. Only now, he wrote, had he found the opportunity to answer Haultain, and even now, he did not have time to answer in depth. It was quite impossible to add to the Territorial grants in the present session because Parliament had already approved them, so the matter would have to wait until the following year. "I quite appreciate the difficulties of your position and would gladly assist you if it were in my power."[22] In later-day slang, this was a "too bad, so sad" dismissal.

There was more to this episode than the mere denial of funds. The way Ottawa delivered its message was as aggravating as the message itself. Time after time during the two decades before the November 1918 Conference, the federal government would act in this cavalier fashion. Sifton was genuinely overextended: that summer, in the West, his ministry was handling scandals in the Yukon goldfields, operating its booming immigration campaign and negotiating a treaty with northwestern Aboriginals. Haultain's administration was an afterthought, and Sifton did not bother to hide this. The premier lived in a world where the residents treated him with dignity. Now, dismissed with an airy wave, he had to perform triage on fiercely competing needs. That brief exchange of correspondence in 1899 was a classic example of why the West distrusted Ottawa.

Sir Wilfrid Laurier's Liberal government could certainly play hardball. In 1890 Manitoba had abolished French as an official language and eliminated funding for Roman Catholic schools. Six years of bitter strife had ensued. A Catholic ratepayer had challenged the legislation at the Judicial Committee of the British Privy Council, which was then Canada's highest court. He had lost. That same court later ruled, however, that Ottawa could reverse any provincial law that deprived a minority of educational rights that had existed at Confederation. Ottawa and the province had groped in vain for a compromise. And then Laurier took power. His emissaries opened negotiations in Winnipeg, and Sifton clinched the deal in November 1896. The province agreed to hire Catholic teachers where numbers warranted, and to allow instruction in French *or any other language*. (That provision for teaching in any other language would cause trouble during the First World War.) A short period of religious instruction was also allowed at the end of the day. The Manitoba Legislature approved the legislation in 1897.

The federal Liberals made sure that their political cousins in Winnipeg got the message: Laurier ran a strong central government that protected minority rights. In February 1898, Manitoba premier Thomas Greenway wrote to Laurier with a familiar grievance. His province's finances were "in a very bad shape, to say the least," and he had delayed the opening of the Legislature in the hope that Ottawa, which controlled his province's land and resources, would help.[23] The premier wanted to build more schools, but Ottawa had not approved his request for an advance from the School Lands Fund, which held money from the sale of Western lands. A resolution to approve an advance had died on the order paper in the previous Parliamentary session.

Sir Wilfrid replied three days later. In velvet tones that scarcely cloaked his iron will, he emphasized the reason for the earlier delay. "As you are aware, there were last year some difficulties about the settlement of the School question," he wrote, "which would have made the support [of] our friends from Quebec . . . very uncertain." He was sure, however, that approval for the advance would now be "comparatively easy" as long as the province behaved with "a fair and generous spirit towards the Catholics."[24] Greenway understood: he would only get advances if he could overcome the objections of his vocal anti-Catholic, anti-francophone voters and implement the compromise.

The premier would have been grimly aware of the irony of his situation. Although he shared Laurier's Liberal allegiance, he had been a strong candidate for the Provincial Rights Party in Manitoba in 1883. As Opposition leader in the Legislature in the mid-1880s, he had mercilessly hounded Conservative premier John Norquay for his failure to wrest more money from the federal Conservative government and for his inability to halt Ottawa's disallowance of provincial railroad charters.

Now the federal Liberals controlled the West's resources, including its school lands. And the Liberal premier was as dependent as Norquay had been on Ottawa's goodwill, including Laurier's substantial caucus of Quebec MPs.

Haultain was well aware of the Laurier government's hard-nosed conduct in the West. But, after Sifton's casual dismissal of his needs, the Territorial premier upped the stakes. On October 7, 1899, in a speech in Yorkton, he called for provincial status. As the *Calgary Weekly Herald* gleefully reported, the Haultain government "had got to the end of its tether." Ottawa was spending $60 million in total in the 1899/1900 fiscal year, which was a comparatively large sum. But it had not found one extra dime for Western grants. It was time to take a "large jump" to secure the fixed per capita subsidies that provinces received.

More important, Haultain demanded control of the lands and the resources. The Territories would *not* settle for a deal like Manitoba's: it would insist on the right to control its lands, minerals and forests, including "a right to every cent, every stick and every straw." Westerners had heard Easterners prattling, "We bought you" or "We own the Territories." Piffle. "Our fathers helped to pay what was the purchase price," was the Calgary newspaper's summary. "Canada never bought the Territories. It paid a million and a half dollars for the extinction of a title." (Haultain was referring to the Hudson's Bay Company charter.) Then the premier added a new twist: "because a trustee paid a bogus claim, that was no reason why the Territories should be mulcted for that amount."[25] In effect, the Territorial premier even put the validity of the charter in doubt, alleging that Ottawa had paid to extinguish a phony claim. The newspaper report likely paraphrased Haultain's elegant prose. But the premier was rebutting the same arguments that would so annoy the Western premiers at the November 1918 Conference.

A campaign was born, and the rhetoric would endure. Other Westerners had espoused the cause of provincehood throughout the 1890s with much bravado but little real hope of success. Newspapers had debated the issue, often with each other. In early 1895, residents from the southwestern Territorial district of Alberta had churned out a pro-provincehood pamphlet, and in September 1898, future prime minister R.B. Bennett, who was then running for a seat in the Territorial Legislature, had even advocated provincial status for Alberta, with Calgary as the capital. (The other parts of the Territories could presumably look out for themselves.)

But until Haultain stopped hinting that money alone would placate him, until he gave that speech in Yorkton in October 1899, the cause was not taken up. The premier was still a reluctant convert—his focus remained on the Territories' financial needs—but he was now engaged. Sifton had only himself to blame. A few months later, Ottawa increased the Territories' annual grant, and created a special fund to repair flooded roads and bridges. The premier was grateful, but not appeased. When the spring session of the Legislature opened in 1900, his government *formally* requested provincial status. Haultain had crossed his Rubicon.

As immigrants continued to pour into the land, they brought the needs of their many different nationalities, languages, cultures and religions. Once settled, often in remote hamlets, newcomers expected less from their governments than their descendants would today. They certainly received less. But they wanted to educate their children—indeed, churches and schools were among the first community institutions that they built—and they chafed at their economic isolation. They needed government help to make new lives, and that included the provision of teachers for their schools. Ukrainian settler Dmytro Romanchych recalled his pioneering years around Dauphin: "one thing perturbed us constantly: our children were growing up illiterate." Manitoba could not find enough teachers for "all the colonies of the various European nationalities which rapidly filled its open spaces."[26] Further west, in the Beaverhill Lake district of central Alberta, it was the same: a Polish-speaking priest performed the first community Mass in a settler's log cabin in 1899, but the first government school would not be built until 1905.[27]

Not all newcomers were farmers. Some men took advantage of the huge turn-of-the-century surge in Western development, snapping up jobs on the railroads or in the mines, and crowding into dodgy accommodations in the camps or the towns. Giovanni Veltri and his brother Vincenzo came from the small Calabrian town of Grimaldi, which young men traditionally left to seek their fortunes abroad. The brothers first worked as labourers on small American railroad branch lines, and by 1897, Vincenzo was a subcontractor for the CPR, while Giovanni was running the improbably titled family construction firm, J.V. Welch Company Ltd. Giovanni travelled the West with aplomb, braving the impossible roads, often in the company of immigrants from other nations. He was self-sufficient. "It was always I who obtained work, signed the contracts and bank loans," he would boast.[28] But he did not put down roots, and would end his life in grumpy retirement in Italy, confined in his "primitive native village."[29] He would miss the polyglot world of Western Canada, where immigrants who could master English and meet the West's economic needs could find good lives.

Others newcomers were not so lucky or so free. In Winnipeg, tensions were sky-high because so many people from so many lands were settling there: immigrants would account for more than fifty-five per cent of the city's population growth in the years before the First World War. Health problems arose because infectious diseases spread quickly in the crowded and impoverished conditions. In 1899 Sifton had improved inspection and quarantine facilities in Halifax to stop ailing immigrants before they boarded trains, but typhoid deaths in Winnipeg were still frequent. Outdoor toilets in the immigrant areas of Winnipeg drained into the ditches along the streets, and families were crammed into single rooms. Mothers, fathers and children often worked for minimal wages, and often in dangerous environments. Slavic and Jewish newcomers in particular usually had little or no capital, and they had to surmount what was then "the added handicap of a foreign language and culture."[30]

The association of disease with foreigners brought out violent reactions in the Anglo-Saxon population. In 1892 Calgary residents had rioted when a Chinese man, who had probably entered Canada as a labourer during the construction of the CPR, contracted smallpox after visiting relatives in Vancouver. In a letter to Sifton in the late 1890s, Haultain denounced federal inspectors in Vancouver for waving through an immigrant who had then carried smallpox into the West, saddling his government with a huge bill for its eradication. Many Anglo-British residents resented the use of *their* lands and resources to aid unwelcome foreigners, whom they dubbed "Sifton's Pets."

Ottawa had little reason to care what Haultain thought. In the federal election of November 7, 1900, the Laurier Liberals handily captured more than fifty per cent of the national vote, including all four seats from the North-West Territories. The Liberals were less fortunate in Manitoba, where the Conservatives captured three seats and Independents took two. The Liberals, including Sifton—who defeated a tough challenge from Sir John A. Macdonald's son, Hugh John Macdonald—secured the remaining two. Sifton had taken no chances with his home turf. He had used special workers to woo specific groups such as francophones, Icelanders, labourers and farmers. To counter charges that Laurier lacked loyalty to the British Crown, he had bombarded English voters "with every 'loyal' statement" that the prime minister had made.[31] Sifton's victory confirmed his ascendancy as the Liberal strongman in the West, and it cemented the clout of his Liberal ally in the Territorial government, treasurer James Ross.

Sir Wilfrid won his second term amid a world in ferment. In South Africa, Canadian troops were fighting on Britain's side against two small Boer republics. The decision to dispatch troops in 1899 had been difficult: many English Canadians had supported the Empire, while French Canadians and recent European immigrants had felt less connection with Britain and its foreign entanglements. Laurier had initially sent a battalion of one thousand avid volunteers—the first time Canadian troops had been officially dispatched to an overseas war. A second contingent of mobile mounted troops had followed in early 1900. Soon afterward, a third unit, which was largely composed of North West Mounted Police officers, had reached South Africa. By late 1900, when the voters confirmed Laurier in office, all hope of an easy victory had vanished.

As the new century opened, the changes seemed dizzying. On January 1, 1901, six self-governing British colonies formed the federation of Australia. On January 10, exhausted drillers in Spindletop, Texas, were re-lowering a drill that they had pulled from a depth of one thousand feet to refurbish, when, to their astonishment, the drill pipe blew out of the ground. Mud gushed into the air, followed by natural gas and then oil. That enormously profitable discovery marked the beginning of the modern North American energy industry that would bring so much wealth to the Canadian West. On January 22, 1901, after more than sixty-three years on the British throne, Queen Victoria died. An era ended.

Eight days later, in the new Edwardian world, Haultain tackled Sifton again. There was a hint of ambivalence in his letter, as he confessed that "financial embarrassments rather than constitutional aspirations" had driven the Legislature to demand provincial status in early 1900. But, he added, there were also sound practical reasons for that request. The Territorial population was growing rapidly, and the Assembly was ready for more responsibility. Ottawa was "apparently incapable" of increasing its grants. Assembly members wanted to administer criminal law and determine land titles. But, most important, they resented their lack of control over the lands and the resources. That control, he wrote, "should be settled at once" before Ottawa gave away more Territorial lands.[32]

Haultain waited for seven weeks. Sifton finally sent a reply that provincial-rights advocates would bitterly cite for decades. The Interior Minister was soothing. He even offered a contorted declaration of sympathy. "I admit that there is very much in the suggestions . . . regarding the necessity of a change in the financial and constitutional position of the Territories." Then came the vague sentence that many optimistic Westerners, including Haultain, pounced upon. "Without at the present moment committing myself to any positive statement," Sifton wrote, "I am prepared to say that the time has arrived when the question of organizing the Territories on the provincial basis ought to be the subject of full consideration."[33] That sounded great, even though it was really a promise to talk, not a promise to act.

Three months later, the Territorial government pushed again, and the pressure came from Sir Clifford's older brother, Arthur. Clifford Sifton had appointed Haultain's ally, treasurer James Ross, as commissioner of the Yukon Territory in early 1901. That move had opened up a cabinet vacancy, and the non-partisan Haultain had scrupulously appointed another Liberal member, Arthur Sifton, as Territorial treasurer. In virtually any other family, Arthur would have been the standout. But, despite his enormous success in later life, his hard-driving sibling always eclipsed him. Arthur is almost an afterthought in biographies of Clifford. "A popular and engaging personality, Arthur seemed to inherit his father's restlessness, impulsiveness and—at least in his youth—talent for near-success," observed historian D.J. Hall. "He was the sort who gaily skipped classes at college, confident of his own ability to pass."[34] By the time Arthur was paying serious attention to politics, Sir Clifford was running the Liberal show in the West.

But the new Territorial treasurer was no pushover. Arthur Sifton was already frustrated with the Territories' status, and he made no effort to hide the strong Western partisanship that he would privately display before the November 1918 Conference when he was a member of the Borden Union government. In June 1901, he tabled his budget—along with some dour reflections. Most of the money came from Ottawa, but it was not enough. Ottawa was apparently going to deliver more money in January 1902, but until the federal cash arrived, the Territorial government was in deficit. Arthur Sifton grumbled that it was impossible to spend money—to earmark funds for specific needs—without the ability to raise that money. The NWT government needed to secure a settlement with Ottawa to change those intolerable circumstances.

Ottawa seemed to get the message. On October 9, 1901, Haultain and Arthur Sifton arrived in Ottawa to meet their financial masters. *The Globe* heralded their coming, citing Clifford Sifton's ambiguous promise of "full consideration" of their case for provincial status.[35] The capital itself was still buzzing with news of the seemingly endless Boer War and other developments. Britain had just launched its first submarine. A letter posted in Toronto at 7 a.m. on September 30 had reached London at noon on October 7—"the quickest delivery on record."[36] The two

Westerners met with Clifford Sifton, and the older brother asked for his younger brother's intercession. But two federal heavyweights, Laurier and Finance Minister William Fielding, were out of town, not to return for two weeks. Haultain headed to Toronto and Buffalo, and during his absence *The Globe* noted that roughly one-third of Territorial deaths were categorized as "without medical attendance or certificates of the cause of death."[37] That statistic belied the premier's claims that he was winning his West.

Ironically, Laurier had been criss-crossing the nation himself, from Halifax to Victoria. The prime minister returned with the conviction that the Territories possessed "immense wealth" and "wonderful possibilities," and he was determined to lure more immigrants onto Western lands.[38] That dream, of course, required continued possession of Western lands and resources. It was an ill omen for Haultain. Undeterred, in late October, seven months after Clifford Sifton's ambiguous letter of sympathy for their cause, the two Westerners met with a high-powered federal delegation in Laurier's Ottawa office. For two hours they argued their case with Laurier, Clifford Sifton, Fielding, Trade Minister Sir Richard Cartwright, Public Works Minister Joseph-Israel Tarte and Territorial MP Walter Scott. Haultain was at his most persuasive, praising the skills of his government to the skeptics, many of whom believed that Ottawa owned the West.

The premier did not dwell on provincehood: rather, he emphasized the need for an increase in his annual grant. Perhaps the sheer urgency of his financial plight diverted him. Perhaps he had lingering qualms about provincehood. Haultain certainly had no chance of winning provincial status and resource control from this particular audience. But he sent the wrong signal for what he thought were the right financial reasons, and he diluted the power of his message. The meeting broke up amicably enough: the courtly Laurier asked Haultain and Arthur Sifton to put their case in writing—so that the entire cabinet could consider it.

Heartened, the Territorial delegation returned to the West. But once again, they had mistaken Ottawa's attention for a promise of action. Six weeks later, in early December 1901, Laurier was taken aback when

Haultain sent a lengthy letter along with draft legislation for province-hood. The premier once again listed the challenges that Laurier had heard so often. Haultain could not raise taxes on struggling settlers who were unable to make a profit from their products because they were so far from markets and shipping terminals. "Nearly every small group of new settlers, united by any tie whatever" was founding new hamlets.[39] Furthermore, the Territories had no capital account—it had no power or authority to borrow—so it had to pay for public works out of its annual grants instead of spreading the expenditures over several years. One statistic (and Haultain packed his letter with statistics) conveyed the severity of the situation: between 1896 and 1901 the number of school districts had escalated from 436 to 649; plans for another thirty-five were underway. The NWT could not tap its rich resources, Haultain grumbled, anticipating the pivotal complaint that would so preoccupy the Prairie premiers at the November 1918 Conference. "All our public revenues go to swell the Consolidated Fund of Canada, *our public domain is exploited for purely federal purposes*, and we are not permitted to draw on the future."[40]

His bill had twenty-four clauses, complete with explanations. The first mistake was clause one, which called for the creation of one large province, encompassing 404,000 square miles of modern-day southern Saskatchewan and Alberta. Although the proposed province would be considerably larger than Ontario, Quebec or British Columbia, Haultain anticipated no administrative difficulties. Other clauses dealt with representation in the House of Commons and the Senate, the selection of judges and the right to regulate all water. Then came the public lands. Haultain claimed all lands that were not already reserved for Aboriginals, persons or corporations, and homesteads—and Ottawa should continue to pick up the tab for immigrant settlement on those homesteads. He wanted all mines, minerals, timber and royalties. He wanted the federal buildings that were used for courts or jails, and all money from the sale or lease of school lands. The Territories also needed the right to sell or lease its resources to raise cash.

The worm of jealousy that had gnawed within each provincial premier's heart since Confederation curled within Haultain's litany. The premier cited Ottawa's per capita spending in the provinces in 1900 to dramatize his grievances. Ontario had received $1.74, Quebec $2.74, British Columbia $9.98 and Manitoba $4.58. In contrast, the desperately needy North-West Territories had received only $2.98—when it required between $6 and $7. Haultain needed more cash, and he wanted that cash to come as a subsidy, not as a grant.

Then came the kicker, the last clause of his letter, which would resonate up to the November 1918 Conference: he wanted Ottawa to pay interest on a sum of one dollar per acre for every acre that Ottawa had handed out, with the exception of land for homesteaders or for Métis claims. This was the pivotal demand: Haultain wanted *continuing compensation* for lands that Ottawa had already allocated, especially for railroads. The Western premiers at the November 1918 Conference would only be echoing Haultain when they asked for their lands and resources—along with the continuation of their subsidies in lieu of those resources—to compensate for Ottawa's past generosity. That demand exasperated Laurier, just as it would annoy the other six premiers at the November 1918 Conference.

The prime minister, who never rushed headlong into major change, and who always considered the regions before he acted, undoubtedly viewed the bill as a provocation. Haultain did not know when to stop. Laurier did not like to say an outright "No," so he pored over the huge package, scribbling his opposition to many clauses in the margin. He was almost certainly the writer who scrawled the word "reserved" beside the demand for control of the public lands and for per capita subsidies. But the prime minister kept his peace while Westerners squabbled about the issues. Should the North-West Territories become one province or two? Would one province be easy for Haultain to administer, or were Western interests already dividing between the Saskatchewan and the Alberta districts? What about special recognition for Catholic settlers, and for those who spoke French? The stakes were as high as the expectations.

Ottawa remained silent. Haultain became impatient. In mid-March of 1902, more than three months after he had sent the draft bill, and more than five months after he had met with the federal ministers, he asked

Laurier for his decision before the opening of the Territorial Legislature. The "great conciliator" dodged what he could not conciliate: he could not answer, he replied, until Sifton had recovered from an illness. Nine days later, the thunderbolt struck. Sifton was back on the job, and the minister presumed that his verdict would not "come in the way of a surprise." After all, he wrote, Haultain "would have gathered our views" at the meeting in Laurier's office.[41] It would not be wise to convert the North-West Territories into a province. The population was sparse, but increasing rapidly, and no one could decide if there should be one province or two. Given those reasons, there was no need to consider Haultain's draft bill. Here again was a brush-off, curt and offhand. If Haultain had reached too far—which he had—Sifton dismissed him far too casually. More fuel to the flame of Western alienation.

Sifton had other plans for the West during those early years of the twentieth century, and they did not include Haultain. With Laurier's backing, the interior minister was grasping at every promising settlement proposal. In 1902, that enigmatic celebrity, Count Paul Oscar Esterhazy, who had brought Hungarian-born settlers into the West in the mid-1880s, secured federal support for the production of an immigration brochure. In July, Esterhazy visited the families that he had lured away from Pennsylvania coal mines and Hungarian peasant plots, and the ensuing pamphlet included his brief introduction along with statements from individual settlers. Ten thousand copies were produced in English, along with 25,000 in Hungarian. Esterhazy was crafty: the majority of the families in the brochure had been among the earliest settlers, so they were far better off than later arrivals. Settler Babijak Janos recounted that he and his family had "arrived here without having a cent of money."[42] Then his narrative did a fast-forward: twelve years later, "after a hard beginning and many sore trials," he was prospering.[43]

There were photos of Sunday worshippers outside their church on the flat prairie; the women wore white dresses and wide-brimmed hats while the men were in sombre suits and white shirts. The message was unmistakable: they had succeeded. Families posed outside log houses with real windows, a second storey, well-crafted outbuildings and herds of livestock. The North-West Territories might not be paradise, but it surely beat a few acres of property and dawn-to-dusk work on someone else's land in Hungary. The pamphlet was superb PR from a man whom some revered as an aristocratic do-gooder, and others reviled as a shameless self-promoter who had turned immigration into a business. Esterhazy, of course, was both. But his pamphlet worked: long after the land reserved for Hungarian settlement had been claimed, newcomers would still approach the colony. Sifton could not have hoped for more.

The Manitoba MP knew the West was very far from won, even in the more settled areas. In the southwestern Manitoba community of Reston, local doctor Alva Burton Chapman was still driving his team of horses to visit patients forty miles away. The roads were not paved, and few were even coated with gravel. Chapman rode through the night, and often through blizzards, to help isolated families. "On his long country drives," an elderly local chronicler would recount, "he often carried a gun for game he might possibly see."[44]

Roads were a luxury in the more remote wilderness. Joseph Prechtl left Hamburg in 1903, bound for the United States, because he "thought success would come easier to a worker [there]." He made his way to Wisconsin, where Canadian brochures promoted free homesteads with pictures of happy farmers in grain fields and ranchers with healthy herds. Prechtl could not resist. He and a German-born compatriot had to walk from the nearest train station to their lands in northern Saskatchewan. On the first night of their trek, Ukrainian peasants sheltered them in their shack, where they slept on hay and were served a breakfast of boiled flour and water paste. They tramped past poplar bluffs, and swamps teeming with muskrats. "We were insight [sic] of the promised land," Prechtl would ruefully recall in 1926, when he was a well-established farmer. "I had forgot one important thing . . . 'that it takes more than romantic surroundings, and fresh air to start a homestead.'"[45] Prechtl had to work for several years on other homesteaders' farms and on the railroads and as a hotel porter before he would have enough money to farm. This new West of the twentieth century still looked a lot like the old frontier.

Haultain governed from Regina, which was scarcely two decades old itself. The Territorial capital was a frontier town, with fewer than three thousand people and few elegant buildings. But Territorial politicians demanded respect. When Sifton dismissed the possibility of province-hood, Haultain snapped that the decision had "come not only as a surprise, but as a deep disappointment."[46] He berated Sifton for his (noncommittal) promise in March 1901 to consider his request, and brushed aside Sifton's reasons for his refusal. He fumed about the disregard of his draft bill. Once again, the premier was forced to raise the issue of money. Since Ottawa had refused to let his government help itself, he could only hope that Sifton would hike his annual grants. It was the same old song. And Haultain could hardly bear to sing it.

The uproar attracted attention. In the House of Commons, Conservative leader Robert Borden asked about the issue—the first hint of his subsequent espousal of the West's resource claims. In Calgary, NWT legislator and future prime minister R.B. Bennett argued that it was time to throw off the federal shackles. Territorial newspapers squabbled among themselves, but generally agreed on one thing: that Ottawa had betrayed the West. The issues had percolated beyond the elites into the English-speaking mainstream. Haultain called an election for May 1902, and his non-partisan supporters romped to victory. In September 1902, Borden went across the West with a pivotal message: "I see no reason why the people of this country should not be intrusted [sic] with the control and management of their own public domain."[47] It was a pivotal signal to Haultain.

Perhaps the premier was misguidedly emboldened by the spirit of the times. In December 1902, the provincial premiers met formally for the first time since their gathering in Quebec City in 1887, but Haultain was not invited. These talks were all about subsidies: the provinces were forming a united front to lean on Laurier for more cash. In 1887 Sir John A. Macdonald had flatly refused to base their subsidies on current census figures, rather than outdated population counts. (Actually, different provinces received subsidies based on different figures.) In 1902 the premiers repeated that demand, which required a constitutional amendment. Laurier formally acknowledged their resolutions, inwardly seething but outwardly polite.

The prime minister knew that this was not a catchy issue. Each province pocketed different types of subsidies: large bulk amounts and per-person subsidies. Ottawa also added an arcane subsidy known as a debt allowance, which was purportedly based on the debts of each province when it joined Confederation. No one had ever rioted in the streets about errors in Ottawa's calculations, or about the size of the debt allowance.[48] But these intricate calculations were the very stuff of Confederation—the stuff that could turn premiers into jealous and manipulative politicians who would begrudge any suggestion that *their* resources should be transferred to the West without huge compensation for them. And Laurier knew this.

Haultain would not back away from his fight for provincehood and resource control, but neither would Clifford Sifton back down. In January 1903, in a daring speech on the NWT Legislature's home ground in Regina, the interior minister argued that the issue required more thought. He might as well have said, "These things take time." Haultain could not resist, and sent another indignant letter to Sifton that reviewed the endless discussions, dismissed Sifton's excuses and itemized his demands. The West wanted equal rights. It wanted control of the lands and resources "in the west, by the west, and for the west." It wanted compensation for the lands and resources that Ottawa had already used "for purely federal purposes."[49] And it wanted an end to the CPR's tax exemption. Gone were the days when Haultain urged provincial status for financial reasons. Now he demanded control for its own sake. Haultain was radicalized, if it were possible to use such a word to describe that normally cool politician.

His judgment slipped. In early February, he boldly resubmitted his draft bill on provincial status to Laurier, stipulating that his subsidies should be raised if Ottawa met the demands of the 1902 interprovincial conference. Laurier forwarded his letter to Sifton, who ignored the call for provincial status but offered to meet to discuss the size of the Territorial grants. Haultain stewed. In March 1903, the premier attended a meeting of the Territorial Conservatives in Moose Jaw, where the party endorsed provincial autonomy. Haultain, who had up to now insisted upon non-partisanship in the Legislature, accepted the role of honorary president of the Territorial Conservative Association. On April 24, 1903, the NWT Legislature again asked Ottawa for provincial status.

Inadvertently, Haultain and his legislators were becoming enmeshed in Ottawa's maze. The premier diligently penned more requests for more money and for provincial status to Laurier. Finance Minister Fielding, in turn, raised the Territories' subsidies. In early June of 1903, the prime minister replied that it would be better to delay provincial status until Ottawa could increase the Territories' representation in the House of Commons. Haultain dismissed this as meaningless. Five weeks later, Sifton wired Haultain, demanding copies of four documents that Ottawa had mislaid; they should be "sent to me here by first mail."⁵⁰ The NWT cabinet clerk cobbled together what he could, although Sifton had apparently furnished incorrect dates and names.

Laurier's secretary then asked Haultain for two more letters. The clerk sent more copies of more documents. In November 1903, the Territorial Assembly once again forwarded the petition for provincial status that it had first adopted in the spring of 1900. Things were becoming farcical. The under-secretary of state in Ottawa insisted that his ministry had no record of the original petition, so the Territorial government sent more copies. On December 9, the under-secretary found the documents from 1900, and apologized. This madcap hunt for documents had consumed five months.

Haultain had lost faith. Everywhere he looked, there were communities of so-called "foreigners" that Sifton had funnelled onto his doorstep. The premier shared the Anglo-British prejudices of his time—he had stopped the printing of French-language Assembly records in 1892. But the polyglot population also flummoxed him: he could barely relate to these struggling peasants, let alone provide them with services. He showed no empathy for them in his letters to Ottawa.

Meanwhile, federal land agents were everywhere. In Qu'Appelle, in modern-day Saskatchewan, roughly a hundred Jewish settlers from Romania, who had survived a devastating outbreak of diphtheria in 1901, now formed the nucleus of a farming community with limited livestock and a few cultivated acres. Their Métis neighbours had kindly shown them how to build log houses with sod roofs and walls sealed with clay. But their official supervisors could not communicate with them, and they spent precious cash on food supplies instead of farm equipment. Their greatest obstacle was the unrelenting hostility of their Anglo-Saxon neighbours, who "despised the newcomers."[51] Prairie life in 1903 was a cycle of unrelenting and often back-breaking chores. The gulf between the immigrants and their premier was unbridgeable.

Politicians saw different messages in the numbers. During the autumn of 1903, Haultain addressed a Conservative rally in Calgary, bemoaning the loss of almost forty million acres of Territorial lands that the Dominion had handed out. Meanwhile, Sifton was boasting about his homestead allotments, which had averaged three to four million Western acres per year during the first years of the new century. The Hudson's Bay Company and the railways were also selling the lands that Ottawa had given them in the last decades of the nineteenth century: they would dispose of ten million acres between 1900 and 1905. By 1904 Ottawa had issued licences to cut almost 95 million board feet of forests in Manitoba, the Territories and the British Columbia railway belt; it had leased 2.3 million acres for grazing; and it had sold more than 86,000 acres of coal lands.[52] Ottawa was doling out the West's lands before Haultain's stricken gaze.

The political differences widened. In a startling editorial in March 1904 the *Calgary Herald* declared that the federal government's conduct was "quite sufficient to raise another rebellion." This time, the revolt would not be led by "a few poorly organized half breeds, but by the stalwart and honest inhabitants of the country."[53] Ottawa had taken in $2.4 million from the sale or lease of resources in Manitoba, the Territories and the

British Columbia railway lands in 1903. But federal grants to the Territorial government did not come close to meeting local needs. The editorial asserted that Ottawa had made a profit of more than $1 million *after expenses* from the West's lands and resources.

Such calculations would become a huge issue in future years, especially at the November 1918 Conference. How much had Ottawa earned from the West's resources, and what were its administrative costs? Ottawa had few revenue sources at the turn of the century: customs and excise duties had constituted almost 79 per cent of federal revenues, and that figure would remain constant until income taxes were imposed during the First World War. No politician, federal or provincial, had a lot of cash, but it was politically too risky to raise taxes. In the early years after Confederation, "even rich and powerful jurisdictions like Ontario [had] derived almost half their revenue" from federal subsidies.[54] That reliance on Ottawa had dwindled in the early twentieth century—because provincial subsidies were based on outdated population figures. By the November 1918 Conference, every province would clamour for more federal cash.

The Territories did not even have the solace of predictable subsidies. Only the Yukon Territory, which Ottawa had carved out of Haultain's domain in 1898, faced the similar challenge of varying federal grants. Haultain wrote repeatedly to Laurier throughout 1903 and 1904, but remarkably, Laurier did not reply. Sir Clifford Sifton's policy was "not to antagonize [Haultain] or take an attitude of opposition."[55] The federal government was lying low. Only after Laurier called an election for November 3, 1904, did he finally write to Haultain, saying that if his government were re-elected, he would "immediately" open negotiations for provincial status, and deal with the issue during the next Parliamentary session.[56]

For Haultain, this vow was too late, and it was too suspect. In a misjudgment that would cost him dearly, the premier campaigned openly for Borden's Conservatives. He argued that the only way to force Ottawa to negotiate provincial status was to trounce Liberal candidates in the West. "Give them a crushing defeat and we will get the rights we demand," he told Regina voters.[57] He lost his bet. The Liberals swept seven of the ten seats in Manitoba, and Sifton won his own riding handily. They took every seat in British Columbia. And, worst of all for Haultain, they took seven of the ten Territorial seats. Laurier was back in power with an enormous majority. The coolly rational Haultain apparently had no Plan B.

True to his word, however, Laurier and key cabinet ministers in early 1905 opened talks on provincial status in Ottawa with Haultain and Public Works Minister George Bulyea. Sir Clifford Sifton was in the United States, recovering from debilitating stress. In his absence, Haultain and Laurier disagreed about virtually everything. Ottawa wanted to create two provinces, if only to dilute the Territorial government's political clout. Haultain continued to press for control of the lands and resources, and payment for the resources that Ottawa had already given away for purely federal purposes.

As the negotiations dragged through the frigid Ottawa winter, Laurier asked Sifton about the resources. The interior minister did not mince words. Any transfer would be "disastrous to the whole Dominion."[58] Immigrants might stay away if a new government administered the lands, and, he added, "the continued progress of Canada for the next five years depends almost entirely" on sustained immigration.[59] Laurier had another proposal. What if Ottawa opted to "hold all lands as a trust, keeping the administration in our own hands and handing over the proceeds annually to the Province"?[60] That way, the conciliator reasoned, Ottawa could safeguard its interests but concede the principle of provincial resource control. That might satisfy Haultain.

Sifton's reply was brisk, and it perfectly captured the case against the Prairie premiers at the November 1918 Conference. The original provinces owned their Crown lands when they entered Canada, the minister affirmed, but "the Dominion owns these lands."[61] Sifton was adamant: if Ottawa allowed Western leaders to view themselves as the owners of their resources, it would "set up an elaborate & untenable fiction."[62] No good would come of it. "You will lay the foundation for a perennial agitation in these new provinces for sure," he warned.[63] Then he added startling advice: "We should say to the Provinces You have no lands & we cannot give them back we shall provide a liberal revenue in lieu of it."[64] In his agitation, the interior minister dropped any semblance of punctuation. But Laurier got the message, and agreed. Haultain was disgruntled, but increasingly powerless.

The most ferocious arguments were about education. In 1892 Haultain had pushed Ottawa to remove the guarantee of separate schools from the North-West Territories Act. As a compromise, the two governments had agreed that the majority of taxpayers in any school district could determine the denominational character of a public school. The minority could establish separate schools, but they would have to pay for them with taxes that they imposed on themselves. In 1901 Haultain had clamped all schools under a central authority, and now Haultain demanded complete control over education. That prospect alarmed Roman Catholics, particularly francophones across the West and in Quebec, and they put enormous pressure on Laurier. The prime minister did not trust the premier's generosity toward minorities, and he did not ask for Sifton's views. Given the religious tensions of the times, an explosion was inevitable.

The two bills to create the new provinces of Saskatchewan and Alberta debuted in the House of Commons on Tuesday, February 21, 1905. Two days later, Sifton returned from the United States. The legislation retained federal control over the lands and resources, and, in lieu of those resources, offered the provinces subsidies that would increase as the population grew. The bill also reaffirmed the principle of separate schools and insisted upon the equitable distribution of public money to minority schools. Sifton erupted. On February 27 he resigned from cabinet, largely because the bill permitted the diversion of public money to sectarian schools.

In the ensuing uproar, Laurier changed the bill so that it largely reaffirmed the existing Territorial system. Sifton was mollified, but he was mortified by Laurier's lack of consultation and would never return to cabinet. During the debate on the bill, Borden expressed his regret at Ottawa's continued control of the public lands. As Britain had discovered, he observed, the residents of any dependency "must be entrusted, and might be safely entrusted" with resource control.[65] He would eventually rue that assertion.

In mid-March of 1905, Haultain wrote a lengthy letter to Laurier, complaining about the educational clauses and quibbling over the size of the subsidies in lieu of resources. Then he added a stunning concession. "I am not unwilling to admit that an immediate income, increasing with population and certain in amount," he declared, "may in the long run prove quite as satisfactory as any probable net income, resulting from local administration of the public domain."[66] Laurier must have read that sentence twice. After all Haultain's complaints about the lack of resource control, he now admitted that the subsidies in lieu of those resources might be worth more to his government than the resources themselves. It was an astonishing admission from a professional politician. Was the canny warrior ambivalent about resources to the end? Perhaps. But it was more likely that Haultain needed federal cash more urgently than he needed the resources. He was always preoccupied with budgetary pressures.

The experience turned Haultain into a fervent advocate for provincial rights. In June 1905 he took the highly unusual step of campaigning against the federal Liberals in two Ontario by-elections. Ontario electors should defeat the Liberal candidates, he said, because Ottawa had interfered with his own government's right to administer education and natural resources. It was a quixotic mission, and the federal Liberals won both seats. Haultain had forgotten his family motto, and lost command of himself. And he had made dangerous enemies.

The Saskatchewan and Alberta bills became law on July 20, 1905. The official inauguration of both provinces was set for early September. Behind the scenes, wheels were turning. In Alberta, Liberal stalwart C.W. Cross warned Alberta MP Peter Talbot that if Laurier selected Haultain as the new premier, "he certainly will be doing it in opposition to his best friends."[67] A few weeks later, Laurier appointed Liberal George Bulyea, who had accompanied Haultain to Ottawa for the autonomy talks, as Alberta's first lieutenant-governor. In August, the provincial liberals selected Alexander Rutherford as party leader, and Bulyea appointed him as interim premier. To act as Saskatchewan's lieutenant-governor, Laurier selected Amédée-Emmanuel Forget, who had been the Territories' lieutenant-governor. Forget then appointed Liberal MP Walter Scott as the new premier.

There was no job for Haultain. Laurier could officially pretend he was not involved in this enormous snub. But the prime minister had decided that "the breach was irreparable" and he would never have any peace with Haultain at the helm.[68] Haultain was now the leader of the Conservative opposition in Saskatchewan, which he transformed into a Provincial Rights Party. He would lose three subsequent elections.

Haultain's private life too would be difficult. In 1902, at the coronation of Edward VII in London, he had renewed his acquaintance with the daughter of his former nemesis, NWT lieutenant-governor Charles Mackintosh. He had been smitten, and in 1906, he secretly married Marion Mackintosh, who was the twice-divorced mother of a young daughter. However, Haultain's bride never lived with him but stayed in Ontario, ostensibly to prevent scandal over her status as a divorcée. The marriage remained secret for thirty-two years, until Marion's death. In 1912 Haultain retired from the Legislature to become chief justice of the Superior Court of Saskatchewan. By 1917 he was chief justice of the Saskatchewan Court of Appeal and chancellor of the University of Saskatchewan. After a second and happier marriage, Haultain would die in Montreal on January 30, 1942.

He left a poignant record of enormous success and unrealized dreams. His contemporaries portrayed him with sympathy as a Western hero, who dreamed of one great province with a non-partisan government that would control its own resources. That monochromatic portrait of an alienated Westerner did not capture the prejudices or misjudgments of this elite lawyer. But the premier worked hard to fashion Canadians in his own image or, at the very least, Canadians who could integrate and talk with each other. His stunning political blunders during his later years as premier were perhaps an understandable response to Sifton's dismissive behaviour. If Sifton populated the West, Haultain taught those newcomers that they had come to a place with its own fierce pride.

Together, as the two men scraped against each other, they brought the West, occasionally kicking and screaming, into the modern world.

There is another photograph of Haultain, which was taken at the banquet to inaugurate Saskatchewan on September 4, 1905. The speakers for this glittering gathering included Laurier, Governor General Earl Grey and Indian Commissioner David Laird. Although Haultain was at the head table, this unofficial Father of Confederation was not asked to address the dignitaries. In the photograph, the former premier is formally attired. His elegant white cuffs protrude from the arms of his jacket. His right arm is resting on the white tablecloth. He looks down, his eyes seemingly closed. There is no sign of triumph on his face.

"A BURDEN ONEROUS TO BEAR":
FRANK OLIVER DOES IT HIS WAY. 1905 TO AUGUST 1911

—

THE NEW MINISTER OF THE INTERIOR was a scrappy newspaper publisher who never backed away from a fight. Frank Oliver had languished on the House of Commons backbenches for nine years when Sir Wilfrid Laurier brought him into cabinet in 1905. The Edmonton MP promptly vowed to dismantle the legacy of his predecessor. Oliver had never concealed his disdain for Sir Clifford Sifton's efforts to recruit Eastern and Central European peasants for Western lands. He agreed that those alien newcomers could grow wheat and buy Central Canada's goods. But they also diluted the West's Anglo-British stock and its Imperialist inclinations. Oliver intended to develop the West with the right kind of folks. No Liberal had ever been "so severe a critic of Sifton's policies as Oliver."[1]

Such prejudice was deeply ingrained. His contemporaries admired Oliver as an upright man, but there was always an edge. As a teenager, Oliver had quarrelled with his father and changed his last name to his mother's maiden name. His formal education had ended with high school, although he later trained as a printer. In 1873, when he was twenty, he had followed the path of many ambitious youths and headed west. He was an early pioneer: he joined ox-cart brigades, rafted across rivers and camped out on the prairies. By 1880, he was cranking out his *Edmonton Bulletin* newspaper on a small press that weighed only two hundred pounds. He understood the hard lives of Anglo-British farmers, and he admired them.

But there was little room in his soul for those he regarded as lesser peoples. He viewed Aboriginals as unworthy landholders; during his six-year stint as interior minister, huge chunks of their reserves would be whisked away, often under pressure. He dismissed Central and Eastern Europeans as undesirables who did not care about their children's education or the well-being of the broader community: their very presence was an impediment to Western growth. In 1901 he had told the House of Commons that Canadians should "look to our own people and to kindred people upon whose industry and loyalty we can depend."[2] Westerners, he said, resented "the idea of settling up the country with people who will be a drag on our civilization and progress . . . and we resent the idea of having the millstone of this Slav population hung around our necks."[3] With his unruly moustache and his deep-set dark eyes, Oliver seemed to bristle with opinions. His ideal settler looked like him.

Why would Laurier select this rough Alberta newspaperman to replace the polished Sifton? Laurier craved Western peace and prosperity, especially after his endless battles with that elegant pit bull, Frederick Haultain.[4] Oliver understood the hard life on the land: he came from an Ontario farm family and had a bond with farmers. Laurier also knew it was far too late to stop the flow of those so-called foreigners into the West, no matter what Oliver did, because Canada needed more workers. But Oliver's public denunciations of past practices might moderate the anger across the West at "Sifton's Pets," those care-worn immigrants who huddled together for support, spoke foreign languages, and retained their old-country garb. As an added bonus, Oliver was a political operator who could take over Sifton's network of patronage appointments and his Prairie election machine.

Perhaps most important, even though Oliver had been a member of the Territorial government for thirteen years, he improbably shared one of Sifton's views: Ottawa should retain control of Western lands and resources to provide free land for newcomers. The Prairie provinces could not be trusted: they might sell those lands. Oliver was no idealist, and one pithy remark said it all. "The interest of a province in the land is in the revenue it can derive from the sale of the lands," he proclaimed during the debate on the bills to create Alberta and Saskatchewan. "The interest of the Dominion in the lands is in the revenue that it can derive from the settler who makes that land productive."[5] If nothing else, this Albertan was succinct. When Alberta was created in 1905, the irreverent *Calgary Eye Opener* newspaper had run a cartoon showing Oliver and Laurier celebrating their retention of resource control. Oliver is singing that Albertans "ever shall be slaves." Laurier is seated at a dinner table. "I don't think these cutlets are good for you, my dear Alberta, so I'll just eat 'em myself," the prime minister purrs as he waves his knife over chops marked *Lands* and *Timber*. "But you can ask a blessing, if you like."[6]

The Prairie premiers knew they could not look to Oliver for control of their lands and their resources. Throughout his six years as interior minister, he would tap those lands and resources to settle newcomers in the West and to fuel an extraordinary economic boom. In the beginning, the new provinces of Alberta and Saskatchewan grudgingly accepted this reality because they had so much else to do. It was hard to build competent administrations to handle their extra duties as provinces. Manitoba was the exception: throughout Oliver's tenure, it would fight a caustic battle with Ottawa over its northern boundary and over federal intrusions onto the swamplands that the province was supposed to control.

Provincehood did not bring harmony to the West—because the federation remained deeply flawed. The provinces were jealous of each other, and all coveted Ottawa's cash flow. A scant year after the creation of Alberta and Saskatchewan, those two provinces would briefly unite to prod Ottawa for additional subsidies for all three Prairie governments. The Prairie provinces would also join the general provincial clamour for higher subsidies at an interprovincial conference in late 1906. But the cries of poverty from British Columbia at that gathering would drown out their complaints. Laurier would mollify the premiers with a general subsidy increase, but he could not satisfy everyone. Indeed, this new crop of Westerner politicians would soon sound uncannily like that ousted provincial-rights advocate Frederick Haultain.

In late 1910, Alberta and Saskatchewan would begin work on the demand for resource control that the Gang of Three would bring to the November 1918 Conference. The two provinces would not work together at first, nor were their demands identical. But, by then, Ottawa's recipe for Western growth had worked all too well, and the Prairie provinces still could not meet the demands for expanded services. They wanted resource control—but Oliver would flatly object. The interior minister had watched that old master Sir Clifford Sifton in action: the grievances of the three Western provinces, occasionally with each other, would rarely preoccupy him. He would play politics, not peacemaker.

By then, the West had changed more than most Victorian-era residents could ever have imagined. The economy was booming. Western farmers were reaping rich crops. Western cities were sprouting and sprawling, with pockets of poverty amid the plenty. Western resource producers could scarcely keep up with the demands of Central Canadian industries. In the last half of the first decade of the twentieth century, the chaos of prosperity would animate the West.

Oliver took over the Ministry of the Interior with relish. After so many futile years on the backbenches, the hyper-partisan politician now had an empire to run, resources to exploit, and people to keep out. In 1906 he introduced a new Immigration Act, which gave the minister and his anointed agents greater leeway to deport newcomers. Cabinet could now prohibit the entry of any specific class of immigrants. New arrivals were legally obliged to have money on hand: in 1908 the federal cabinet would specify that newcomers should have $50 on hand in winter and $25 in other seasons. This stipulation was designed to whisk away the welcome mat from penniless Ukrainian peasants, but by then, Ukrainians already established in Canada were often able to supply their country-men with cash. The minister also cancelled Ottawa's secretive contract with an Amsterdam-based company that recruited farmers from conti-nental Europe and Scandinavia. The unhappy Sifton failed to convince Laurier to overrule Oliver, and maintain that contract. Instead, Oliver increased the number of immigration agents in Britain. In case anyone missed his message, Oliver told the *Calgary Albertan* that Ottawa was "not pushing Continental immigration at all."[7]

Western lands and resources were so rich, however, that Oliver could not turn back time. The resource boom was astounding. In 1901 the Ministry of the Interior had stopped selling coal seams without charging royalties. It was now levying a royalty of ten cents per ton. The output was huge: between 1901 and March 1907, Ottawa sold an estimated 514,000 acres of Albertan land to coal-mining firms.[8] In Manitoba the ministry was leasing out sprawling tracts of timber, and those leases often encroached on the swampy terrain that the provincial government in principle controlled. New railway branch lines snaked across the West. In mid-decade, the Canadian Northern Railway opened access to Saskatoon for the Ukrainians at Fish Creek as well as access to Winnipeg for the Doukhobors at Verigin. Westerners were snapping up farm machinery, consumer goods and construction materials, and building grain elevators. The West's lands and resources were driving national growth, and the Prairies crackled with possibility.

But prosperity brought its challenges to all the provinces. Thousands of people were streaming into the cities; hundreds of new factories were operating. Such global trends were unsettling in a federation designed to meet the needs of an earlier era. Federal subsidies had not grown with provincial responsibilities, and the provinces jealously eyed Ottawa's revenues, demanding a bigger cut. Have-Not provinces in the Maritimes resented thriving provinces, including those in the West, because they were magnets for their ambitious youths. Prairie provinces envied settled regions that did not have to build their infrastructure from scratch, and British Columbia squawked about its distance from Central Canadian markets. The grievances that would engulf Westerners at the November 1918 Conference now plagued Laurier and his strong central government.

The prime minister did get one break: the new premiers of Alberta and Saskatchewan were initially too busy to fret about resource control. In Saskatchewan, Premier Walter Scott was struggling to fill the shoes of Frederick Haultain, who was now fiercely espousing provincial rights from the opposition benches. Scott was a haunted soul who lived in fear that his electors would learn of his illegitimate birth on a rural Ontario farm (during his lifetime, they never did). In photos, Scott appears lean and tentative; his wide black moustache looms over the merest hint of a smile. Despite a lifelong struggle with depression, however, Scott was surprisingly adept at governing. Plunked into power in 1905, the former Liberal MP assembled a cabinet, cobbled together a platform and faced the electorate with the slogan "Peace, Progress and Prosperity." He scoffed at Haultain's attacks on Ottawa, and his Liberals won handily. Optimism was in the air.

Scott had a lot on his plate. His province needed a competent civil service, infrastructure, higher education, and a better grain-elevator system. It also required a new Legislative building to house those civil servants and the newly elected politicians. Scott was a Laurier loyalist, who followed "the Liberal Party line" on Ottawa's control of the West's lands and resources.[9] He even thwarted Haultain's campaign to refer the laws that created Saskatchewan and Alberta to the Judicial Committee of the British Privy Council to examine their legality. To snatch the initiative from Haultain, Scott unexpectedly asked the Legislature to endorse a motion calling for a referral. It was a ploy—because the decision to make the referral was left to Laurier, who naturally did nothing. Even if Scott had wanted to rock Frank Oliver's boat over resource control during those early years, he could never have found the time.

Scott's Alberta counterpart, Alexander Rutherford, was even less likely to cause trouble. Born on a dairy farm near Ottawa, the son of prosperous Scottish immigrants, Rutherford had worked as a lawyer with an Ottawa-area firm for nine years before he rode the Canadian Pacific Railway across the West in search of a new life. He liked the vibrancy of South Edmonton, and the dry air eased his bronchitis. In 1895, with his unwilling family in tow, he had moved west, bought land and opened a law office. He was a staunch Liberal, a devout Baptist, a successful lawyer and a shrewd business investor. In photographs, he appears dignified, mildly pompous and so complacent that, in earlier decades, he could easily have been a member of Upper Canada's cliquish Family Compact.

When Ottawa created Alberta in 1905, Rutherford was not Laurier's first choice for premier, nor did Lieutenant-Governor George Bulyea want him. Oliver simply did not trust him: the federal Liberals wanted someone who was politically astute. But the local Liberal elites espoused his candidacy, and the provincial Liberals selected Rutherford as party leader in August 1905. It seemed a safe choice. "Though he was neither dynamic nor innovative," historian Howard Palmer notes bleakly, "he was viewed as someone with few enemies."[10] Rutherford won a landslide electoral victory in November 1905, partly because his opponent, Conservative leader R.B. Bennett, worked as a corporate lawyer for the despised Canadian Pacific Railway.

But politicians needed more than goodwill to survive Alberta politics toward the end of the twentieth century's first decade. Rutherford was well aware of Laurier's lukewarm attitude toward him. Knowing he might need the prime minister's help someday, he curried favour. In early 1906, after the Scott government passed that devious motion to examine the legality of the acts that created Saskatchewan and Alberta, Rutherford disavowed any support for that phony challenge. He told Laurier, "the Alberta people are very well satisfied" with the act; he dismissed Haultain's complaint that it interfered with provincial powers over education, and he vowed that Alberta would never drag Ottawa to the Judicial Committee of the British Privy Council in London, Canada's highest court.[11] A few months later, when the premier asked Laurier to force the Bell Telephone Company to submit to municipal regulation and provincial control, he made sure to add: "We would not care to submit any resolution that would in any way embarrass your government."[12] Laurier was undoubtedly grateful: this man seemed far more amenable than the fractious Haultain.

Rutherford's accomplishments were respectable. He created one of Canada's first government-owned telephone systems, expanded the public school system, set up a public library organization and established the University of Alberta. To appease militant farmers, he subsidized agricultural societies, creameries and sugar-beet producers. He poured money into capital projects, including railways, which would eventually be his undoing. His contemporaries regarded him as an "unlikely politician, an unexpected leader, and an unusual Premier." His supporters respected his "hard work, honesty, and administrative skills."[13]

But the establishment lawyer had his blind spots. Although Rutherford passed legislation for an eight-hour workday in the coal mines along with workmen's compensation, he was hesitant to tackle the treacherous conditions. Mine bosses usually allocated the most dangerous tasks to foreign workers. Labour militancy escalated, foreshadowing the disruption of the war effort. But the premier was harried and otherwise occupied. Like Scott, he had neither the time nor the inclination to pursue any demands for resource control.

The frontier communities and trading posts of the last century were now towns and cities that were growing faster than rural areas.[14] Spidery networks of sidewalks, paved roads, water mains, sewers and street railways crisscrossed communities. That great historian J.M.S. Careless captured the challenges of time and space: western cities were compressing a century of Eastern urban growth into a few decades; they sprawled because they "felt they had all God's room to grow in"; now they were confronting the transportation and communication problems of too much distance.[15]

Calgary had become a lively commercial centre for agriculture and ranching. Businessmen in imposing brick offices struck deals to exploit Western resources. Winnipeg was the hub for continental railways and the distribution centre for Western grain headed for markets in Central Canada and abroad. Stone pavement coated streets where mud had once spattered Louis Riel and his Métis. Ritzy brick and stone buildings had replaced wooden shacks. Thousands of passengers clambered each day onto clattering trolleys. Wealthier residents were taking advantage of the new bridges, and moving into the South and West Ends. New immigrants clustered in the city's North End, where their lives were one long struggle. Mothers and children had to work to ensure that their families had enough to survive. Infant mortality rates were high; many victims died from gastrointestinal diseases. But local boosters now touted Winnipeg as the Chicago of the North.[16]

Those city dwellers could now reach for the luxuries of civilization and the efficiencies of progress. In Regina, Premier Walter Scott established the first provincial museum in the three Prairie provinces in 1906. Ottawa knocked down Regina's Knox Presbyterian Church and built the elegant Prince Edward Building to serve as the post office (a Beaux-Arts landmark that would become a city heritage property). Only a few blocks away, the Lake of the Woods Milling Company constructed a massive grain elevator with vertical bins that could pour grain directly into railway cars.

The Laurier boom was in full swing. By mid-decade, roughly two-thirds of all Canadian homestead entries were in Saskatchewan.[17] In Alberta, the number of farms would double between 1906 and 1911, and the acreage for field crops would triple.[18] Manitoba was "on the upswing of the greatest boom" it had ever seen: wheat production would climb from 54.5 million bushels in 1906 to 60.2 million in 1911.[19] Those farmers bought steel plows, shipped their crops on the railroad, and demanded the latest in equipment and domestic conveniences. The West's lands and resources were fuelling a bonanza. Despite a brief downturn in 1907, manufacturing industries, service firms, hydroelectric utilities and the financial sector exploded.

Oliver may have wanted Anglo-British farmers, but industrialists needed labourers who could mine the coal, harvest the Western crops, toil in the foundries and operate the plants, and they did not care about Oliver's ethnic preferences. Nations throughout North and South America were competing for workers. Corporate leaders argued that unrestricted immigration was good for the nation, and Oliver's federal cabinet colleagues were on their side. Large businesses and private agencies recruited Italians, Bulgarians, Russians, Ukrainians and Poles. Between 1907 and 1914, the railways alone would hire 50,000 to 70,000 workers, most of them from Southern Europe, to build two new transcontinental lines including the Canadian Northern track, a second CPR line, and an extensive network of branch lines.[20]

Unions, on the other hand, complained that the arrival of those contract labourers violated Ottawa's undertaking to attract only domestics and farm workers. Their opposition was fierce: newcomers were undercutting wages, doing little to resist unsafe working conditions and slipping into their members' jobs. The Laurier government disregarded them. "No actions were taken to significantly impede" the entry of those immigrants.[21] Overruled in cabinet and ignored in corporate boardrooms, Oliver could no more stop that tide than cancel the icy Prairie winters.

Not everyone prospered, even among the settlers of the fine stock that Oliver so admired. Genes were no substitute for farming expertise. In 1906 American Sarah Ellen Roberts trekked with her husband and three children to a homestead near Talbot, on the eastern edge of central Alberta. She and her husband were both over fifty. Her husband, a physician, had abandoned his medical practice in Illinois when his doctors prescribed an outdoor life to remedy a liver ailment. But the vastness of the outdoors overpowered them. En route to their homestead, Sarah wrote, they passed a house in which no one would dwell "because the man who had lived in it became discouraged and committed suicide."[22] When they reached their land, the family gathered poplar branches for fuel, excavated a well, huddled in a tent to wait out fierce storms, and painstakingly assembled a log hut. To earn desperately needed cash, Sarah's middle son found farm work in the haying season and her eldest son helped out in a general store.

At first, Roberts could not cope with the empty space. When she discerned far-off herds of cattle on the otherwise endless prairie, she wrote, "I felt as though I were absolutely alone in the world, and my sense of littleness and helplessness overwhelmed me."[23] Even her adjustment tells a tale of loneliness: "Since then I have, of course, learned to think of anyone who lives within five miles or six miles of us as a 'neighbor.'"[24] Homesteading was "not a life for weaklings."[25]

Profits from the Robertses' grain sales barely exceeded their expenses. In the summer of 1909, they would hunt frantically for anti-toxins to survive diphtheria. To subsist during the harsh winters, her husband resumed his medical practice. Eventually, it all became too much, and in 1913 the Robertses sold their oxen, horses, machinery and household goods, and moved back to the United States. Oliver's dream of luring Anglo-British pioneers to the land could not succeed for those whose life was easier in their homelands. Eastern and Central and Southern Europeans came and stayed because their lives at home were often unbearable. They had to succeed, because they could not contemplate failure.

Just about everyone in twentieth-century Canada wanted a taste of the better life. Premiers across the nation were fending off demands for paved roads, hospitals where patients did not get sicker, and better schools— and they wanted to raid Ottawa's piggy-bank. Laurier was running out of excuses. The prime minister had deftly shelved provincial requests for a subsidy increase after the interprovincial conference of 1902, but four years later, the provinces were fed up with federal dawdling. In January 1906, Prince Edward Island premier Arthur Peters, the scion of a prominent local Liberal family, told Laurier that he absolutely must do something about subsidies. "I am extremely sorry to have to write thus at the beginning of a new year, but there is no better time to make good resolutions."[26] A few days later, another Laurier ally, Quebec Liberal premier Lomer Gouin, forwarded a letter about subsidies that Peters had sent him, adding a pointed remark. The Prince Edward Island premier was "at least as impatient as we [are]," Gouin observed, underlining the word "impatient."[27]

If Laurier wanted to keep his Liberal friends, including Westerners Scott and Rutherford, while soothing disgruntled Conservatives, he had little choice. As his biographer André Pratte has pointed out, the prime minister usually espoused what he called the "sunny ways" of conciliation within his disparate federation.[28] He recognized that he had to find more cash, and reach an agreement. At his invitation, the premiers convened on Monday, October 8, 1906, in Ottawa. Laurier greeted them in the Railway Committee Room on Parliament Hill, offered to discuss their eventual resolutions, and left them to their talks.

Little did he know how time-consuming, tumultuous, dramatic and difficult the upcoming six days would be. Every resentment that would eventually sideswipe the Western premiers at the November 1918 Conference emerged during those talks. To Laurier's relief, Alberta and Saskatchewan did not ask for the right to control their lands and resources. Compared to their peers, those two Western provinces created so little agitation that Interior Minister Oliver was not even a member of the federal delegation. The only exception to Prairie goodwill was Manitoba's prickly Conservative premier Rodmond Roblin, but his complaints about Ottawa's refusal to extend his province's northern boundaries and about federal misuse of its swamplands were not on the official agenda. Laurier largely ignored him: he could rarely bring himself to practise his conciliatory ways with stridently partisan Westerners.

The nine premiers had arithmetic on their side, and numbers could be formidable political weapons. After two days of talks, on Wednesday, October 10, Ontario premier James Whitney summed up their case. In 1867 the Fathers of Confederation had agreed that Ottawa would collect customs and excise revenues. In return, it would distribute fixed per-person subsidies to the provinces. Now, almost four decades later, their respective populations had grown, and those subsidies had generally remained fixed. In 1868 the provinces had received roughly twenty-four per cent of the customs and excise revenues; in 1905, less than nine per cent. Whitney was forthright. Laurier had brushed off the provinces in 1902 but "under our system it can never be too late to apply a remedy, where one is called for."[29]

So far, so good. The premiers had apparently got their act together. After Whitney's presentation, they essentially endorsed the resolutions from their 1902 conference. They wanted higher per capita payments for everyone, although individual provinces could still press special claims. Whitney and Quebec premier Gouin, who was the conference chairman, strolled to Laurier's office to present their tidy list.

But there was a malcontent in their midst, and his disruptive interventions foretold the troubles of the November 1918 Conference. British Columbia premier Richard McBride did not want to float along with the crowd. He dumped a lengthy memorandum, which demanded exceptional treatment, on his peers. The per capita cost of running his government, he argued, was five times higher than their average costs, and it had been that way for thirty years. His province's landscape was daunting: arable land was limited; the cost of building roads over the mountains was sky-high; and the price tag for services to communities within those mountain chains was astronomical.

He resentfully compared his subsidies with those of the other provinces, including Manitoba. He overreacted: even though Saskatchewan and Alberta had only existed for fourteen months, he insisted that his subsidies were "far too little" in relation to theirs.[30] He called for a three-man commission to investigate B.C.'s financial woes. He irritated everyone.

McBride was only thirty-five, and he was ambitious. He had been on the job since 1903—after the lieutenant-governor had dismissed his predecessor for glaring irregularities in the tendering of public contracts. The Conservative lawyer was the first premier born in the province, and the fourth during the first three years of the new century. His twelve years in power would only occasionally prove that "for once, someone had got something right."[31] McBride was usually affable, charming, a star on the social circuit, and used to getting his way. He was also profoundly racist: his government passed legislation, which Ottawa later disallowed, to choke off Asian immigration, and it ignored valid Aboriginal land claims. He did not know how to build alliances with his peers, including his fellow Westerners, and he never bothered to reach out to them.

His conference partners dutifully wrestled with his demands for more cash for two days, and the premiers even met with Laurier to discuss the situation. The prime minister rejected McBride's call for a three-man commission, but he did accept British Columbia's eligibility for special treatment "owing to the vastness of her territory, to its mountainous character, and the sparseness of her population."[32] He urged McBride to co-operate with his fellow premiers: they were looking for a compromise that would satisfy him, but leave enough federal money in the kitty for their own needs.

The prime minister could have saved his breath. Eight of the nine premiers agreed to ask for an extra $100,000 per year for ten years for British Columbia. But McBride would have none of it. On Friday afternoon, after five days of meetings, he declared that the conference had no place considering his special claims: his dispute was between Ottawa and Victoria. This was the tactic that Roblin had adopted in his fight to expand Manitoba's boundaries—which was one reason why that issue was not on the agenda. The Western premiers would eventually espouse the same approach at the November 1918 Conference as well, but it did no more good for McBride than it would do for Roblin or the Western premiers. This was the blight of the federation: if one province pried something out of Ottawa, all the rest usually wanted something, too.

McBride's declaration did not remove his demands from the Conference agenda, but two provinces were particularly annoyed. On Saturday morning, in their first historic alliance, Alberta's Premier Rutherford and Saskatchewan's Premier Scott, their feathers ruffled, proposed a defiant addition to the Conference resolutions: Ottawa should earmark $50,000 per year for ten years for each of the three Prairie Provinces because of their "very exceptional conditions of settlement."[33] Although the two Liberal premiers were not close to Roblin, they included Manitoba in their demands. It was the first intimation of an improbable Prairie alliance, and a rebuke to McBride. It was also the last straw for the impolitic British Columbia premier. He stalked out of the Railway Committee room. The other premiers could not agree on the request from Scott and Rutherford, but they unanimously recommended that Ottawa send an extra $100,000 a year for ten years to British Columbia. It was a generous deed, albeit with someone else's money.

That did not end the drama. From his perch outside the conference chamber, McBride scrawled a message to chairman Gouin. He outlined his case, once again, and then demanded even more money: his subsidies, he wrote, should not be based on the number of residents, but on the cost of delivering services to those residents. After all, British Columbia's problems were "of a permanent character and [could] never be overcome"—an assertion that would surprise future generations.[34] He would boycott the proceedings in protest. Gouin shared this news with the other premiers. They instructed Gouin to reply, and they took a clean shot along the way. "The Conference recognized the claim of British Columbia for exceptional treatment," wrote Gouin. "I am specially directed to add that it was passed *after* all the data presented by you had been fully considered."[35] They were tired of McBride's habit of repeatedly telling them what they already knew.

McBride's temper tantrum had done little good in Ottawa, and maybe some harm. The British Columbia premier eventually returned to the conference chamber, took his seat, and presumably said nothing. But he boycotted the afternoon session with Laurier. He had lost whatever allies he might have had, including the Prairie premiers. That afternoon, Sir Wilfrid outlined a preliminary scheme for a subsidy increase, but he reserved judgment on British Columbia's special needs. McBride was hailed as a hero when he went home, which was surely his intention. Four months later, in early 1907, he would win an election with the slogans "Better Terms" and "Fight Ottawa."

What McBride's dramatics had accomplished was to push the Prairie provinces closer together. But the October 1906 Conference also raised another barrier to Western resource control. Previously, Ottawa had sent a hodge-podge of allowances and subsidies to the provinces every year. The system was a mess. The Constitution Act of 1867 had actually spelled out the per-person subsidy amounts for the original four provinces, and other provinces had secured different amounts when they joined Confederation. Worse, Ottawa had also tinkered with obscure formulas, increasing the debt allowances for Ontario and Quebec in 1884, and raising Manitoba's subsidy in lieu of resources in 1885. The differences in subsidy levels confounded everyone.

But any changes to the basic subsidy would require a Constitutional amendment, which Britain had to approve. Laurier's Constitution Act of 1907 spelled out a new formula with higher specific numbers: the subsidy would increase as provincial populations grew; the maximum amount would take effect when the population in any province reached 1.5 million. Laurier also granted that extra $100,000 per year for ten years to British Columbia. Higher subsidies would purportedly allow the premiers to tackle the wrenching problems of their industrial age.

No good deed goes unpunished. In 1907 McBride actually travelled to London to lobby for changes in that legislation, and he charmed the under-secretary of state for the colonies, Sir Winston Churchill. (Canada could not amend its own Constitution until 1982.) At the premier's behest, Britain removed from the legislation a declaration that the subsidy levels were "a final and unalterable settlement"—but the phrase remained in the schedule attached to the bill.[36] It was a very small victory: those new subsidy levels could never have survived anyway in an inflationary world. The triumph did not mellow the implacable premier, however, and in the 1908 federal election, McBride would campaign fiercely for Robert Borden's Conservatives—arguing that Laurier was sending too many Asian immigrants and not enough money to his province.

But there was a trickier problem in the Constitution Act of 1907 that no one envisaged. What would happen if the population of any region remained stagnant, but the cost of providing better services escalated? The Maritimes would soon complain about that very difficulty. At the November 1918 Conference, Maritime premiers would refuse to countenance Western demands for resource control until their special concerns were resolved, and Laurier's efforts to standardize the formula would backfire on the Western premiers. The history of federalism is littered with ironies.

Oliver's Western empire was expanding—with immigrants that he did not want. They clustered around the coal mines or the sawmills, accepted the brutal jobs on the railroads, or crammed into urban tenements. Most newcomers from Continental Europe could not farm because they had little cash and earlier arrivals had already taken up the best Western land. There were exceptions. In the mid-1900s, a small cluster of Italian families staked out a farming community that they dubbed "Naples," sixty miles northwest of Edmonton. The first settlers found themselves in such thick forests that there was no room to pitch a tent. They often went hungry, and they had to walk to Edmonton for supplies, carrying home bags of sugar and flour. When they eventually had wheat to sell, it was a ten-day journey to get it to market. Such pioneers were rare, however, and few Italians came to Western Canada in the early twentieth century "who were not poor labourers."[37] Newcomers might aspire to own land, but they needed cash first.

These immigrants were different from earlier arrivals. In 1907 unskilled labourers constituted thirty-one per cent of all immigration. That figure would reach more than forty per cent on the brink of the First World War. The trend unsettled Oliver, enchanted as he was with life on the farm. But even Oliver had to recognize the ways of his changing world. Despite his public pronouncements, the Department of the Interior soon resumed the practice of paying a bonus to European agents who recruited carters, railway builders and rail-bed repairmen, miners, gardeners and farmers from such areas as southern and central Europe. The numbers were astonishing: 143,000 people immigrated in 1908, almost 174,000 in 1909, more than 286,000 in 1910, and 331,000 in 1911. That was a torrent of people, especially for a nation that in 1911 would have a population of only 7.2 million people. To his chagrin, Oliver was now the target of caustic cartoons, which charged that *his* immigrants were paupers and criminals.

Those immigrants flocked to the Prairies, and they did the hard jobs. Mine owners packed their workers into camps, where they were isolated because they could not speak English. To survive, the newcomers cherished the very customs that Oliver disdained. Polish priest Anthony Sylla would recall Christmas celebrations in Canmore, Alberta. One Christmas Eve, after Midnight Mass, a group of devout Polish miners arrived at his door singing carols. One celebrant had a homemade replica of the holy family in the stable strapped to his back. When the miner placed it on the ground and removed the cover, the priest saw tiny figures, including kneeling shepherds and a devil urging King Herod to murder the infant Jesus. The men lit candles and sang as the craftsman used a handle attached to a string to move the figures surrounding the family. The priest was deeply touched.

Those gruff miners cherished another custom that would have driven Interior Minister Oliver to despair. During Christmas holidays, young Polish-born men went from door to door with a "goat": that is, one of the youths was dressed as a goat and walked on all fours. They offered to sell the goat, and when no one wanted to buy it, they announced that they would slaughter the goat—despite its pleas for mercy. As the priest would recall, the men sharpened a knife and held the goat's head. Animal blood flowed from a concealed sac. "Very good acting," Sylla would remember, adding that the young people then drank, danced and sang traditional songs.[38]

Even British newcomers cherished their communities and traditions. In Winnipeg, they clustered in the West End, near the CPR shops, where they could rely on each other for jobs and lodging. But the ability to speak English was no guarantee of any easy life. Nearly sixty per cent of the British-born single women who settled in Winnipeg between 1900 and the outbreak of war were domestic servants, and forty-six per cent of the men held unskilled jobs. But their ethnic background did open more doors. Winnipeg boarding houses welcomed them as lodgers who might raise their own status. Hostels sheltered timid newcomers; one elderly Scottish woman would remember the Girls' Home of Welcome as an "Old Country place."[39]

These British newcomers set up societies, clubs and lodges where the homesick could find solace. They dominated the workforce in key firms. The T. Eaton Company recruited skilled seamstresses from London, and those workers then found jobs for friends and family. The shop foremen for the Canadian Pacific Railroad were usually British, and they "routinely discriminated" in favour of their fellow countrymen when hiring.[40] Churches, especially the Anglican parishes, opened social centres. Today, it might seem odd that British newcomers would stick together when Frank Oliver assumed that they would slip into the Canadian mainstream. But their neighbourhood groups created a "competitive advantage" in the job markets.[41] Envious outsiders saw only a privileged clique in a prosperous blue-collar Winnipeg neighbourhood.

Manitoba was thriving at the crossroads of the national boom, but Conservative premier Sir Rodmond Roblin could see nothing but grievances. The map of Canada illustrated his first complaint, for which he blamed Ottawa. When Laurier had carved the provinces of Alberta and Saskatchewan out of the North-West Territories in 1905, he had drawn their northern borders along the 60th parallel. (In 1906, Ottawa officially changed the name of the North-West Territories to the Northwest

Territories.) Manitoba's northern boundary was less than half as far north, and the province had no outlet onto Hudson Bay. Prior to the October 1906 Conference, Roblin had pushed Laurier to extend the boundary, and to raise Manitoba's subsidies in lieu of resources. When Laurier had decided to consult Roblin's neighbours, Saskatchewan and Ontario, Roblin had countered that the issue was solely between Manitoba and Ottawa. That pique had kept the issue off the agenda at the October 1906 Conference, but Roblin persisted.

This Western premier did not charm Laurier, and there is not much evidence that he even tried. The former Manitoba grain merchant was a dynamo, blunt and politically astute. He had courage: he had stood up for francophone education rights against the Liberal government's plans to restrict them in 1892—at the cost of his own seat. But his flaws—"a certain pomposity of speech and manner, a self-confidence which verged on arrogance"—could be unsettling.[42] Perhaps worse in Laurier's eyes, the partisan premier had a formidable political machine, which he swung behind Conservative leader Borden during federal elections. Roblin was a scrappy soul. After his hopes for federal concessions were dashed at the October 1906 Conference, he called an election for March 1907, denouncing the injustice of Ottawa's boundary limits, and won handily. His downfall, enmeshed in scandal, would be well in the future.

His quarrel with Ottawa drifted along until early 1909, when the Legislature demanded the extension of the province's boundaries and increased subsidies in lieu of resources "based upon, and at least equal to" those that Saskatchewan and Alberta were pocketing.[43] Manitoba had first received those subsidies in the 1880s, and the amounts had not kept pace with inflation. Roblin asserted that his province was extremely ill-treated. Laurier was less than sympathetic, however, and presented a frugal counterproposal.

By the fall of 1909, the prime minister and the premier were openly bickering. Laurier offered to extend the boundary to the line that marked the northern edge of Alberta and Saskatchewan, which would vastly increase the province's land mass. But he offered only a token sum of $10,000 a year as compensation for Ottawa's use of the land and resources within that area. Roblin wanted far more money because those lands were rich in minerals. He scoffed at Laurier's offer. "I do not think," he wrote, "you could have been serious when you suggested it."[44] His graceless tone was startling. He also advised the prime minister to proceed with legislation to extend the boundary, with the reservation that the law would only go into effect if his Legislature approved it. This was a novel approach, to which the exasperated Laurier replied: "You tell us that we may ask Parliament to legislate, but you reserve to yourselves the right to reject such legislation . . . it cannot be seriously contemplated to ask Parliament to enact legislation which would not be final."[45]

The face-off was irresolvable because neither leader would yield. Laurier viewed the October 1906 Conference as the final settlement of the subsidy issue, and he refused to contemplate a major increase in other subsidies such as payments in lieu of resources. The Liberal prime minister was also irritated: he would not go out of his way for the brash Conservative, and ill will prevailed. In October 1910, Roblin threatened to put the federal and provincial proposals "directly to the people of this Province" in a referendum.[46] Laurier did not take the bait, and merely offered to resume negotiations. Roblin, in turn, agreed to participate. But the talks remained deadlocked.

The animosity ran deep because Roblin had resurrected another grievance over resource control. In 1885, at the behest of the doomed Manitoba premier John Norquay, Ottawa had transferred the swamplands to his government. But what was a swampland? Did the province cease to control the lands if they were drained? After more than twenty years, officials had not even completed a survey to categorize the land. Roblin was incensed, and he was also looking for a fight. In November 1906, the premier had penned a diatribe. Oliver's Interior Ministry had reclaimed swamplands, it had kept land within provincial timber plots, and it was pocketing dues. It was isolating out tiny plots of swampland for the province, and keeping the good surrounding land. Roblin claimed that Manitoba should have received between eight and ten million acres. Instead, the records showed that because of Ottawa's shoddy administration, it was owed two million acres—but controlled only 1.25 million acres.[47] The premier wanted an extra 768,374 acres and $53,991.13. Four years later, nothing had happened. As long as the fiercely partisan Frank Oliver remained in the Ministry of the Interior, nothing ever would.

Oliver wielded immense power over the West. His responsibilities included the Indian Affairs Department, which he represented in cabinet. That conflict of interest received virtually no attention at the time, but it would spawn a thicket of legal claims in future decades. The Aboriginals were in Ottawa's way: even though the federal government was responsible for Indians and their lands, it was careless with their lives.[48] Ministry officials pestered Aboriginals to surrender land that they considered surplus to the band's needs, and they allowed speculators to snap up indigenous land before settlers arrived. Extortionate officials told hungry Aboriginals that they would not get equipment or supplies "unless their reserve lands were surrendered and sold."[49] Some senior officials benefited personally from the sales. From 1896 to 1911, twenty-one per cent of the lands reserved for Prairie First Nations were "surrendered to the Crown to make way for western expansion and an influx of immigrants."[50]

The bulk of those land transfers took place during Oliver's tenure. But his predecessor, Sifton, was not blameless: he had put a single deputy minister in charge of both Indian Affairs and the Interior Ministry, and he had severely curbed the growth of the Indian Affairs budget. To his credit, however, during his time in power, he had informed Edmonton MP Oliver— who wanted to open up the Cree reserve at Stony Plain to settlers—that Ottawa could not grab land belonging to Aboriginals without their consent, and it had to spend the proceeds from any sale on them. Oliver had fewer scruples. On his watch, Aboriginals would "come under every inducement and pressure to sell their lands and become assimilated."[51]

That was the darker side to Ottawa's resource control. Indeed, no one could ever maintain that the federal Liberals were idealistic. Laurier and his government had made a hard-nosed decision to continue the offer of "free" land to attract settlers who would stimulate the nation's economic growth, and thereby boost the Liberals' popularity. Resource control was also useful for patronage. The Liberals awarded positions such as

resource administrator to loyalists; they expected Interior Ministry officials such as homestead inspectors to double as party workers during
elections.[52] Ownership had its privileges, and Oliver was exercising them.

His empire hummed with such promise that he was not overly disturbed when Alberta premier Rutherford landed in the soup because of
his aggressive promotion of railroad construction. The premier could
not resist the siren song of Edmonton businessmen who wanted railroads around the rapids of the Athabasca River to Fort McMurray's Oil
Sands—which would at that time prove too difficult to exploit. In 1909
Rutherford removed railways from the control of Public Works Minister
William Henry Cushing, and created a Department of Railways, which
he decided to administer himself. The premier guaranteed the bonds for
three railways, including the new Alberta and Great Waterways Railway.
That guarantee was a bonanza for the A&GWR promoters: $20,000 per
mile for 350 miles. With the province's backing, the bond issue was naturally oversubscribed, and the railroad promoters tucked $7.4 million into
three bank accounts.

Their good luck, however, became the premier's downfall. Members
of Rutherford's own caucus, along with Conservative leader R.B. Bennett,
peppered the hapless premier with questions about that ultra-generous
financing. On February 14, 1910, Public Works Minister Cushing resigned
in protest. In a devastating letter to Rutherford, he condemned the premier's guarantee of those A&GWR bonds. "This transaction, put through
without my knowledge or consent, is, in my judgment, such that I cannot
with sincerity of heart and honesty of purpose defend before the electorate of this province," Cushing wrote. Then he added a zinger: "You have
utterly failed to protect the interests of the people."[53]

It took almost two weeks for news of Cushing's resignation to become
public, and the premier must have dreaded the revelation. His fears were
justified. The voters were fiercely critical, Rutherford's damage control
was inept, and more ministers resigned. Interior Minister Oliver, who had
never trusted Rutherford, supported the rebels—though he would eventually regret the loss of a politician who refused to rock Ottawa's boat.
Rutherford's hold on power was slipping, and in March, the premier

established a Royal Commission to inquire into the mess. Caucus insurgents kept plotting. In May, Lieutenant-Governor George Bulyea and the president of the provincial Liberal association, Senator Peter Talbot, turned on him and he resigned.

Six months later, that Royal Commission would clear him of wrongdoing, but two of the three judges would lambaste him for his lack of judgment: he had been negligent and gullible. Rutherford would linger on the backbenches, where he would stew impotently until he lost his seat in the 1913 Alberta election. According to historian Patricia Roome, he had always nurtured "two other passions: religious service and financial speculation," which had probably absorbed him during his railway negotiations, to the detriment of his taxpayers.[54] But his private real estate deals would also provide wealth, including a Banff cottage and a Strathcona mansion near the University of Alberta, where he would sit in the university senate. Rutherford would remain bitter: in 1921, the once-devout Liberal would campaign for the provincial Conservatives. His graceful mansion, Rutherford House, would become a provincial historic site.

In May 1910, however, his most notable political legacy was chaos. The man who had few enemies in 1905 had accumulated plenty by 1910, and many in his own caucus were seething. None of those Liberal members had the clout to unite the factions after Rutherford's departure, and Conservative leader Bennett was a formidable and damaging opponent. Lieutenant-Governor Bulyea and Liberal leader Talbot, in consultation with Laurier, finally spotted Rutherford's replacement among the judiciary.

Their choice was none other than Arthur Sifton, that loyal Liberal and fervent proponent of Western rights, who had left Haultain's cabinet in 1903 to become chief justice of the North-West Territories Supreme Court. Four years later, Sifton had become chief justice of the Alberta Supreme Court, and he liked the job. But when the Liberals dangled the position of premier, the prospect was too alluring. He accepted and became Alberta's second premier. By late June 1910, he had appointed a cabinet, won a by-election in a riding east of Edmonton, and joined Laurier's summer tour of the West. The new premier spoke at every Alberta stop, impressing the prime minister.

Prosperity reigned. Fifty miles west of Estevan in Saskatchewan, young Jewish farmers from Galicia were thriving. Israel Hoffer had saved enough money to purchase a team of horses, along with a cow and calf. He had survived on flour and corn meal, made his own butter and cheese, and eaten wild duck eggs. He had lived in a sod hut, which had collapsed, and then in a shack with no roof. He had cleared stones. But by 1910 his crops were flourishing. He would eventually own "one of the best organized and best operated farms" in Southern Saskatchewan.[55]

Such optimism was bracing.

The boom in the West even caught the attention of the Maritime premiers. Although their manufacturers were struggling with their distance from markets and the small scale of their operations, the premiers fixated on the Maritimes' dwindling political clout, which reflected their stagnant population growth. After the 1891 census, the three provinces had collectively lost four seats in the House of Commons. A decade later, another four seats had vanished because populations in other regions such as the West were growing. Ironically, while the Western provinces were complaining about the cost of providing services to more people, the Maritimes were jealous because their brightest youths were heading west.

By the early 1900s, as their proportion of the Canadian population declined, Maritimers felt the need to shore up their influence. Prince Edward Island had only four seats, and wanted to go back to the six seats that it had held on its entry into Confederation. New Brunswick, in turn, insisted that Ottawa should not include Westerners when it calculated how to distribute seats: New Brunswick's representation should be based solely on its proportion of the population of the original four provinces. That suggestion would have artificially maintained the number of Ontario seats over the heated objections of Quebec, so Laurier would not consider it. New Brunswick and Prince Edward Island had taken their case for more seats to the Supreme Court of Canada in 1903, and then to the

Judicial Committee of the British Privy Council in 1905. They had lost. But the issue continued to simmer, and bitter advocates of Maritime Rights would thwart the West at the November 1918 Conference.

By 1908, when New Brunswick premier John Hazen—whose Conservatives had just swept into power—called for a "united Acadia" at a Halifax public gathering, the first signs of a Maritime Rights movement were appearing. In 1909, the Methodist newspaper, the *Wesleyan*, urged politicians and journalists to unite in the fight for fair Maritime representation. Easterners drew a bitter connection between their diminished clout and the growing number of MPs from the West. In 1910, desperate for a solution, Maritimers offered to settle for "a guarantee of their existing allotment [of seats] as a basic minimum."[56] To their dismay, Western MPs objected vehemently, claiming that special status for the Maritimes would damage the balance of power with the Rest of Canada.

Laurier looked for a diplomatic dodge. He expressed his sympathy for the Maritimes' situation, but shrewdly suggested that the three premiers should first look to their counterparts for a solution. The trio took the bait, because they were hoping to find allies. At their instigation, representatives from seven provinces gathered on Parliament Hill on Friday, December 9, 1910. The records from that interprovincial conference are brief, but they tell a revealing tale. The premiers discussed Maritime representation in the House of Commons "at some length" in the morning, adjourned for lunch, and then met again at 2:45 p.m. Quebec Premier Lomer Gouin's resistance was pivotal: he refused to consent "to any decrease in the proportion of [Quebec] MPs" in the House of Commons.[57] One hour and forty-five minutes later, at 4:30 p.m., Ontario Premier Whitney "declared the Conference adjourned to a future day."[58] So much for Maritime aspirations.

The conference had been a debacle, and the Maritime premiers went home empty-handed. Their campaign was now stalled but the regional positions that would define the November 1918 Conference were already taking shape. A month after that stalemate, in January 1911, Saint John MP J.W. Daniels told the House of Commons that his voters were people from good stock, the descendants of the early French, English, Irish and

Scottish settlers. Western MPs, he sniped, mostly represented "Galicians, Lithuanians, Ruthenians, Buckowinians and possibly Dukhobors [sic], all new comers [sic], a great many exceedingly illiterate, not able to speak the language of the country and unacquainted with its constitution."[59] Such incendiary rhetoric did little for inter-regional harmony.

The federal government, however, was on a roll. Maritime complaints were foiled, and Laurier had neatly dodged the blame. Central Canadian industries were humming. British Columbia was in such good financial shape that the brash McBride guaranteed the interest on the bonds of one railroad, and subsidized another. Despite Manitoba premier Roblin's fulminations, the economy of the West was thriving and for a while the federation was relatively calm.

But the Alberta premier kept alive the saga of federal–provincial disagreement over resources. A veteran of the failed campaign at the turn of the century to obtain control of the North-West Territories' natural resources, Arthur Sifton lacked his predecessor's ardent loyalty and his reluctance to rock the federal boat. At that time, the elder Sifton had been treasurer in Haultain's cabinet. Now, after a prod from his still-divided caucus, he revived a modest version of that campaign. In November 1910, a Rutherford loyalist demanded that Sifton secure control of the natural resources "of a purely local concern," because Alberta needed the revenue.[60] Sifton assured the Legislature that his cabinet was already considering the issue, and the insurgent withdrew his resolution. On December 1, 1910, at Sifton's behest, the Alberta Legislature endorsed his efforts to open negotiations with Ottawa for the "control of all such natural resources as are of a purely local concern."[61]

In Saskatchewan, Premier Walter Scott was also chafing at Ottawa's limitations. As well as asking for control over all natural resources of a purely local concern, his Liberal government also wanted control over all lands in the northern part of the province that were not required for immigrants. As well, they wanted jurisdiction over the school lands, along with any money in trust from the sale of those lands. The Westerners were back at Ottawa's door.

In January 1911, Arthur Sifton took his message on the road to the Canadian Club in Toronto. His speech lasted only twelve minutes, but that normally glacial politician, whose nickname was "The Sphinx," entranced the staid Toronto *Globe*. "Somewhat of a western whirlwind struck the Canadian Club yesterday," the paper recounted, praising Sifton's "breezy eloquence." The premier conceded that Ottawa should probably retain control of any vacant homestead lands, so that immigration and settlement could proceed in one seamless process. But the province should manage its own natural resources; Sifton specifically mentioned hard coal, lignite, timber and waterpower. He eerily anticipated the conflicts that would come up at the November 1918 Conference. "There must be one solid plane for every Province," he told his rapt audience. "The prairie Provinces have no quarrel with the eastern Provinces, there is not jealousy over political power, but a desire to live with you in unity. . . . The people of Alberta and Saskatchewan and Manitoba must be placed upon an equality."[62] He presumed to speak for all three provinces.

Two months later, in March 1911, the Alberta premier sent a remarkable thirteen-page letter to Laurier that outlined his case for resource control and his province's financial predicament. It would become a pivotal missive in Western history—and Western militancy. Laurier must have shaken his head in consternation: Arthur Sifton was picking up where former NWT premier Haultain had unhappily left off in 1905. Now, Sifton was arguing the same case, with the same skill as Haultain.

The Premier's arguments were powerful, and they would recur in the speeches of later Westerners like a refrain. He traced his province's growth from a "state of semi-feudalism" to *almost* full status.[63] He acknowledged that Ottawa had the legal right to determine the West's powers. Sifton even made a remarkable admission: "In the light of the events of the last forty years," Ottawa's decision to retain resource control in 1870 "may be conceded to have been justified . . . and may be held to have been the only possible one."[64] Every Canadian, including Westerners, had initially benefited from Ottawa's control, because the Dominion could afford to give away Western lands to new settlers. If Manitoba and the Territories had

controlled the resources in those early days, the cost of surveys and settlement would have fallen upon "scattered communities, undeveloped and
sparsely populated."[65] That would have hurt everyone.

But that was emphatically in the past. "The time has arrived in the history of Western Canada," Sifton wrote, "when the reasons cited before
no longer apply."[66] Now, Alberta's population was skyrocketing. The
Provincial government was paying for roads, bridges and telephone lines,
but the newcomers could not generate enough economic growth to
cover that tab. Once again, the green eyes of jealousy appeared. The
Alberta government had spent $10.15 per person in 1910, while Ontario
had spent only $3.50 in 1909 and Quebec had spent only $3.46 in 1910.
Alberta had not asked for these obligations: Frank Oliver's immigration
officers had created them. Sifton needed cash. The Dominion's exploita-
tion of the West's lands and resources was placing "a burden onerous to
bear" on his government. "[It] should be relieved or placed where the
benefits fall, if serious financial distress is to be prevented."[67]

Sifton then outlined the demands that constituted his province's
bottom line; they would only grow more insistent and ambitious over
the coming decades. Ottawa had used 100 million acres of the Northwest
Territories' land for free homesteads and a national railway system.
Now—with key exceptions such as land earmarked for homesteads—
Alberta wanted control over its lands, mines, minerals, timber and any
royalties collected after a fixed date, as well as any waters that Ottawa
administered under the Irrigation Act.[68] Sifton added a bitter arithmeti-
cal shot: Ottawa's revenues from Western resources were $2.86 million
in 1909–10, while its administrative costs were only $600,000. If Alberta's
demands were met, the Dominion could still cover its immigration costs
as well as rescuing his Provincial Treasury. Alberta officials could easily
do Ottawa's job—despite Ottawa's claim that it could administer the
resources more efficiently and cheaply.

Laurier was annoyed. Huge numbers of immigrants were still pour-
ing in and the current system was comfortable and convenient. It would
be difficult to decide which lands were still needed for settlement. But
the prime minister could hardly ignore a thirteen-page letter from

Alberta's Liberal premier. The two politicians had informal discussions throughout the summer of 1911, including at least one face-to-face encounter in early August. During that pivotal meeting, Laurier verbally outlined Ottawa's conditions for any transfer. However, he did not draft his formal written summary of those conditions until after the meeting, and he would not send that summary until it no longer mattered.

His response was scarcely a triumph of diplomacy. Laurier clearly worked on the letter: several drafts have survived with varying dates and handwritten notations. His tone was cool, almost aloof. He reviewed the constitutional history. In 1870 Ottawa had retained jurisdiction over Manitoba's public lands, because they were "largely prairie lands, immediately fit for settlement and tillage, and likely to be sought by immigrants." That same principle had applied to Alberta and Saskatchewan in 1905.[69] Now, he noted, Sifton wanted control over *some* of the lands that were not set aside for settlers, and over the resources that lay beneath those lands.

Laurier expressed pleasure that Sifton had not asked for complete control over the entirety of the provincial lands, but he still took a swipe at the premier. Sifton's request would alter the way that subsidies in lieu of resources were calculated, so "those financial terms would have to be revised."[70] This was a polite threat that it would be then necessary to slice Alberta's subsidies, which no Western province could afford. The prime minister blandly added that the topic might be "a fair subject of discussion" at a later date, but he would not commit himself.[71] Laurier did not send this letter immediately: he was waiting for Interior Minister Oliver to return to Ottawa for consultations.

Arthur Sifton headed home. En route, in Winnipeg, he spotted a headline: "Laurier Fearful of Western Sentiment Offers Resources." When journalists confronted him, he confirmed that Ottawa was contemplating a transfer of resource control.[72] A day later, Laurier reproached Sifton: he was "distressed" to read the Premier's remarks when Ottawa was not ready to make an official statement. "The premature announcement which you made will a little complicate matters," he wrote in the tones of a strict schoolmaster, "though I believe that in the end we can come to a satisfactory conclusion."[73] Taken aback, Sifton replied with a curt, scribbled note from Edmonton on August 18, 1911. He was "a little surprised" to receive Laurier's letter because he was only responding to a newspaper headline, which he enclosed.[74] He had put a good spin on that dour story of impending Liberal doom: "It was by good luck that I succeeded in getting a fair version out."[75] Anyway, the initial report was the "result of information presumably from your office."[76] The premier even offered to supply the name of the reporter who had filed the story from Ottawa.

They could not know it, but those two politicians were wasting each other's time. Laurier called an election for September 21, 1911, which he fought on the issue of free trade with the United States. The West and the Maritimes largely supported that initiative; Ontario and Quebec flocked to the Conservatives under Sir Robert Borden, who stoutly opposed it. Laurier lost. In early October, probably as a result of bureaucratic inertia, Laurier's formal response on resource control was finally sent to Sifton. Laurier's letter noted that Frank Oliver had agreed with his decision to pare subsidies in lieu of resources if Alberta received control over any resources. Ottawa's ultimate position at the November 1918 Conference was already hardening.

Oliver would hold his Edmonton seat in that 1911 election. As Western political boss, he had helped to ensure that Saskatchewan and Alberta voters remained behind his party. But his days as a Western kingpin were over. He would remain in Parliament until his defeat in the 1917 federal election, and in 1921 would lose another bid for re-election. In 1923 he became a member of the Board of Railway Commissioners. Ten years later, while handling board issues in Ottawa, he took ill; he died at the Ottawa Civic Hospital in March 1933.

The Edmonton MP would leave an ironic legacy. By 1911, although the number of British immigrants had increased, the number of newcomers from Central, Eastern and Southern Europe had grown even more. When Oliver left the Ministry of the Interior, ethnic Canadians made up an astonishing nine per cent of the country's population. Most of them had settled in his Western empire.[77] The Prairie provinces watched the giveaway of their lands and resources, pinched pennies, and bided their time.

BORDEN DALLIES:
THE BIRTH OF THE GANG OF THREE.
SEPTEMBER 1911 TO AUGUST 1914

—

Sir Wilfrid Laurier had gambled his party's fate on the promise of free trade with the Americans, and he had lost. While Western Liberals mourned his defeat, they hoped that the new Conservative prime minister would step in to solve their problems. Sir Robert Borden had staunchly proclaimed their right to administer their resources when he had barnstormed the West in 1902. He had set no conditions on the transfer. When Laurier had withheld resource control from Alberta and Saskatchewan in 1905, Borden had accused him of not trusting the people "as I would be willing to trust the people, in this regard."[1] When the dapper Conservative lawyer from Nova Scotia took office on October 10, 1911, Western premiers expected he would keep his word.

Perhaps Alberta premier Arthur Sifton and Saskatchewan premier Walter Scott had forgotten they were Liberals, and early twentieth-century politics was an extremely rough game. The dignified Borden may have professed his dislike for politics, but he played it with clumsy resignation and able henchmen. The new prime minister knew which premiers were his political friends. Manitoba premier Rodmond Roblin had thrown his powerful machine behind Borden during the election. So had British Columbia premier Richard McBride. Borden owed them gratitude. But he owed nothing to Sifton or Scott.

They had supported Laurier on the campaign trail, and partly because of their opposition, Borden had taken only one of the seven seats in Alberta, and only one of the ten seats in Saskatchewan. Borden would have to placate the farmers in those two provinces that had swung behind Laurier's promise of free trade—and cheaper American farm machinery. But the prime minister would not do anything that would make the Liberal premiers look good to their electorates. Borden was not nasty to Liberal partisans—he left that chore to underlings—but he was never nice.

A better politician might have tackled the regional tensions that were disrupting the federation and thwarting the West's bid for resource control. Many Central Canadians still asserted that they had bought the West and owned the resources. Borden tried to placate those voters. In 1907, at a Tory rally in Halifax, he had promised to restore the public lands to the Prairie provinces "upon fair and reasonable terms."[2] He had also vowed to manage the "public domain so that [the] increment arising therefrom shall inure to the people."[3] That could mean everything. Or nothing. The only certainty was that the Prairie premiers were ignoring reality: Borden was now detached from their cause because the other regions would not countenance any transfer of resource control without compensation to them. Such negotiations would be too much trouble.

The prime minister had no gift for domestic politics. He would reward his badgering friends, one by one, until the onset of the First World War, with no overall plan for harmony or balance. He simply tried to give them whatever they wanted. Over those same years, he would largely ignore the Liberal governments in Alberta and Saskatchewan, and when pressed, would disdainfully brush them off. It would be on his watch that politics edged out practicality as the determining factor in withholding resource control from the Western premiers.

The prime minister would finally discover the limits of party loyalty in October 1913 during a futile meeting on resource control with the three Prairie premiers. Borden would prepare for that gathering with secret briefing notes that outlined why he should wriggle out of his promises to transfer resource control, and how he could shift the blame for that decision to the Liberals. He did not expect that the improbable— indeed, surely the impossible—might happen.

The events of those crucial peacetime years, from the fall of 1911 to

August 1914, would stoke every resentment that the Western premiers would encounter at the November 1918 Conference. Borden's partisan favours would only deepen the jealousy in neglected provinces such as Nova Scotia and New Brunswick, and beneficiaries of his largesse, such as Roblin, would have short memories. Dismay at Borden's neglect would even disrupt the operations of Parliament and derail his international agenda. A skilled statesman abroad, the prime minister seemingly could not tap his diplomatic skills at home to make the federation work. His disinterest and distraction, coupled with the machinations of his henchmen, would do much to foster Western alienation during those halcyon days before the world went to war.

Borden took office during the boom years, while Western provincial governments were expanding their services and their skills. The Prairies teemed with life. Winnipeg was thriving after a decade of breathtaking growth, and its population had tripled since 1901. Four hundred industries churned out $50 million in products each year, and twenty-seven railroad lines stretched into the city. Winnipeg was the "metropolis of western Canada," with its imposing Grain Exchange and banks, its flourmills, breweries and slaughterhouses, and its transportation network.[4] New arrivals could easily find jobs amid the industrial tumult.

The city was at the height of its pre-war influence and affluence. Although many vehicles were still horse-drawn, by 1912 there were more than two thousand registered automobiles in Winnipeg alone. Those automobiles brought "excessive speed, accidents, and even some fatalities," but they also symbolized wealth and cachet.[5] The social elite held afternoon teas, which the newspapers hungrily chronicled. The Royal Alexandra Hotel, the pride of the Canadian Pacific Railway chain, was adding guest rooms, a ballroom and a large banquet room. The wealthiest were building palatial homes south of the Assiniboine River with attached stables and full staff. Multi-storey office towers spiked along the main streets.

The *Canadian Annual Review* even published a special twenty-page segment on Winnipeg and its "certainty of greatness." The prose was

breathless. Winnipeg was "the capital of a Province where public prosperity and individual opportunity are manifest."[6] Such bravado would
soon seem poignant: Winnipeg was edging toward an economic turn-
down and the collapse of its real estate market. But, in those early years
after Borden's victory, the city was the prosperous intersection of the
Canadian East and West.

Good times prevailed across the Prairies. In Saskatchewan, a diminu-
tive Englishman, Seager Wheeler, who had emigrated from the Isle of
Wight, watched in awe as seeds from a federal agriculturalist ripened on
his farm north of Saskatoon in the autumn of 1911. This new Marquis
wheat matured a week earlier than the standard Red Fife variety—and
the yield in bushels per acre was astonishing. Wheeler entered grain
samples in the World's Premier Wheat Competition in New York and
won the top prize of $1,000. The new Wheat King of the World promptly
paid off his mortgage with the Canadian Pacific Railway, which had
sponsored the contest. Hardy wheat would be a godsend for the entire
economy. It would bring "big profits down the line" for the railway and
shipping companies, and it would spur immigration. The Prairies' reputation for grain growing would soar from "questionable status . . . to its
recognition as the British world's breadbasket."[7]

In Alberta, Sifton's government created a full-page immigration promotion, "Free Homes for Settlers," with photographs of the premier
and his cabinet. That slick pitch boasted about Alberta's agricultural
output, its "advantageous climate" and the "stream of immigration"
from the United States and the Homeland, which was a coded assurance
to those who disliked foreigners. The "deep, rich black soil" could be
cultivated "at comparatively small cost." Although the province was rich
in resources, "the great mass of her people will live upon the land, and
agriculture will be the basis of her permanent greatness." Immigrants
should not fear loneliness and hardship. Settlement was so rapid "that
pioneering is shorn of its desolation. . . . The pioneer will always have
neighbors."[8] The fine print added that roads and schools would follow
settlement "in due course." Alberta was depicted as heaven on earth.

To the west, in British Columbia, Premier Richard McBride was

revelling in this good life. He travelled to England in 1911 for the coronation of King George V. He would return there in 1912, and renew his friendship with Winston Churchill. Furthermore, after a lengthy audience with
the king, the captivating Westerner would be knighted. Meanwhile at
home, his province's economy was flourishing, and the premier was busily
forging railroads across the mountains and up the wild Western coast.
McBride's love for railroads would eventually trump his common sense
and strain his taxpayers' pocketbooks—and would ultimately be his
undoing. For now, however, in those forward-looking years before the war,
with his ally Borden in power, Sir Richard was on a decidedly heady surge.

Borden barely had time to settle into his new office before his provincial
counterparts called in their debts. On October 31, 1911, Manitoba premier
Roblin pounced. He sent an extraordinary sixty-two-page memorandum
that called for a boundary extension and a subsidy increase. In his
covering letter, Roblin played that destructive game of comparing
Ottawa's treatment of his province with the fortunes of other provinces.
Saskatchewan and Alberta received higher subsidies in lieu of resources;
Ontario and Quebec controlled their lands, timber, mines and minerals.
Roblin wanted his subsidies to equal those of his Prairie colleagues, or
he wanted resource control. Past inequality had already sparked a "feeling of unrest akin to irritation." Just imagine, Roblin wrote, "how unjust,
unreasonable and inequitable" Manitoba's situation would be without
remedy.[9] The premier's tone was vaguely menacing: the prime minister
should consider the "injurious effect" of doing nothing. Roblin did not
need to remind Borden that he had just delivered eight of the province's
ten seats to the Tories.

A mere week after Roblin's missive, on November 6, British Columbia
premier McBride sent a fifteen-page shopping list of more than a dozen
demands. His machine had delivered all seven of British Columbia's
seats to Borden, and now the premier wanted payback. His wish list
included a halt to Asian immigration, control over the railway belt and
the Peace River block, and the appointment of a commission to examine
British Columbia's need for higher subsidies. In a covering letter, McBride

called for "an early adjustment" of his claims.[10] His first whirlwind messages were sent to Borden's temporary home at the Russell House Hotel in Ottawa: they were a sign of the nuisance ahead. McBride would become the most adamant of Borden's supplicants. He would deploy guile and charm and stubborn persistence to pester Borden through peacetime and, shockingly, into wartime.

Borden endured doggedly. As *The Globe* once sourly observed, he was "incapable of inspiring enthusiasm."[11] He was a man of his class and time—and he was also a bit of a snob. Whenever he visited Europe, he carefully recorded his elite social connections. During a trip in 1912, he was agog at the elegance of a state ball in London. "It was most brilliant in every way," he would later write. "I particularly remember the wonderful display of gold plate."[12] He noted encounters with Lord Salisbury, Lord Kitchener and Lord Robert Cecil. He chatted about his delightful meetings with Foreign Secretary Sir Edward Grey and the secretary of state for the colonies, Viscount Lewis Vernon Harcourt. It was "busy work," hurrying between appointments, hobnobbing with the British upper class, and he revelled in every hectic minute of it.

On that same trip, however, he met with a delegation of London suffragettes, and peered down his long nose at them. "The secretary, who spoke at some length, was rather aggressive and bumptious and did not at all impress me," he sniffed. "Another young woman, with the face of a martyr, spoke with great feeling and effect. . . . I informed them that we would be guided by reason and our best judgment." The suffragettes should not imagine that any coercive measures "would have the slightest influence upon us."[13] The interview was quickly closed. Borden rarely recorded the appearance of his male acquaintances, but women who wanted to participate in politics were lesser souls. He could not conceal his disdain.

Unfortunately, the prime minister adopted the same patronizing attitude whenever Liberal premiers asked favours. In his *Memoirs*, Borden wrote that a friend of Sir John A. Macdonald had revealed the Old Chieftain's maxims: a political leader should never allow his personal feelings to affect his selection of colleagues; and he should never quarrel so violently with a political opponent that the two could not work together. To this, Borden added his own principle: "A political leader should use discretion in attack and should not always hit with all the strength at his command."[14] The prime minister forgot a fourth principle: leaders should adopt Sir Wilfrid Laurier's "sunny ways" of compromise. Although Borden had plausible reasons for keeping control of the West's lands and resources, he did not use honeyed tones when he explained them. He did not even bother to promise "eventual" change. Provincial bickering exasperated him.

Another thing: he could not always physically cope with the day-to-day hassles of office. He was prone to nervous stress: he would push himself to the limit and then collapse, often with serious outbreaks of boils in that pre-antibiotic age. Although he would write earnest letters to supporters, he disliked the grubby business of politicking. He also hated the constant importuning. As he once confided to his diary: "Countless interviews with people who have been hovering like vultures."[15] In photographs, Borden usually appears serious, his greying hair combed away from a centre part, his thick moustache neatly trimmed, his mild eyes fixed on a better time. He looks like a responsible person who could be trusted to handle the complexities of diplomacy. What the photographs cannot show is the political tin ear for domestic affairs that would shape his legacy.

The Western Liberal premiers addressed their first pleas to Borden in late 1911. They had no luck. On November 8, 1911, two days after McBride had sent his imperious shopping list, Alberta premier Sifton forwarded the thirteen-page letter on resource control that he had originally sent to Laurier, along with the Assembly's 1910 call for control.[16] Borden ignored him.

That same day, Saskatchewan premier Scott protested the hints that Ottawa would soon extend Manitoba's boundary. He was outraged. When Regina was the capital of the Northwest Territories, he noted, its jurisdiction had extended over the northern land that Borden had now apparently earmarked for Manitoba. That land should go to Saskatchewan, he insisted, so that his grain growers could have access to Hudson Bay.[17] Furthermore, Scott claimed that Laurier had agreed to consider his claim.

Five days later, Scott sent another list of issues to Borden: he wanted control both over his resources and over his school-lands trust funds. He requested a meeting. On November 14, 1911, Borden promised to consider—eventually—Scott's objections to the Manitoba boundary extension, and requested more information. Scott's second letter received no reply. Six weeks later, Acting Premier James Calder—who now routinely represented Scott after a severe bout of pneumonia and stress in 1906–07 had forced the premier to spend half of each year in a warmer climate—renewed Scott's request for a meeting.

In a two-sentence letter, Borden curtly refused. "It will be quite impossible to fix any date for a conference before the end of the present session," he wrote impatiently on January 9, 1912. "We have been obliged to defer by reason of absolute necessity, consideration of many matters which are of the highest importance."[18] Calder was stung. The prime minister was not winning friends in the West.

As for the relentless McBride, perhaps he had already driven the prime minister to distraction. The British Columbia premier pestered Borden and his ministers with telegrams and letters. It was surely no accident that the diary entry about hovering vultures was dated just after another impossible demand from McBride. But Borden did go out of his way to pay his political debts. In late December 1911, at Borden's request, the mercurial Sam Hughes, federal minister of the militia, had investigated one of McBride's problems with uncharacteristic diplomacy. The province was building a university on the tip of Vancouver's Point Grey, even though the Ministry of Militia and Defence controlled that land. Hughes concluded that although Ottawa's title was ironclad, it would "seriously inconvenience and discredit" McBride if Ottawa asserted its ownership.[19] So he proposed that McBride should transfer other parcels of land to his ministry as compensation.

In January 1912, the prime minister generously offered to swap lands to settle the Point Grey dispute and to investigate British Columbia's case for special subsidies. Laurier had previously turned down McBride's earlier request for a three-man commission on subsidies, but Borden now accepted that proposal. Ottawa would name one commissioner, the provincial government could appoint the second, and the two commissioners would select the third. If they could not agree, the British secretary of state for the colonies, Viscount Harcourt, would select the third.

McBride responded with alacrity. He was "greatly obliged" for Borden's offer to establish a "Better Terms" Commission.[20] But he rejected any compromise on Point Grey: he simply wanted to keep the land. He also asked the prime minister to address three more issues soon so that he could report on his progress to the Legislature. All this, in a terse, almost peremptory one-page letter. It was hard not to sympathize with Borden: over the next few years, McBride would send a stack of correspondence.

The new prime minister wanted to dispense with his election obligations as expeditiously as possible. He listened to the pleas of Prince Edward Island's Conservative premier John Alexander Mathieson, and awarded an extra annual subsidy of $100,000 to his struggling government. He also had a debt to Quebecers, who had elected a surprising twenty-six Tories, up from twelve in the 1908 federal election. Although the Quebec government was Liberal, Quebec nationalists under former Liberal MP Henri Bourassa had swung behind Borden's party. He was also indebted to Ontario's Conservative premier James Whitney, who had helped to deliver seventy-one of Ontario's eighty-six seats. To repay that support, Borden was carving huge sections out of the Northwest Territories to extend the boundaries of both provinces. He was simultaneously pushing Manitoba's boundary into the Northwest Territories, and mulling its demand for higher subsidies and compensation for the misuse of its swamplands. When Federal finance minister Thomas White scoffed at Manitoba's bookkeeping and demanded an audit of its swampland expenditures, Borden demurred. He was not inclined to penny-pinch his allies.

He went to remarkable lengths for his friends. When Premier Whitney asked what would happen if his province needed to push its Temiskaming and Northern Ontario Railway to a port on Hudson Bay near the mouth of the Nelson River, Borden actually brokered a compromise. Manitoba agreed to let Ontario operate a railroad within a five-mile corridor across its lands; it would even exempt that railroad from taxation.

But with Saskatchewan the prime minister would not discuss the Manitoba boundary extension. He was well aware of its objections, if only because the Deputy Premier would not stop writing. In late January, Calder reminded Borden that he had not replied to Saskatchewan's request for a meeting on resource control—even though he was extending Manitoba's boundaries. Borden did not answer. On March 6, 1912, when Calder learned that Ontario was getting access to Hudson Bay ports, he protested "in the strongest possible manner," and warned that his voters would remember the inequity "for all time to come."[21] It was an ill-advised telegram.

A day later, he received a crisp reply, in which Borden pointed out that Ottawa controlled the public lands of the Northwest Territories, including the land that it was granting to Manitoba. Ontario had merely been granted a five-mile-wide strip of land for its state-owned railway. If Saskatchewan wanted to construct a state-owned railway to the shores of Hudson Bay, Ottawa would consider that request, too. But Borden was now up-to-date on this dispute: Laurier had already rejected Saskatchewan's claims. His decision was firm: the new boundary would run straight north from the existing Saskatchewan–Manitoba border.

Calder, however, would not contemplate retreat. Saskatchewan had never considered Laurier's rejection to be final. Anyway, he sniped, since the prime minister was handing out land to Manitoba, it was evident that he should transfer resource control to Saskatchewan. Local grain production was growing so fast that a provincial railroad would eventually need access to Hudson Bay. "At the very least," Borden should establish Saskatchewan's right-of-way through Manitoba immediately.[22] Once again, the prime minister paid no attention.

But Borden would remember the objections of his Maritime allies to the boundary extensions, and he would exploit them. In late March 1912, New Brunswick's fledgling Conservative premier James Flemming listed a compendium of complaints. If the areas of the other provinces grew, Maritime representation in the House of Commons might have to shrink. Borden, he complained, had granted an extra $100,000 a year to Prince Edward Island before the other provinces had the "opportunity of being heard."[23] His province was picking up the tab to train teachers who then moved west.[24] Flemming had a bottom line: before Borden doled out any more subsidies or Dominion lands, he should "make some reasonable arrangement" with Nova Scotia and New Brunswick.[25] Competitive federalism continued to nibble away at national harmony.

Flemming had no impact on the boundary bills. As he was surely aware, his letter arrived far too late to affect that decision. But the New Brunswick premier was putting the Maritime position on the record: Borden, he implied, should consult every province before he transferred any more federal lands and resources or raised individual subsidies. For Borden, it was a highly convenient letter. He would rely on such Maritime objections when the Gang of Three demanded resource control at the November 1918 Conference. The Easterners would not disappoint him— they would insist that their special claims trumped the West's requests.

The boundary bills zipped through the House of Commons, as Opposition MPs had far more delectable political fish to fry. Borden's government had hired so-called executioners to investigate the role that Liberal patronage appointees had played in the last election. One "executioner" had probed the actions of an Intercolonial Railway superintendent who had instructed an engineer to blow his whistle repeatedly during an open-air Conservative rally. The executioner had then billed $2,500 for his services, which the auditor general refused to sanction.[26] The Liberals pounced on this attempted outrage on the public purse.

Amid that uproar, the boundary bills received Royal Assent on April 1, 1912—less than six months after Borden had taken office. The transfers were massive: Ottawa almost doubled Quebec's land mass, pushed Ontario into the northwest, and brought Manitoba's northern boundary to the level of Saskatchewan and Alberta. Manitoba returned control over its swamplands to Ottawa. Borden, in turn, increased Manitoba's subsidies in lieu of resources from $100,000 per year to more than $550,000. That was well above Laurier's offer of an extra $10,000 per year to compensate for federal resource control over the expanded turf.

It was also a mark of the attention that Borden accorded his supporters.

Alberta's Arthur Sifton and Saskatchewan's Walter Scott could only look on enviously. That saying on the other side of Alice's looking glass had come true: "The rule is, jam tomorrow and jam yesterday—but never jam to-day."[27] The Liberal premiers were disgruntled, but there was little they could do.

The inequity rankled. But the notion of inequity was ingrained in the West during those pre-war years, when science had supposedly proven that many newcomers came from inferior genetic stock. The premiers were oblivious to the ironies of their world, and in mid-March 1912— perhaps because it was an election year—the Scott government passed a law that epitomized the worst of the times. It stipulated that no white female could reside or work in any business that an Oriental person owned. Offenders faced a fine of up to $100, or up to two months in jail. In late May 1912, the inevitable happened: naturalized Canadian citizen Quong Wing, better known as "Charlie," was convicted for employing two white waitresses at his restaurant in Moose Jaw. His fine was five dollars.

The incident might have ended there, but Quong Wing was insulted, and other British colonies and allies were paying attention to his dismay. In the summer of 1912, while Borden was socializing with the elites in London, the British Colonial Office sent a devious memorandum to Borden at his Savoy Hotel. The governments of India and Japan fiercely resented the legal discrimination. Why not alter the legislation so that anyone who wanted to employ a white woman needed a "licence to be obtained from some executive authority?"[28] There was no need to be explicit about race when the executive authority could simply deny a licence. Ottawa sent word of Britain's displeasure to Scott, who in turn amended the law to ensure that it applied only to employers of Chinese ancestry.

Meanwhile, as a hard-working citizen, Quong Wing appealed to the Saskatchewan Court of Appeal. He lost. In 1914 he would also lose at the Supreme Court of Canada. Chief Justice Sir Charles Fitzpatrick would stoutly defend the law. "The Chinaman is not deprived of the right to employ others," he would write, "but the classes from which he may select his employees are limited."[29] There would always be a reason. In 1919 the Saskatchewan government would amend the law again to delete all references to race. Instead, it would adopt the British Colonial Office's ploy: it would empower municipalities to issue a licence for the right to employ women. Remarkably, it would not repeal that law until 1969. Prairie governments did not make everyone feel at home on the range.

In Regina, Premier Scott was still smarting from Borden's evident slight. His occasional ally, Alberta premier Sifton, was distracted, caught in a legal fight over his predecessor's naïve guarantee of the bonds, along with interest payments of five per cent, of the faltering Alberta and Great Waterways Railway. Sifton was struggling to seize the $7.4 million proceeds from that bond sale from the A&GW Railway's bank accounts to cover his government's losses. (After a protracted legal battle, he would fail.) So Scott was on his own when he went after Borden, once again. In early May of 1912, Saskatchewan resubmitted its request for control of the school lands in a pre-election ploy that Borden simply ignored. He hoped to get rid of Scott in the July 11, 1912 provincial election.

The relationship between Regina and Ottawa reached a new low during that campaign. Scott would later claim that federal agencies had supported Conservatives who were running for the Provincial Rights Party. Federal ministers had even linked a potential Tory victory with the fate of Saskatchewan's request for resource control.[30] Such "notorious" intervention—"in some cases illegal use as has been judicially established"—set an extremely nasty precedent.[31] Scott's opponent agreed. The dignified Sir Frederick Haultain, who had resolutely refused to align his Provincial Rights Party with the Conservative Party, considered the election "to have been the most vicious in his campaign experience, mainly because of the increased activity of the two old parties."[32] Although local Liberal newspapers charged that Haultain was under the thumb of Borden's conniving interior minister, Robert Rogers, the former Territorial premier thoroughly disapproved of the conduct of *both* mainstream parties.

Scott's government was far from blameless. When a cyclone struck Regina on June 30, 1912, killing thirty people and destroying five hundred buildings, Scott waited until two days before the election before unveiling a massive aid package. He showcased his efficiency. "Plans for new buildings on the old sites were being made while the debris was cleared away," wrote historian J.F.C. Wright.[33] The premier won his third term. His victory was so decisive that the disappointed Haultain resigned, and the few remaining members of his caucus now sat as Conservatives.

A month later, Scott was back at Borden's door. During the provincial election, both Haultain and Scott had demanded jurisdiction over the lands and resources. Now, with his new mandate, Scott protested the prime minister's refusal "to open negotiations leading to the transfer of the resources." Without that control, his government's operations were "considerably inconvenienced."[34] It was time to talk.

Borden stalled, arguing that he needed to consult his interior minister. Scott waited, impatiently. In October, the prime minister shuffled Robert Rogers into the Public Works portfolio, and put Manitoba doctor William James Roche in the Interior Ministry. But the prime minister still did not contact Scott. On December 21, 1912, more than four months after his first post-election letter, the Saskatchewan premier wrote again. The Minister of the Interior had been back in Ottawa for nine or ten weeks.[35] Surely Borden could now schedule a conference.

Scott fairly spluttered with frustration. "Your Government is now well entered into its second year in office," he wrote, "without even a first step being taken towards the transfer of resources which was promised the western provinces."[36] Agreed, the prime minister was busy. But Saskatchewan residents, Scott pointedly observed, "cannot forget that within three months after taking office you found it possible to negotiate and conclude a new arrangement" with Manitoba.[37] That deal, he added, stood "in striking contrast" to his treatment of Saskatchewan.[38] In the diplomacy of the time, his one-page letter was both guilt trip and electoral threat.

Borden was now in the position of Laurier: by his reckoning, Ottawa still needed the West's resources for settlement and economic growth.

Scott's request was inconvenient—and it was also risky, since the Maritimes vehemently opposed any transfer. The Liberal political machines in Alberta and Saskatchewan were strong, and the transfer of resource control would bolster them. Borden had many reasons to renege on his earlier promises, but he did not want to take the blame. He stalled, and perused a federal report on Saskatchewan's case. That crucial document, which was later buried in the National Archives, showed that Borden had absolutely no intention of keeping his election vows.

The prime minister did not play overt hardball, but his regional henchmen did. This federal report was a sarcastic summary of Scott's varying positions on resource control from 1901 to 1912. The undated and unsigned memorandum asserted that the Scott government had waited until the Laurier Liberals had lost power before it "declared itself favourable to a policy [the transfer of all resources] which it previously denounced as pernicious and fraught with danger."[39] The unknown author, who was possibly Public Works Minister Rogers or his successor, Roche, did not recognize that the Prairie governments had grown in competence. Instead, this political operative contrasted Scott's current demand for control over all lands and resources with his more limited request in early 1911 for control over all resources in the northern region and those of a "purely local character" in the rest of the province. Scott was playing politics, the writer scoffed: he was a "political weather-cock."[40] The absence of any diplomacy or tact in the memorandum hinted at the troubles ahead.

It would have been relatively easy to mollify the exasperated premier, but Borden barely tried. Instead, he chastised Scott: there was, he replied, no connection between Saskatchewan's request for resource control and Manitoba's boundary extension; he had merely rectified "what seemed to us a serious injustice" to Manitoba.[41] Borden added testily that it would be useless to arrange a meeting with Saskatchewan because the issue of resource control involved all three Prairie provinces. He hoped "in the early future" to arrange a meeting with the three governments. He then did nothing. In response, the Saskatchewan Legislature passed another futile resolution urging Ottawa to get cracking.

The prime minister had no time for Western Liberals and their resource requests when the superpowers were arming and the British Empire was menaced. In the fall of 1912 he already sensed that war in Europe was dangerously close. He was resolved to assist Great Britain in its naval build-up, and to have a voice in the Empire's defence and foreign affairs. But his inability to practise diplomacy in Canadian circles would thwart his efforts to shine abroad during those pre-war years.

He walked into a minefield that was partly of his own making. In 1909, the British government had asked Canada for money and reinforcements for its massive shipbuilding program. When French-Canadian nationalists had objected, Laurier had opted to create a domestic navy, which could be put under British control in wartime. (That provision had driven Quebec nationalists into Borden's camp during the 1911 election.) Britain, in turn, had transferred two elderly cruisers to Canada. The Royal Canadian Navy was afloat. When the Conservatives took power, however, Borden had changed course. In March 1912, he had halted the Liberals' naval program, leaving the service in no man's land and infuriating the Opposition, especially the Liberal-dominated Senate.

Partisan politics were rife on Parliament Hill, and Liberals were up in arms everywhere. When the House of Commons opened on November 21, 1912, MPs fought over a bill to provide provincial governments with money for highway construction. Although the bill allowed the minister of railways to determine the amount for each province, Borden asked for trust: the grants, he asserted, would be doled out on the same population basis as federal subsidies. Cynical Liberal MPs did not believe him; nor did the Western provincial governments. A similar bill had died in the previous session. At the urging of many partisans, the Liberal-dominated Senate would effectively kill this replacement in the early spring of 1913.

That fate may seem inexplicable. But, as Saskatchewan premier Scott later explained to Ontario premier Whitney, the Liberals did not trust the Conservatives to play fair. Suppose the Liberals slipped back into power in Ottawa: "All Provinces with local Liberal Governments in any year or period of years might be in receipt of federal highways money and work," Scott wrote, "while all Provinces with Conservative Governments were receiving none, and vice versa."[42] Borden was reaping the legacy of his dismissive behaviour toward Scott and Sifton.

The distrust was corrosive and it paralyzed Parliament. When the prime minister finally unveiled his own naval program in December 1912, it was to a very divided House. It was perhaps an omen that, when he tried to sit down during an interruption from the Speaker, he forgot that he had pushed his chair into the aisle. He tumbled to the ground and broke his glasses. At the time, however, he was triumphant. Canada, he declared, would provide $35 million to the British government to construct three Dreadnought battleships.

The plan would boost the Dominion's international profile, he maintained, as much as it would assist Britain. If Canada shared in the responsibility for the defence of the Empire, Britain could no longer "assume sole responsibility" for its foreign policy.[43] Canada would give to get. The big ships would be built in the United Kingdom, which would order smaller vessels from Canadian shipbuilders. Canadian ministers would be permitted to attend all meetings of the Committee on Imperial Defence in London, and Canada would have the right to recall the Dreadnoughts if it wanted to establish a Canadian unit of the Royal Navy. In the meantime, Canadians could serve on those ships. Borden would later recall that, when he finished, MPs and the crowd in the galleries sang, "Rule Britannia" and "God Save the King."[44] It would be his last moment of triumph on the issue.

The Naval Aid Bill consumed him. By early March of 1913, the debate was running twenty-four hours a day, six days a week. MPs attended the House of Commons in shifts, and one evening, Borden found himself going home at 3:45 a.m. and returning at 9:45 a.m. On two days, debate "became so violent as to occasion apprehension of personal conflict"; near midnight on another evening, there were "scenes of great disorder."[45] The stress afflicted Borden: he broke out in carbuncles, and twice left his sickbed to go to the Commons with a bandaged neck. "I became greatly exhausted by the intense strain of the long debate, the night hours and the tumultuous scenes."[46] The discussions dragged into mid-May 1913 until, after closure, the bill finally received House of Commons approval. At the end of May, Laurier having rejected several possible compromises, the Liberal-dominated Senate defeated it. Borden virtually collapsed.

That summer of 1913 passed almost uneventfully. Immigrants were still streaming into Canada. More than 400,000 arrived in 1913, an all-time record that included almost 160,000 British newcomers, along with roughly 28,000 Italians and 29,000 Russians. Canada still symbolized a better life or, at the very least, better wages. An Italian emigration official came upon a Calabrian youth in the bush, four hours by westward train from Port Arthur on Lake Superior. To the official's amazement, the young man, who had been in Canada for only three months, working as a section hand on the Canadian Pacific Railway, had already sent home to his mother the equivalent of six months of wages in his Italian village.[47] Such youths were often migrants, who came for a few years to earn money and then went home. After several turbulent trans-Atlantic crossings, however, many decided to stay.

Immigrants were pushing ever farther into the Prairie northlands. In 1912 a small group of Polish Protestants had trudged into Athabasca in northern Alberta. As a minority within a minority, their adjustment was especially hard. They joined the English-language Anglican Church because they did not have enough members to build their own church. They cleared the bush, enduring a tough life. Farmer Karol Góra once confronted a huge black bear, and when he ran home to get his rifle, his wife insisted on joining the pursuit. They did not find the bear, but they shot a moose: they were a good team. When Góra left home every winter to earn extra cash, his wife took care of the livestock and the family. She even "had to be a doctor, sometimes, as there were no doctors or medical facilities" in that remote area.[48] Another settler, Pawel Kawulok, once swam across the wild Athabasca River while tugging a raft that carried his family's food and his own clothes. But the water was too rough for him to pull the raft to shore, and he had to return home in his "birthday suit."[49] Those wild lands in the northern Prairies were still not fully settled: they remained Ottawa's challenge, and the pioneer's ordeal.

Many settlers, however, could finally glimpse an end to their hard times. In Saskatchewan in 1913, during his fourth summer in Canada, Dutch farmer Willem de Gelder was clearing another twenty acres on his 320-acre homestead. Day after day, his horses tugged a plow through the stony turf and de Gelder heaped the rocks into a huge pile. He was a well-educated man, the son of an upper-class banker, who had arrived in Canada in 1910. His neighbours never did find out why he had left Holland, although he did attend a public lecture on Canada before he set out. He never allowed anyone to photograph him. But his surviving letters to an unknown Dutch recipient captured the "drudgery of the settler's hard fight to survive and succeed."[50]

His first summer had been "miserable, wretched enough to discourage the most courageous."[51] In 1913, he wrote that gadflies like wasps bit the heels of frantic cattle, "and nothing can hold them."[52] He recorded a drought that was followed by seven days of torrential rain. He lamented the dull blades on plows. His life was still filled with trials, but there were now small moments of triumph. He purchased the new Marquis wheat seeds that had "won first prize in New York."[53] He measured out his life in boulders. "It's a very satisfying job," de Gelder wrote. "Every stone you've picked and piled up won't bother you again."[54]

That same summer, Borden came to the realization that he had to tackle provincial complaints before the next Parliamentary session also dissolved in acrimony. In mid-June 1913, he asked Ontario premier Whitney to convene an interprovincial conference. He promised to consider the demands that had accumulated over the last two years, and he offered to meet with individual premiers before the formal gathering. At his request, New Brunswick premier James Flemming agreed to ask his Maritime counterparts if they would meet with the Western Provinces before the Conference. (This meeting apparently did not occur.) At the end of June, desperate for a break, the prime minister left for the picturesque United Empire Loyalist retreat of St. Andrews by-the-Sea in New Brunswick.

Borden would remember his month away with mixed feelings. "We had a very pleasant time but I was continually bombarded with letters and telegrams so that half the benefit of a much needed rest was lost."[55]

Not surprisingly, Borden found himself dealing, once again, with British Columbia premier McBride. The bureaucratic wheels had turned slowly since Borden had offered to establish a three-man commission to study McBride's case for special subsidies, as McBride tried to heap more issues on the commission's agenda. Both leaders had needed to consult their colleagues, and it had taken a year merely to finalize the commission's terms. By mid-1913, each government had named its appointee but, predictably, the two appointees could not agree on the third member, who would be the chairman. Borden had to approach the British secretary of state for the colonies, Viscount Harcourt, who was swamped with duties and busily pursuing young people of both sexes, for an informal list of possible candidates. (Harcourt would eventually kill himself in 1922 when his fellow aristocrats learned of his attempted seduction of a twelve-year-old boy.)

The prime minister knew British Columbia was already preparing its case for the commission, and he had enough political sense to worry. From his roost in the Algonquin Hotel in St. Andrews by-the-Sea, he wrote to his minister without portfolio, George Perley, fretting that federal officials were not matching McBride's zeal. What was Ottawa's position on British Columbia's special circumstances? "I am afraid nothing has been done," he confided. "Will you please ascertain and stir them up?"[56] Borden did not want to delay the commission. He just wanted to get it underway, and out of his way.

The Prairie premiers also had Borden's attention, if not his goodwill. That summer, the prime minister asked his new solicitor general, Arthur Meighen, who represented Portage la Prairie in Manitoba, to discuss the issue of resource control with Public Works Minister Rogers and Interior Minister Roche, who were also Westerners. Significantly, the prime minister also forwarded a letter to Rogers that he had received from former Saskatchewan MLA Archibald Gillis, who had represented the Provincial Rights Party until his defeat in that rancorous 1912 election. Gillis was still bitter, and he seemingly forgot all about provincial rights. Westerners would be happy, he told Borden, if the prime minister did *not* keep his election promise to transfer resource control. Premier Scott had driven the provincial debt from zero in 1905 to nearly $20 million—"with very little to show for it." He could not be trusted with the cash. Gillis's parting advice was blunt: "under no circumstances should the lands be transferred."[57] It was an ill-natured screed, in keeping with the times, but Borden took the advice seriously enough to draw it to his Western ministers' attention.

The prime minister also put his tough operatives to work. Three remarkable memorandums, which have been preserved in the National Archives, indicate how dutifully he prepared for the Western premiers when he had little choice. Those reports also show that, no matter what Borden said publicly, he had no intention of surrendering the West's resource wealth. He would adopt a deeply cynical position, marshalling his version of the facts against the West.

In the first memorandum, "Suggestions for preparation," which included marginal notes in Borden's handwriting, the prime minister called for a detailed eighteen-point report on the public lands in the three Prairie provinces and British Columbia. He wanted: statistics on immigration and the boost that immigrants gave the economy; a breakdown on the acreage earmarked for schools, parks, railway firms and the Hudson's Bay Company; the amount of land available for homesteading; and a detailed breakdown of each province's subsidies. Most significantly, he wanted to see "a discussion of the claims of the Maritime Provinces for compensation in case the natural resources of the Prairie Provinces should be transferred."[58] The issue might have seemed tedious, but Borden knew enough to ask the right questions.

The very existence of the second report shows how much the Borden government had politicized resource control. Even the title was ominous: "Memorandum as to presentation of a brief touching the attitude of the Liberal Party and especially Western Liberals in respect of the administration of the natural resources of the three prairie Provinces."[59] It asked for the position of Laurier and "other prominent liberals including especially Mr. Walter Scott" on resource control prior to 1905, as well as their positions during the debate on the Saskatchewan and Alberta autonomy bills in 1905, when Scott had been an MP. It also asked what the Alberta and Saskatchewan governments had said about the issue since 1905. Finally, it requested a review of the prime minister's own statements on the issue since 1907 when he had promised to set "fair and reasonable" terms for any transfer. Significantly, the memo did not ask for his position prior to 1907 when he had supported unconditional transfer. This memorandum was not dated or signed, but it was tucked among the briefing papers for the October 1913 Interprovincial Conference. The prime minister was readying his escape.

The final memorandum in those archival files was perhaps the most disturbing. Ottawa had tussled with Western governments for years over resource control, and at times, the exchanges had been acrimonious and sarcastic—and occasionally, even threatening. But the sheer nastiness of the "Memorandum Re Western Conditions" marked a new low. There was no signature and no date on the memorandum, except for a scrawled "October 1913?" notation. But it was clearly the work of a highly partisan Conservative operative. And it was brutal.

The writer started with a snide assertion: the two Liberal governments in Alberta and Saskatchewan "have got into a position of great temporary strength, partly by a profligate and reckless policy in dealing with the needs of the moment, partly by a clever use of the Radical element."[60] Both provinces had guaranteed railway bonds of at least $25 million, which was "a most reckless and dangerous" amount. Those pledges were hampering their ability to borrow. If Borden transferred resource control, he would be boosting the strength of vulnerable political opponents. But if he did not transfer resource control, those opponents would allege that he had broken his word. Ottawa "will be in a difficult position if it does not early take steps to forestall *these people*."[61]

There was a way out. Borden had to assail the popularity of Sifton and Scott. He should show that the Liberals were unfit to govern: Ottawa was simply protecting provincial taxpayers when it kept the resource revenues out of their hands. Borden should talk about the West's real problems such as high freight rates and the railways, which had left huge areas of the countryside "utterly unserved, to the positive hardship of the people settled in them."[62] If Borden could change the topic, the writer added, he would "extricate his Government from any awkwardness in refusing to entrust the public domain to men who would squander it and build up a tremendously effective political machine with it."[63]

This was early spin, and it was distasteful. Borden should distract Western voters by attacking their Provincial governments, so that those voters would not notice his failure to keep his own promises. The policy issues around the transfer of resource control were scarcely mentioned. Political advantage was everything, even if it meant thwarting provincial aspirations and enforcing provincial inequality. It was no wonder that Parliament was deadlocked.

Borden almost certainly read this deviously useful advice on how to wriggle out of his promise. It suited his circumstances. Huge numbers of immigrants were still arriving, and any negotiations to transfer resource control might delay their settlement. The Maritime provinces were also an expensive impediment to change: it would take a far better politician than Borden to figure out how to buy them off.

The highly partisan climate trumped principles. It is fair to speculate that Public Works Minister Robert Rogers, who ran the Conservative machine in the West, was the author of that explosive memorandum. Western Liberals had come to detest his partisan tendencies, and the dislike was mutual. Rogers's opinionated ways would eventually lead to his being dropped from cabinet, but many Conservatives shared his political views. The bottom line was clear: any transfer of resource control would boost Western Liberal governments. It was a non-starter.

While his minions plotted, Borden dolefully visited his constituents and the regional party bosses in his native Nova Scotia. Predictably, the ward heelers called for more patronage appointments. "My engagements were incessant," he recorded.[64] He loathed the whole gritty business, and he was just not interested. That same month, as Borden had anticipated, British Columbia unveiled its formal case for special subsidies. In October the province forwarded a corrected copy of that submission. By the end of that month, the prime minister was reluctantly ready for his close-up.

The 1913 Interprovincial Conference finally opened at 11 a.m. on Monday, October 27, in Ottawa. Ever the dutiful host, Borden welcomed the nine premiers and their delegations to the Railway Committee Room of the Senate. Then he scuttled back to the safety of his Parliamentary haunts. Officially, Ontario premier James Whitney and Quebec premier Lomer Gouin had summoned their peers to pick up where an earlier conference in December 1910 had disastrously left off: the nine premiers would once again consider the Maritimes' demand that the number of their MPs should increase to reflect their proportional representation at Confederation. Alberta premier Sifton tartly declared that the West was already shortchanged: it should have received twenty more seats after the 1911 census, but the federal government had failed to enact redistribution. Worse, the West's population had increased by 600,000 since that census, "really entitling them to twenty more."[65] Sifton was not winning allies. Naturally, the discussions reached a deadlock—again—that very day.

Abashed, the premiers fell back on something they could all agree upon: Ottawa should raise their subsidies. On Tuesday, October 28, they even settled on the amount of that subsidy increase—an extra ten per cent of Ottawa's customs and excise duties, which would roughly double their current amount. At 5 p.m. Borden trudged to the Railway Committee Room, and the premiers presented their requests. As the conference minutes dryly report, Borden "saw no objection to the Provinces coming at stated intervals, —say, every ten years—to discuss" their financial needs, "if circumstances warranted it."[66] It was a firm rebuff: ten years had not yet passed since the last subsidy increase in 1907. So much for that provincial ploy.

That same day, as Borden expected, Alberta premier Sifton and Saskatchewan premier Scott formally requested a meeting with him on resource control. "The matter has been pending a comparatively long time," they wrote, "and during more than two years past no progress has been made towards a settlement."[67] Borden promptly invited Manitoba premier Roblin to join any discussions. It was virtually a command performance: "They [Sifton and Scott] suggested (and I think it most desirable) that you should be present."[68] He had every reason to assume that Conservative Roblin would be his ally: after all, he had increased Manitoba's land mass and subsidies only eighteen months before. He also invited Interior Minister Roche to the upcoming gathering. On the evening of October 28, Borden hosted a dinner at the Château Laurier Hotel for the provincial delegates, many of whom who were nursing bruised feelings.

On Wednesday, when the Conference resumed, the three Maritime provinces took another run at their colleagues, outlining the grievances that would do so much to thwart the Western premiers at the November 1918 Conference. At Confederation, they had been self-governing colonies, "something more" than the mere sum of their inhabitants.[69] In those days, they had never imagined that the growth of the West would diminish their status. Their taxpayers had contributed to the purchase of Rupert's Land and the cost of driving the railway across the prairies. Their sons and daughters had left home to settle the frontier. When Ottawa had expanded provincial boundaries in 1912, the Maritimes had lost their stake in that valuable Northwest Territories terrain. Then came their *cri de coeur*: "There is surely something wrong with a system of representation which penalizes some Provinces for the sacrifices they have made."[70] Prince Edward Island added a passionate special plea for the return of the two seats that it had lost since 1873. (It now had four seats.) The discussions again stalled.

As the conference limped toward adjournment, the three Prairie premiers gathered in Borden's East Block office to discuss resource control. As Borden would later recall, he set three conditions on any transfer. The increase in federal subsidies for Alberta and Saskatchewan in 1907 and for Manitoba in 1912 had been gestures "in lieu of natural resources and as compensation therefore."[71] If the Westerners wanted their resources, they could not expect to keep those subsidies in lieu of resources. Second, the trio had to agree to retain the "present provisions" for virtually free homesteads.[72] Finally, they could not hatch schemes that would interfere with the "continued flow of desirable immigration."[73] In effect, unlike their peers, the three Prairie provinces would have to accept federal limitations on their resource control.

The meeting was such a catastrophe that it would unite the fervently Conservative Roblin with Liberal premiers Scott and Sifton in the Gang of Three, though Roblin would never publicly explain his decision to join that unlikely alliance. The premier was not on particularly friendly terms with Scott or Sifton: neither of them had attended a huge dinner in Roblin's honour, with a cross-Canada guest list, in Winnipeg in April 1912. But, throughout his career, the Manitoba premier had espoused provincial equality with the same vigour that he had brought to defending Canada's ties with Great Britain. Borden's three conditions had violated those principles, and Roblin's outrage overcame his political affiliations.

Predictably, Borden did not even mention that interprovincial conference in his *Memoirs*. He merely noted that, by the end of October 1913, he was "on the verge of exhaustion."[74] Liberal premiers had that effect on him.

It took only two months for the Western alliance to solidify. On December 22, 1913, the three Prairie premiers consolidated their demands into a one-page proposal. Sifton sent it to Borden along with a covering letter, which pointed out that all three premiers had actually signed it. Indeed, their signatures floated below their one-page resolution. The missive began with a convoluted snarl: "After having an interview with you in regard to the questions in respect of which the Prairie Provinces have received different treatment from the other Provinces of Canada . . . "[75]

The three Premiers then condensed their request into one remarkably clunky sentence: "It has been agreed between us to make to you, on behalf of said Provinces, the proposal that the financial terms already arranged between the Provinces and the Dominion as compensation for lands should stand as compensation for lands already alienated for the general benefit of Canada, and that all the lands remaining within the boundaries of the respective Provinces, with all natural resources included, be transferred to the said Province, the

Provinces accepting respectively the responsibility of administering the same."[76]

That clumsy sentence would become the single demand of the three Prairie provinces for the rest of the decade, the one they would take into the November 1918 conference. The prose was awkward but mercifully brief. The premiers wanted their subsidies to continue, as compensation for the resources and lands that Ottawa had already commandeered. They also wanted complete control over their remaining lands and resources, including the lands earmarked for homesteading. So much for Borden's conditions. Unfortunately, to their peers, the Gang of Three's demands would simply seem greedy.

Now what? In January, Borden sent copies of the Gang of Three's missive to the Maritime premiers. Their opposition was vehement, and almost instantaneous. On January 19, only eleven days after Nova Scotia premier George Murray received Borden's letter, he replied: the Western Provinces did *not* own their resources. Ottawa had deliberately retained control because "the Dominion . . . had to spend so liberally for immigration."[77] Furthermore, Western provinces received "very generous" subsidies, especially when compared with the Maritimes' subsidies.[78]

Murray added tolerantly that the matter was of no concern to him, because it was between Ottawa and the West. But—and there would always be a "but" over the next few years—Murray did bring forward his own caveats. Western subsidies were far from stingy. It was not "fair and reasonable" to request the lands *and* the money because that would "produce wide-spread discontent."[79] If Ottawa did decide to transfer the resources or to increase the West's subsidies, "such action should only be taken as part of a general re-adjustment [and] all the Provinces should have an opportunity to participate."[80] Strike one: more money for everyone, or nothing for the West.

The other Maritime premiers were equally adamant. On February 2, 1914, Prince Edward Island premier John Mathieson argued that Saskatchewan and Alberta deserved no compensation for the land that Ottawa had used before they became provinces in 1905: this claim was "not well founded" because the Dominion had picked up the tab for the

West's settlement, and it had ensured that the two provinces were debt-free at their creation.[81] The original provinces had been forced to pay themselves for the "public works and services" that Ottawa had freely provided for the Territories.[82] If the Western provinces retained their subsidies as well as receiving their resources, "it would give rise to a new or additional claim for compensation from every other Province in Canada."[83] That was strike two for the West.

Strike three came fifteen days later when New Brunswick premier James Flemming insisted that the three Western provinces pocketed large subsidies—so they did not have "any just or reasonable claim" to the lands.[84] Then he sang his own song of jealousy: Ottawa had used huge parcels of the Northwest Territories for boundary extensions in 1912, and it was sending regular payments to the Western provinces from the trust fund that held the proceeds from the sale of school lands. This was all done, Flemming claimed, at the Maritimes' expense: "The immense area of the public domain north of Quebec, north of Ontario and on the Western prairies is *our common property*."[85] Ottawa was free to transfer the resources, the premier observed coolly, but he added that it would not be right "to do so without some compensation being made to the Maritime Provinces, whose geographical position is such that they can receive no territorial expansion."[86]

The trap that Borden's henchmen had devised was sprung. Decades later, in the 1930s, when Borden penned his *Memoirs*, he would recall a satisfying exchange in the House of Commons with Alberta Liberal MP William Buchanan on February 24, 1914. Buchanan had read aloud the Gang of Three's letter, and then accused Borden of breaking his promises on resource control. The prime minister had clearly been waiting for this question and he pounced. He cited Buchanan's own opposition to any transfer in 1905, along with the earlier opposition of Western Liberals to a complete transfer. He cited Laurier's warning in 1911 that any transfer would mean subsidy cuts. He harrumphed that the Liberals were hypocrites.

Borden piously claimed that he had wanted to discuss the issue at the October 1913 Interprovincial Conference, but that this had been impossible because of "the well-known opposition of the Maritime Provinces to any such proposal, except upon compensation to other provinces."[87] The concerns of both regions "ought to be considered and dealt with at one and the same time."[88] Anyway, the Western Provinces had ignored his conditions, he wrote: surely the Gang of Three's proposal was not "one that they really expected us to entertain."[89]

Two weeks later, on March 5, 1914, the prime minister finally got around to replying to his three petitioners. He had raised three conditions at their meeting in October, he observed, and the premiers had largely ignored them. Borden was condescending. "I shall be glad to discuss the subject with you at a further interview," he proclaimed, "or to receive from you a further communication embodying a new proposal or suggesting a mode of inquiry as to the matters above alluded to and other relevant considerations."[90] That was gobbledygook. The three premiers surely fumed as they deciphered it. In effect, the prime minister had volleyed the political problem back into their court: they should come up with a new proposal. Now he sat back. No one could possibly blame him for this highly convenient stalemate, which would persist into the November 1918 Conference.

Perhaps the only remaining wolf at Borden's door was the irrepressible McBride. In 1909, without informing his cabinet, McBride had guaranteed the bonds of the Canadian Northern Pacific Railway, which was building a third line from the Yellowhead Pass on the Alberta border to Vancouver. As a condition of the deal, true to form, the premier had insisted that the railroad could not employ Asians. In 1912, he had pushed through another deal, chartering the Pacific Great Eastern Railway to run from North Vancouver to Fort George, backed by more government guarantees. He had ignored "early signs of a sputtering economic engine."[91] Now he needed more money to fend off a potential catastrophe.

Borden's agreement to create a special three-man commission had neutralized the British Columbia premier during the 1913 Interprovincial Conference. McBride had been relatively quiet at that gathering because he was sure Ottawa would soon meet his needs. Borden had bought short-term peace, but in the longer term, the unresolved fate of this spe-cial commission would disrupt the November 1918 Conference—because Borden's pre-war promise to McBride would go nowhere.

How could that happen? Bureaucracies moved with excruciating slowness. In November 1913, after the October conference, Ottawa offi-cially asked Colonial Secretary Viscount Harcourt to appoint the third member of the commission. Harcourt asked for more information. Meanwhile, in late December of 1913, federal bureaucrats complained that McBride was still sneaking new issues into his legal case for special subsidies. In May 1914, after a visit to Ottawa, McBride agreed to amend his case, and on June 20, McBride told Borden that he was "most anx-ious" to make progress on the commission.[92] A week later, in a late June telegram, he argued that Borden should push Harcourt to appoint the third commissioner so that the commission could start work in August. Then in mid-July, Borden told McBride that the minister of justice still objected to portions of his claim.

In late July, Borden retreated to Ontario's Muskoka region for four weeks of rest. He golfed and he swam, and he chatted with friends. On July 28, he learned that Germany and Russia were quietly mobilizing, so on July 31, he left for Ottawa, and summoned his cabinet. He cabled a promise of firm Canadian support to the British government shortly thereafter. At 8:55 p.m. on August 4, "the momentous telegram arrived announcing that war had been declared," he later recalled. "Although the events of the past few days had quite prepared us for this result, it came at the last as a shock."[93] With Britain's declaration of war, Canada was automatically at war. Despite that shock, the news had not come as a bolt out of the blue: in January 1914, Borden had asked each department to prepare contingency plans in the event of war. Now those plans would be put into effect.

On that same terrible day, McBride assured Borden that federal and provincial officials had finally agreed on the commission's mandate, and he suggested that the two commissioners could proceed without their third member. No one noticed. Sir Richard could not restrain himself: even though he had been partly responsible for the delays by slipping new issues into his old case, he could not give up. On September 22, 1914, as Canada hurriedly assembled its 31,000-man Canadian Expeditionary Force, which would embark for England on October 3, McBride once again wrote to Borden. His first sentence alone was outrageous. "Now that the first fever of the war is over and conditions are becoming more or less normal," he wrote, "I want if possible to have an expeditious commencement in the Better Terms Commission." He asked for Ottawa's legal case, and he wanted to know Harcourt's choice for the third commissioner. "The war may continue a year or two years," he added, so the commission should no more be delayed than "the ordinary courts of law."[94] In a coolly rational universe, he may have been right— but it is tempting to imagine how the viscount would have reacted if Borden had forwarded McBride's demand. The prime minister wisely did not.

The debate on natural resources was stalled. Borden and the two Western Liberal premiers were now worried about the fate of the Canadian Northern Railway, which would be vital for wartime ship-ments. In exchange for a huge grant, Ottawa had acquired $7 million of that faltering transcontinental railroad's stock in 1913. On the brink of war, it had guaranteed more bonds. The Grand Trunk Pacific Railway had also appealed for aid. That railroad, Borden observed dourly, was "like the Old Man of the Sea . . . continually upon our shoulders and would not be taken off."[95] In June 1914, Ottawa had guaranteed bonds of up to $16 million for the Grand Trunk. The government had little choice: the railroads were the arteries of the West.

Ottawa and the Western provinces needed each other now. There could be no time-consuming negotiations over resource control when the British Empire required those very resources for the war effort. Ottawa and the West had to curtail that dispute—although it would reappear briefly when the conflict persisted. They had another war to fight.

7

THE WEST'S BAD WAR:
STALEMATE FOR THE GANG OF THREE.
AUGUST 1914 TO EARLY MARCH 1918

—

THE GANG OF THREE did not dissolve in wartime. But those two Liberal premiers and their scrappy Conservative colleague were mightily distracted. Their governments tottered after years of high-rolling with the railroad barons and high-spending on projects and partisans. Their Anglo-British residents turned on their Central and Eastern Europeans neighbours, whom they saw as potential enemies, and who had to grapple with that rejection as well as their own divided loyalties. Amid the unrest, the three Prairie provinces struggled to fulfill their wartime mission as resource providers. Their wheat fed the Allies. Their coal fuelled the machinery and munitions plants. It seemed almost unpatriotic to argue about which level of government controlled the lands and the resources when their soldiers were dying in the trenches.

The Prairie premiers walked a tightrope: they could not denounce Ottawa in wartime, but they had to remind Prime Minister Sir Robert Borden that they would not leave the scene in peacetime.[1] Federal steward-ship remained an irritant: oil and natural gas firms, which had secured federal leases, arrived without notice on farmers' lands, set up rigs, and drilled. Farmers, who did not own the sub-surface rights to their land, could only stand by as the interlopers struck it rich in *their* cow pastures. Throughout the war, as the importance of oil and gas to the military machine became blindingly obvious, large energy firms circled the West like hawks, hounding Ottawa for leases to large swaths of Western land. The Prairie premiers wanted that control—and those royalties—for themselves.

Western demands remained constant throughout the war, even though the entire cast of the Gang of Three would be replaced. Manitoba premier Sir Rodmond Roblin would go down in a blaze of scandal in May 1915, embittered and shattered. Within months, his Conservatives would lose at the polls. Saskatchewan premier Walter Scott would resign in October 1916: although he had survived scandals, his fraying nerves and frequent absences would provoke a caucus uprising. Alberta premier Arthur Sifton would muddle through scandals over government-backed railroad bonds but, in the fall of 1917, he would join Borden's wartime unity cabinet. In the meantime, British Columbia cabinet ministers, weary of Premier Richard McBride's erratic one-man brand, would force him to resign.

The West came through for the nation in wartime, contributing more than its share of soldiers and resources. Huge numbers of men enlisted; of those who returned, many had sustained terrible injuries to their bodies and minds. Women rolled bandages and knitted socks and tended livestock and operated factory machinery. Those so-called foreigners dragged coal to the surface in treacherous mines. The Prairie premiers raised the topic of resource control only twice during those early wartime years. Their timing was terrible. Borden rebuffed and rebuked them, but they remained resolute.

In early 1918, with the end in sight, Ottawa convened an informal federal–provincial gathering to talk about food production, farm workers, the fate of the veterans and postwar immigration. The Prairie premiers resolved to secure resource control before Borden could settle more immigrants and the returning soldiers on their lands. They had proven themselves in wartime. Surely they deserved their reward after everything that they had done during and before the war. That February 1918 Conference would be a tough lesson in power plays. Borden would outfox them, and the Rest of Canada—especially the disgruntled Maritime premiers—would thwart them. The stress of the proceedings would help to put one participant in his coffin. The federation would remain dysfunctional.

Those years of war abroad would be dangerous to the peace at home. From the first burst of patriotism to the realization that the conflict would be nasty, brutish and long, it would have been hard to find a Westerner who had a good war.

From the very beginning, from those early days in August 1914, Western Canadian society was horribly divided. As the patriotic calls to enlist grew deafening, Anglo-British Westerners were ardent. There were more unmarried young men in the West than in other regions, and many were recent immigrants from Britain. They flooded into the enlistment offices, anxious to serve in what would surely be a brief war leading to a glorious peace. When the first contingent of Canadian troops finally sailed for England in October 1914, almost two-thirds of its 31,000 recruits had been born in the United Kingdom. More than fifteen per cent of Western males between the ages of fifteen and forty-four volunteered in those early weeks. The comparable number among Ontario males was roughly fourteen per cent, among Quebec males five per cent, and among male Maritimers ten per cent.[2] Westerners were already doing more than their share. The first hint of many bruising regional rivalries could be discerned in those dry statistics.

As their Anglo-Canadian neighbours marched off to war, newcomers from Central and Eastern Europe braced for trouble. On August 22, Ottawa enacted the draconian War Measures Act, which conferred vast powers to protect the "security, defence, peace, order and welfare of Canada." Ottawa could now censor all communications, seize property and approve regulations for "arrest, detention, exclusion and deportation."[3] Residents who were born in Germany or the Austro-Hungarian Empire were now enemy aliens if they were not already naturalized Canadians. Many Ukrainian-born farmers, who had answered Clifford Sifton's siren call for homesteaders at the turn of the century, had neglected to swear formal allegiance as British subjects. Without that magic piece of paper, which roughly fifty per cent of Ukrainian-born residents, among others, did not have in 1914, they were pariahs.

The effect was electrifying. Sifton's Pets were now regarded as traitors who had to register with federal authorities and carry identification cards. Many Anglo-Canadians had never accepted those newcomers, including those hard-working farmers who had lived in their midst for almost a generation; they even turned on the naturalized Canadians. Ironically, their derision extended to those German settlers whom former interior minister Frank Oliver had wooed, less than a decade earlier, as ideal immigrants of good genetic stock.

Ethnic communities across the West were beset. In Winnipeg, employers laid off their German and Austro-Hungarian workers. Others assaulted German-speaking citizens, damaged the German Society of Winnipeg headquarters, and attacked the German and Austrian consulates. Their victims were often torn between two worlds. Some Germans were still too deeply rooted in Europe "to be influenced by the attractions of Canadian nationalism," and loyalty to Canada seemed to come with "too much of Anglo-Saxon ascendancy and British imperialism."[4] Derogatory references to "Huns" only deepened their estrangement. They kept a low profile, insisted that they were Russians or Poles, and prayed for a quick end to the conflict. They would be disappointed.

The West would become a land of "Us Versus Them." More than 8,500 aliens would be interned in camps across the nation for everything from suspicious activities to attempts to leave Canada. By the end of the war, more than 400,000 people—including Germans, Turks, Austrians, Hungarians, Poles, Romanians and Ukrainians—would be card-carrying aliens.

By early 1915, Calgary was essentially a military outpost. Nearly forty thousand eligible residents enlisted here, and the ornate Grand Theatre now opened its doors to the soldiers. In January, the entire 50th Battalion happily watched a performance of *Baby Mine*, an American farce about a compulsive fibber.[5] That gift would become a poignant memory at the Front. Some of those unlucky men would soon be detached from their battalion, and transferred to England as reinforcements for other hard-hit army groups. In late April, nine hundred members of the 50th Battalion returned to the theatre for *My Tango Maid*, and sang lustily along with the chorus.[6] Those troops would find eventually themselves in the trenches in Flanders and France.

When the Canadian Expeditionary Force finally reached the Front in early April 1915, the soldiers burrowed into trenches along the Ypres Salient in Belgium. It was an unfortunate location: German troops occupied the higher ground around three sides of that Allied position. Two weeks after the Canadians arrived, the Germans released a four-mile-wide cloud of chlorine gas that wafted over nearby French troops. The greenish-yellow mist killed five thousand troops within ten minutes as the gas combined with water in the lungs to form acid that destroyed the lining of the lungs. Soldiers writhed in shock, gasping for breath, vomiting slime, and slowly asphyxiating. Two days later, the Germans released more gas, which drifted over the Canadian front line, killing more men. That cloud was smaller, and the Canadians were better pre-pared because army doctors had advised them to hold urine-soaked cloths over their nose and mouth, to partly neutralize the toxic chemi-cals. Somehow, the Canadians held their position.

The collateral damage on the home front was almost as bad. The Allies maintained that the Germans had violated the Hague Convention of 1899, which banned the use of projectiles carrying poison gas. In Toronto, the news was emblazoned on the front page of *The Globe* under the headline "Germans Are Poisoning the Soldiers of Canada."[7] On the same page was a Canadian soldier's harrowing account of the effects of German gas. The German government insisted that it had not used projectiles to deliver the gas: it had simply let the prevailing wind carry it. That legalistic quib-ble disgusted the Allies, although they would later use gas themselves. The Germans in Western Canada were now derided as barbarians.

Fortunately, their Anglo-Canadian neighbours were distracted. The scan-dal that would take down the most unlikely member of the Gang of Three would transfix Westerners for weeks. Conservative Premier Rodmond Roblin had won his fourth election on the brink of the First World War, albeit with a diminished majority. Although the confrontational politician had accrued more enemies with each campaign, he remained emblematic of the pioneering Old West. Born to United Empire Loyalist parents of Dutch descent in southern Ontario, he had read MP James Trow's harrowing account of his journey across the West in the mid-1870s, and left home to seek his fortune.

Roblin was a self-made man like former interior minister Frank Oliver: he had been a sawyer in a Winnipeg mill, a pressman, a fur trader, and the owner of a general store. By the time he entered provincial politics in 1888, he was the proprietor of a flourishing business that bought local wheat for export. He was often on the edge of impropriety: in the Legislature, he pushed for the construction of a rail line to a town where he owned land and purchased wheat; he also promoted the interests of the Manitoba Central Railway—even though he was a director of the firm. When cornered, he could be fierce. "It cannot be suggested that Sir Rodmond was ever involved in an election contest that was not lurid, hard-fought and without political quarter," his fawning biographer Hugh Robert Ross acknowledged in 1936. "His very presence seemed to add these inflammable materials."[8]

The pugnacious Roblin was his own worst enemy. As the premier strutted into war, vowing unflinching loyalty to Great Britain, a flock of enemies marched behind him, including suffragettes, temperance proponents, Orangemen who opposed any teaching of the Roman Catholic religion and bilingual school instruction, and his Liberal opposition. It was a mark of the era's fierce partisanship that the Liberal machine of Saskatchewan premier Scott had swung behind Roblin's opponents in the July 1914 election, despite their mutual membership in the Gang of Three.

Roblin was a polarizing presence. His detractors assailed him as arrogant and condescending. Later generations would remember his assertion to the Political Equality League, including women's rights activist Nellie McClung, that women should not get the vote because they would abandon their homes and children. McClung retaliated with a mock Parliament that satirized him. His supporters viewed him as a resolute if often highly aggressive leader. He stuck to his principles: he refused to bend to Protestant objections to Catholic and French-language education rights. His continuing participation in the Gang of Three in wartime remained a provocation. Borden knew that Roblin might mute his campaign for resource control during the conflict, but he would never abandon it. As long as Manitoba did not have resource control, Borden would not have peace in peacetime.

Roblin's downfall was swift, once the sordid details of government contracts to construct a new Legislature building emerged. In 1911, when Winnipeg was at the booming crossroads of Eastern and Western Canada, Roblin had proudly announced an architectural competition to design that building. The very announcement had symbolized Manitoba's vitality, and the Roblin government's strength. Construction of the palatial edifice had started in 1913, using superb Tyndall stone and the finest marble. But wartime shortages would delay the opening until July 15, 1920, which would be the fiftieth anniversary of the province's founding in the aftermath of the Riel Resistance. Future generations would view the Legislature as a "priceless monument" that would be almost impossible to duplicate because "the craftspeople would be difficult to find and the cost would be prohibitive."[9] In 1915, however, it would be Roblin's undoing.

His troubles erupted in March of 1915 when the Liberals on the public accounts committee opened an inquiry into public works projects, including expensive changes to the plans for the new Legislature. The initial findings were unsettling, and the Conservatives scrambled to control the damage. In the Legislature, in late March, the Tories on that committee called for approval of their majority report, which justified the changes to the construction plans. The leader of the Liberal minority on that committee, Albert Blellock Hudson, asserted that the contractor had defrauded the Province. He called for a Royal Commission to probe his suspicions, and the debate raged into the night.

On the morning of April 1, Roblin asked Lieutenant-Governor Sir Douglas Cameron to prorogue the Legislature, before the Liberals could do any more damage to his government's reputation. To his dismay, he discovered that the Opposition had beaten him to Cameron's door. On the evening of March 31, while the Legislature was lost in debate, Liberal leader Tobias Crawford Norris had presented Cameron with a caucus petition that called for a Royal Commission. Cameron spent a deeply troubled night. When Roblin arrived, the lieutenant-governor delivered an ultimatum: he would only agree to prorogue the Legislature if Roblin appointed a Royal Commission to examine the charges. Roblin reluctantly went along with the scheme. He stalked back to the Legislature to deliver what would be his last speech there.

The political damage to the premier and his cronies would be lethal. The Royal Commission launched its hearings in late April, and within days Roblin invited Liberal leader Norris to his home: his government had decided to resign. Only a new *Liberal* administration could "make the adjustments which had to be made between the province and the contractor."[10] On May 12, 1915, Roblin officially stepped down. A day later, auctioneer T.C. Norris formed his Liberal government; Hudson, who would later serve on the Supreme Court of Canada, became his attorney general.

The Royal Commission soon concluded that the contractor had indeed defrauded the taxpayers. But even worse, large portions of those purported cost overruns had gone to the Provincial Conservative Association. Perhaps most damaging of all, the Royal Commission learned that, between October 18, 1914 and January 1, 1915, Roblin and one of his ministers had "destroyed an order in council and all the paperwork on one of the contracts" with the builder.[11] The premier had also signed one of those fraudulent contracts. Eventually, as the charges multiplied like Prairie gophers, the Norris government would appoint more investigative commissions.

Manitobans were outraged. The scandal coincided with gruesome news from the ongoing battle at Ypres—in which the Winnipeg mayor's son, Douglas Waugh, was badly injured. The local newspapers carried shocking letters from survivors. One injured man described the death of a friend who had been standing beside him, and his own stunned walk "through a landscape that was constantly shelled and littered with dead horses and men."[12] Westerners compared such heroism with the Roblin government's grubby antics, and on August 6, 1915, Norris won a huge majority at the polls. As *The Globe* headlined: "Manitoba Is for Norris and Clean Government."[13] He would ultimately become the Gang of Three's most vehement member. Three weeks later, former premier Roblin and three of his ministers were charged with conspiracy to defraud the Crown. The jury could not agree on a verdict, however, and a new trial was ordered. In June 1917, those charges would be dismissed, partly because of Roblin's ill health. But the taint of scandal would linger. The Conservatives would not return to power in Manitoba until 1958, under the leadership of Sir Rodmond's grandson, Duff Roblin. Sir Rodmond would die in Hot Springs, Arkansas, in 1937, a casualty of the political sword he had wielded for so long.

On the Pacific Coast, after four years of nagging Borden, British Columbia premier McBride was making no progress with his quest for extra subsidies. Mere weeks after the declaration of war, McBride had impetuously urged his political ally to proceed with the special commission to examine his province's case. Although Borden was usually scrupulously polite to McBride, he had not replied. Unabashed, the Conservative premier had written again in January 1915, as German Zeppelins were bombing mainland Britain and the war was spilling into the Caucasus and the Middle East. He had urged Borden to ensure that the commissioners would tour the province "as soon as summer comes." He also demanded Ottawa's position on his statement of claims, which was "submitted some time ago."[14]

In March 1915, the prime minister finally responded to his oblivious petitioner. He was mild: "I must plead the tremendous pressure created by the war and by the work of the present session as my excuse for delaying a reply."[15] But he was also firm: there were irrelevant grievances—such as a demand for the return of the unused railway lands—in British Columbia's claims. Ever the gentleman to his fellow Conservatives, Borden added that he *might* find "an opportunity of discussing the subject" if McBride appeared in Ottawa.[16] It was a crumb of civility that McBride would seize.

As with Roblin, however, McBride's days in office were numbered. The premier had always supported railroads, chartering lines through his rugged territory and guaranteeing company bonds. He had been carefree and careless. When the war started, two additional transcontinental lines were under construction, and the Pacific Great Eastern Railway was struggling to connect North Vancouver with Fort George. Inevitably, in 1915, the Pacific Great Eastern was unable to meet a bond payment. Over the objections of his more sensible cabinet ministers, McBride advanced $7 million to the troubled firm. Then he disappeared to Ottawa in a vain effort to convince Borden to right his province's perceived wrongs. He went on to London to renew his social contacts, call upon skeptical financiers, and visit Canadian troops. When he returned home that fall, he was ill, and his cabinet colleagues were disenchanted. He resigned in December 1915 and, seemingly undaunted, promptly left for London as the province's agent-general.

His immediate political legacy would be disastrous. His successor, William John Bowser, lost power less than a year after McBride's departure for Europe, and Bowser's Liberal successor, Harlan Carey Brewster, would open an investigation into skullduggery in the Pacific Great Eastern Railway agreement. That probe would conclude that the province's agreement with the railroad had been improper and illegal, payments had been made for incomplete work, and portions of those payments had been diverted to Conservative campaign funds. McBride's complicity would never be proven, but his reputation would be tarnished. He would die of kidney disease in London on August 6, 1917, when he was only forty-six. The planned commission into British Columbia's special claims would lapse into wartime oblivion. But British Columbia would revive McBride's claims at the November 1918 Conference, with disastrous consequences for the Gang of Three.

With Roblin's resignation in mid-May 1915, all three Prairie provinces had Liberal premiers in power. They could not contemplate a forceful bid for resource control while the nation coped with trench warfare and daily casualty lists. But in late November of 1915, almost four months after Norris had trounced the Manitoba Conservatives, the Gang of Three finally replied to Borden's letter of March 5, 1914, in which the prime minister had requested an alternative proposal for the transfer of resource control. Borden had laid out three conditions for the premiers:

their provincial subsidies *in lieu of resources* could not continue unabated; they could not hinder immigration; and they had to provide free home-steads for aspiring settlers. He had set out his views in peacetime, only five months before the war, but it seemed as if a generation had elapsed. How could the Gang of Three respond?

The premiers' renewed attempt was ill-advised. Roblin would have hesitated to pester Borden in wartime because he was a fervent British Imperialist. Without his partisan restraint, the Gang of Three's missive was almost rude. They reminded Borden that their demands were based on two principles, which "were enunciated by yourself" in October 1913: the Western provinces had the right to control their resources, and they had a right to compensation for the lands and resources that Ottawa had already given away "for the general benefit of Canada."[17] (They had appar-ently extracted those principles from their disastrous meeting with the prime minister after the Interprovincial Conference of 1913.) Borden's con-ditions were simply "collateral matters and incidental to those principles," which could be tackled during negotiations.[18] Borden had rejected their proposal for resource control and the continuation of their subsidies in lieu of control. Now they dared him to make a counter-proposal, promis-ing "most careful consideration."[19] Borden held back. Why would he bother to respond to three Western premiers, all Liberals, in wartime—especially when resource control had become a heavily politicized issue?

Canada marked the onset of 1916 with grim tidings. On New Year's Eve, from his sickbed, Borden announced that he was doubling overseas Canadian forces to 500,000 men. That decision was taken "in token of Canada's unflinching resolve to crown the justice of our cause with victory and an abiding peace."[20] The governor general appealed for funds for the families of men who were heading to the Western Front. *The Globe* recalled its own insouciant prediction of January 1, 1915: *if* the war were to continue for another year, Germany would be fighting alone, ringed by foes. Now, on January 1, 1916, *The Globe* admitted that the Germans were fighting "everywhere except in Alsace . . . upon the soil of the allied nations," while Austria-Hungary and Turkey had "shown greater powers of endurance than were then believed to be possible."[21]

Remarkably undeterred by this situation, the Gang of Three tried again to get Borden's attention. In early February 1916, the three premiers demanded control over their school lands, along with the trust fund that held money from the sale of those lands. The three premiers collected the interest from that trust fund every year, and they now maintained that they could do a better job of managing the money. They also wanted to tap the capital in the fund. Any delay in building schools could "retard the growth of national unity," but they could not afford to borrow money.[22]

Their timing was disastrous. Borden was beset with problems, at home and abroad. He had been ill for three weeks in January of 1916 with sciatica, neuritis and lumbago. "I was in no physical condition to carry on my work," he would recall.[23] In early February, a terrible fire on Parliament Hill devastated a large portion of the Centre Block. The Reading Room was destroyed, and Borden escaped "by crawling along the corridors on his hands and knees."[24] The Gang of Three's demands were infuriating. The smoke had barely ceased to curl into the frigid Ottawa air when Borden delegated this Western bugbear: he asked four cabinet ministers to report on the Gang of Three's demand for resource control. Less than a week later, they offered a draft response, which Borden sent with a few minor changes on March 10, 1916.

His two-page letter was tough: he repeated the wearying details of their previous correspondence and his three conditions, and maintained that nothing had changed. He would not deal with their request until they dealt with his conditions. So far, they had addressed only one condition, and that demand for resource control along with continued subsidies "could scarcely be said to constitute a reasonable basis for discussion."[25] He awaited their revised proposal. The Gang of Three was back where it had started in December 1913.

Five weeks later, the prime minister tackled their request for control over their school lands and their trust fund. Here he was scathing. The premiers wanted access to the trust fund, which now contained $9.8 million, so that they could quickly build more schools. But surely it would be more prudent to draw upon the fund slowly, so that money would be available in future years? In 1915, Ottawa's net cost to administer the school lands and the fund itself had been only $35,303.09. Surely the three premiers were not claiming that they could handle those school lands more cheaply? Anyway, in case they had not noticed, Canada was at war. "The present time," Borden concluded, "is not opportune to further press for transfer of this fund."[26] The three premiers had their answer. As trains laden with coal and wheat chugged out of the West, both sides were hardening their positions.

Borden's letters to the Gang of Three coincided with an ongoing crisis in Saskatchewan: Walter Scott's Liberal government was under siege. In mid-February 1916, a Conservative legislator claimed to have proof that members of Scott's government had accepted bribes. Deputy Premier James Calder countered that Borden's hyper-partisan public works minister Robert Rogers from Manitoba had orchestrated a plot. Politicians in both provinces were still reeling from the scandals that had toppled Roblin. Scott first assured himself that Manitoba history would not repeat itself: Saskatchewan lieutenant-governor Richard Lake would not unilaterally force him to establish a Royal Commission into the charges. He had time to strategize.

The premier was frazzled. The voters scarcely noticed when he extended the vote to women three days after those charges first surfaced. Then he waded into the never-ending debate over his government's support for Roman Catholic schools, including French-language and German-language separate schools. On February 25, 1916, when Scott was defending that support in the Legislature, he concluded his speech with a brutal personal attack on his own Presbyterian pastor. The Reverend Murdock MacKinnon had long opposed taxpayer support for Catholic schools. Now Scott dismissed him as an "intellectual and spiritual leper." His caucus reeled. The premier "had crossed the line . . . . he [had] caused more damage than he realized."[27]

The premier left for the Bahamas. Through a volley of telegraphs, he and Deputy Premier Calder decided to establish three Royal Commissions into the alleged misdoings in road contracts, construction contracts for buildings such as the Regina jail, and the issuance of liquor licences. In August, the Royal Commission into the distribution of liquor licences cleared Scott and his cabinet of bribery charges, but the strain had been too much. His physical and mental health was fragile, and his annual absences of up to six months from the Legislature had become disconcerting. His colleagues were fed up. Finally, the long-suffering Calder consulted Liberal senator James Ross, who had worked with North-West Territories premier Frederick Haultain during that long-ago struggle for provincial status. The two decided that Scott should resign. Ross brought the resignation letter to his old friend as he lay in a Philadelphia hospital. By the time Scott reluctantly signed the letter, his colleagues had already decided upon his replacement. The resignation took effect on October 20, 1916.

His replacement could handle trouble. Deputy Premier Calder had turned down the top job because he wanted to run in the next federal election (he was also smarting from the scandals). Instead, key Liberals settled on Saskatchewan MP William M. Martin. An experienced lawyer untainted by scandal, Martin was a devout Presbyterian who agreed with Pastor MacKinnon's view "that the existence of a dual school system was unfortunate."[28] More important, Martin had fervently supported Saskatchewan's demand for resource control when Borden extended the boundaries of Manitoba in 1912. The Gang of Three had its new Saskatchewan member.

All three Royal Commissions would eventually clear Scott and his cabinet, although several public servants and Legislature members would be convicted of fraud and the acceptance of bribes.[29] But Scott would never really recover: although he would work sporadically as a journalist after he left the Legislature, his depression would be virtually unrelenting. In 1938, he died alone in an Ontario psychiatric institution.

The times were desperate. While Saskatchewan wrestled with its scandals, Canadians watched the heroism and the horror abroad. On July 1, 1916, as the three Saskatchewan Royal Commissions worked their way through sordid corruption charges, British troops, along with some French troops on the southern flank, attacked along a twenty-five-mile front that spanned both banks of the Somme River, south of the Belgian–French border. It was a bloodbath: more than 57,000 British soldiers were killed, wounded or missing on the first day alone. The plucky Newfoundland Regiment—Newfoundland joined Canada only in 1949—was virtually annihilated. In late August 1916, Canadian troops were shifted from Flanders to the Somme. They dug in along the front line, and when the next major Somme offensive started on September 15, the Canadian Corps pushed ahead to secure the village of Courcelette, repulsing German counterattacks, and consolidating its position.

Over the next few weeks, three Canadian divisions would inch forward, with small territorial gains and huge manpower losses. On September 17, 1916, Lance Corporal Harold Fleetwood Robins from Lethbridge, Alberta, was hit with gunshot in the back. The twenty-seven-year-old British-born soldier, who had made telephones before he enlisted in November 1915, died later that day, almost assuredly painfully. Ten days later, Private James Thomas Alcock was reported missing in action. By the end of the day, his buddies found his remains. The twenty-eight-year-old, also British-born, had been working as a ranch hand in Milk River, Alberta, when he enlisted in September 1915. His body now lies beneath a stark white headstone among the approximately 1,000 Canadian graves in Adanac Military Cemetery, southwest of Courcelette. (Adanac is Canada spelled backwards.)

There were so many stories of futile sacrifice. On October 9, 1916, as the Manitoba Regiment struggled to cross German wire fences, piper James Clelland Richardson strolled alongside that barrier, back and forth, playing his bagpipes. As his citation for bravery would declare: "Inspired by his splendid example, the company rushed the wire with such fury and determination that the obstacle was overcome and the position captured."[30] Later that day, as Richardson accompanied a wounded buddy back from the front line, he realized that he had forgotten his bagpipes. "Although strongly urged not to do so, he insisted on returning to recover his pipes. He has never been seen since."[31] By the time the Somme offensive ended in mid-November, when the battlefield turned to bog, the Allied line had moved forward only seven miles. There were more than 600,000 Allied casualties in this disastrous offensive, including more than 24,000 Canadians.

Across the Prairies, the war was a sadly communal effort. In Neerlandia, eighty miles northwest of Edmonton, a handful of Dutch settlers were raising cattle, hogs and poultry; the cattle were herded to the nearest railway station for shipment. Pioneers had founded this isolated Christian Reformed Church community in 1912, after a week-long oxcart ride from Edmonton. The settlers had cleared "land covered with heavy timber and dotted with swamps, creeks and small ponds."[32] By wartime, they were self-sufficient, shooting game and harvesting family gardens. Each pioneer was "his own butcher, veterinarian, blacksmith, shoemaker and carpenter since none of these services were within 20 miles of their settlement."[33] But those hardy Dutch immigrants from the United States and the Netherlands were now contributing to the nation's food supplies.

In northern Saskatchewan, German immigrant Joseph Prechtl was growing wheat and selling produce, and he was finally making money after a string of peacetime losses. His family lived frugally because prices for everything from sugar to coffee were soaring. The family was also lying low as Anglo-Canadian Westerners denounced the "Huns." Prechtl's son Richard would recall that his family survived because their closest neighbours, the Garnetts, "stood back from the prejudices of war" and maintained their friendship.[34] But "propaganda and the casualty list brought the tragedy of war into every home."[35] The Prechtl family's problems would get worse after the war when the veterans returned home, and compared their lives with the apparently fortunate fate of those who had stayed on the farm. But the German immigrant's butter, eggs, wheat and hogs were vital for the home front throughout wartime.

In 1917, as Canada limped into its third full year of war, the most serious domestic problem was the labour shortage. Women had been working overseas as nurses since the start of the war: more than 2,500 women would eventually serve in England, France and the Eastern Mediterranean in the Royal Canadian Army Medical Corps. In 1917, at home, more than 30,000 women were now working in munitions plants, roughly 5,000 were civil servants and thousands more held office jobs.[36] Others were harvesting crops, and working as streetcar conductors and telegraph messengers.

By now, the suffragettes' cause was compelling. Given the long tradition of female farm labour, it was only natural that the Prairie provinces were the first to extend the vote. Manitoba premier Norris had done so in January 1916, eight months after he had replaced Roblin. (He had also cancelled provincial support for bilingual schools, erasing the compromise that Laurier had forged in the mid-1890s.) Saskatchewan's Scott and Alberta's Sifton had soon followed Manitoba's lead. Other provinces and Ottawa were now under pressure to emulate the Prairies and recognize women's contributions.

Despite the employment of female workers, however, there were never enough people to fill the jobs. In the spring of 1917, common sense finally prevailed: Canada reached out to its under-utilized labour force, those so-called "enemy aliens" from Eastern and Central Europe. The federal government had already released some prisoners from its detention camps to work on contract for mining and railway companies. (Not coincidentally, the release of those prisoners reduced the cost of operating the camps.)[37] Now many more foreign-born workers, who had been dismissed as undesirables at the start of the war, were able to find jobs. It was an ironic reversal. In 1915, Anglo-Canadian miners at the Crow's Nest Pass Coal Company had threatened to strike unless German- and Ukrainian-born employees were dismissed.[38] Now those so-called aliens had been welcomed back to the dangerous pits in Fernie, British Columbia. Those workers would soon join unions for mutual safety and better wages, as the allure of socialism outweighed the attraction of their traditional ethnic organizations.

Worker demands provoked consternation and resistance. Although Ottawa controlled Prairie resources and oversaw mine sites, the turmoil spread across Western societies. The Prairie premiers were now dealing with a new war on their doorsteps. On April 1, 1917, Lethbridge miners, who were largely former citizens of the Austro-Hungarian Empire, went on strike. They charged that their employers were exploiting their tenuous position in the labour force by ignoring unsafe conditions and underpaying them. Those were risky tactics for foreigners in wartime. Many journalists and anti-labour politicians "cast their job actions as the work of seditious enemy aliens."[39] Those miners held out for three months, until the Borden government took over their United Mine Workers of America local, and increased their wages. Such strikes embittered many Anglo-Canadians, however, and troubled high-level authorities.

By mid-1917, Ottawa was closely monitoring the showdowns. Interior Ministry officials wanted the Royal North-West Mounted Police to maintain order in Dominion parks because a "large number of aliens [were] employed in the coal mines" in those parks.[40] On July 10, 1917, Borden asked British Columbia premier Harlan Brewster if he would accept the RNWMP in areas where the mining union local operated in his province. Brewster welcomed any police presence during wartime near the mines "in the event of developments, which might endanger life or property."[41] The RNWMP also kept a close watch on mine workers in Jasper and Banff. The distinction between suspicion and paranoia blurred, and on December 31, 1917, a former solicitor for that United Mine Workers local even warned Ottawa that foreign-born workers might make "embarrassing use of trades unionism in the interests of the enemy[sic]."[42]

Ignorance compounded the backlash. In late July 1914, before the declaration of war, Ukrainian Catholic Bishop Nykyta Budka had dispatched a foolish pastoral letter from his Winnipeg headquarters, urging Ukrainians who still had military reserve obligations to return home. Although the bishop would soon retract that advice, "the damage had nevertheless been done."[43] Throughout the war, Ukrainians vainly explained that Russia and Austria had forcibly divided their nation in the eighteenth century. Such distinctions were lost on the authorities. In 1915, Ottawa had set up a detention camp below Castle Mountain in Banff National Park for those so-called Austrians, who were really Ukrainians and Poles and who generally opposed the Austro-Hungarian Empire's rule. Most would recall that internment as unremittingly grim. They cleared the bush during the day, subsisted on poor diets, and huddled in chilly barracks at night. "They found it hard to be under such strict regulations," Polish priest Anthony Sylla would recount, "and to be fenced in by the high barbed wire."[44] Ottawa finally closed that camp in August 1917, when the men were conditionally released to take industrial and mining jobs.

But the shame of that "alien" designation lingered. Polish-born Waclaw Fridel had a job in the coal mines during the war, but he had to report once a month to federal authorities in Wabamun, Alberta, which was seven miles away. Officials had threatened to send him to a special labour camp in British Columbia if he did not appear regularly. Once, he failed to report for three months, because of personal and work difficulties. On the night before he finally trudged into Wabamun, braced for the worst, Fridel looked up at the midnight sky, and prayed. "I said to myself, 'Oh, God, good God, is there anybody in the world who could talk to the angels on behalf of us Poles? Why am I supposed to go there? What for? I am not guilty of anything, I do not owe anyone anything. Austria is not my country.'" More than half a century later, he would remember that moment "as if it were today . . . Even now it brings tears to my eyes."[45] A sympathetic clerk backdated the stamps on his papers, and Fridel returned to work, hacking out Western coal for the war effort.[46]

Amid the labour strife and the incessant demands of the war machine, one truth was now clear: Western resources were becoming more valuable with every passing day. Energy companies were homing in on Ottawa. In late July 1917, the Shell Transport Company Ltd. of London sent a brash eight-page proposal to Interior Minister William James Roche. The firm wanted exclusive oil and natural gas rights over an enormous swath of the West for the duration of the war and for five years after Armistice. Shell even dictated its terms. If Ottawa granted rights over the 328,000-square-mile parcel, exploration would start when the weather permitted. The company would not pay taxes for the first fifteen years of exploration, and royalties would only commence on January 1, 1930 at three cents per barrel. Meanwhile, the federal subsidy for crude oil production should continue. Shell also wanted right-of-way over all Crown lands for pipelines, telegraph and telephone lines, railways and highways; not to mention land for factories, storehouses, refineries and reservoirs.

Almost a century later, the letter seems astonishingly patronizing, as company officials assumed that they were making an offer that Ottawa could not refuse. "The tremendous importance of oil as a fuel and as a power producer cannot be exaggerated," Shell executive R.N. Benjamin wrote through his attorney Adam T. Shillington.[47] Because Shell was a company with a good reputation, solid technical skills and enormous capital, there could be "no possible criticism" if the rights were granted. Only "a Government or a great corporation" had the ability to exploit that challenging Western land.[48] Roche circulated the letter to his cabinet colleagues, but the federal government prudently did not reply. It also did not consult the Western premiers, who would have instantly resurrected the campaign of their Gang of Three.

Ottawa also concealed the proposal from the embittered Western farmers who would have hit the ceiling if more drill rigs had appeared in their wheat fields. The sight of those contraptions had been so provocative even before the war that the United Farmers of Alberta had warned Borden: farmers were "more important to the community in general than oil lease speculators, or even oil drillers."[49] The farmers had demanded the opportunity to lease the sub-surface rights to their land before speculators could grab them. They had also demanded a share of the royalties along with compensation for any property damage. Now, in wartime, they were even more determined. The right to exploit Western resources had become a very hot potato.

Borden had so many crises on his hands. On Easter Monday in 1917, all four Canadian divisions had attacked German troops on Vimy Ridge, a fortified two-hundred-foot escarpment in northeastern France. It had been a hazardous venture. As Private Donald Fraser from the 31st Alberta Battalion would record in his secret diary, the soldiers had struggled through sleet and snow across a mass of shell holes, passing the walking wounded and the corpses of their mates. Fraser had seen a fellow soldier "struck on the head by a piece of shrapnel which knocked his brains out. They were lying two feet away and resembled the roes of a fish."[50] Fraser would survive, but there would be 10,500 Canadian casualties, including 3,500 dead. By Thursday, April 12, 1917, after four days of intense fighting, the Canadians had secured the entire ridge in a spectacularly successful advance. It was a turning point in the war: the Germans could no longer shell Allied positions from those heights with impunity.

Borden was in London, at sessions of the Imperial War Conference and the Imperial War Cabinet—where Britain's self-governing dominions sat as equals—when news of Canada's "splendid victory" reached him.[51] The prime minister was cheered, but he also knew that voluntary enlistment could no longer keep pace with the slaughter at the Front. On May 18, 1917, soon after his return to Ottawa, he announced compulsory military conscription. Canada would need 50,000 to 100,000 more soldiers. It was a pivotal decision in an already divided nation. "Never before," historian John English noted, "had the state compelled young men to fight in a war beyond Canada's shores."[52]

Borden was endangering the ever-fragile pact of Confederation. Across English Canada, Anglo-Canadians called for a coalition government to shepherd the country through the crisis, and Borden asked Liberal Leader Sir Wilfrid Laurier to join that government. Although Laurier regarded the conflict as a just war, he was torn: in French Canada, especially in Quebec, there was almost universal opposition to conscription; some Western farmers also objected to the drafting of young men who might help with the harvest. Laurier refused Borden's offer, and Western Liberals at a convention in Winnipeg in August initially supported him. But, behind the scenes, virtually daily throughout that summer of 1917, the prime minister was patching together his Union government. The discussions were so critical that Borden even offered to resign as leader so as to attract Liberals. Meanwhile, legislation to conscript unmarried male citizens between the ages of twenty and forty-four for the duration of the war inched its way through Parliament, finally passing in late August.

The stress of that summer reduced Borden to "nervous prostration, arising from fatigue, anxiety and strain."[53] In early September, the prime minister played his trump card: his government introduced the War-time Elections Act, which disenfranchised anyone of enemy-alien birth who had not been naturalized before March 31, 1902. That same bill granted the right to vote to women who were the widows, wives, mothers, daughters or sisters of soldiers. The legislation, which passed on September 20, removed the vote from many Prairie residents born in Central and Eastern Europe—who might have been inclined to oppose conscription. Liberals who joined Borden's government were now safe from voter reprisal.

On September 21, the prime minister left for nine days at a fishing camp in the Laurentian Mountains. A day later, he heard that prominent Liberals would now support his Union cause. The negotiations had been trying, and Borden had been "exasperated beyond measure by the hesitation and dallying," and by their doubts about his leadership.[54] But he now had the backing of Saskatchewan Railways Minister James Alexander Calder and that prominent Manitoba-based president of the United Grain Growers, Thomas Alexander Crerar. Most spectacularly, the prime minister had snared the third original member of the Gang of Three, Alberta premier Arthur Sifton, thanks to the intercession of Arthur's younger brother, former Liberal interior minister Clifford Sifton.

Why did Arthur Sifton leave Edmonton? The premier was worn out from the strain of wartime governance, and his financial misjudgments were coming home to roost. Despite the dire experience of his predecessor with railroad bonds, Sifton had succumbed to the same lure in 1913, and guaranteed $37.1 million in bonds to support the construction of 2,300 miles of railway line. By 1917, the Canadian Northern and the Grand Trunk Pacific railways were in serious financial trouble, as was the Alberta and Great Waterways Railway and the Edmonton, Dunvegan and British Columbia Railway. The provincial Conservatives were snapping at Sifton's heels.

Despite those attacks, Sifton had won his third majority in June, partly thanks to his slick political machine. Two months later, in August 1917, Borden once again bailed out the Canadian Northern: Ottawa could not afford to let the railroads go under in wartime; nor would it write off its huge equity stake. (In 1919, Ottawa would take over that railroad along with the bankrupt Grand Trunk.) The pressure on Arthur Sifton was lifting, but those guarantees dogged him. He was also a fervent British patriot: he had supported the war, and now he supported conscription, despite the strain it would put on available manpower for the farms and the mines. He also viewed a coalition as the best way to win the war. His replacement was Public Works Minister Charles Stewart, who owned a farm implement business and a real estate operation.

In mid-October 1917, Borden unveiled his cabinet, including his contingent of powerful Western Liberals. That strong advocate for farmers, Thomas Crerar, was now agriculture minister with responsibility for the Wartime Food Control Board. Saskatchewan's former deputy premier, James Calder, was now minister of immigration and colonization. Arthur Sifton was minister of customs. All three were prominent advocates of Western concerns; Sifton and Calder were veterans of those tough battles against Borden for resource control.

One man was missing in this new cabinet, and few mourned his absence: former public works minister Robert Rogers, that hyper-partisan Manitoban who had been under suspicion during the investigations into the Roblin government (a *federal* Royal Commission would eventually clear him) and who had bitterly opposed any Union government. When Rogers had warned Borden that his government was headed toward destruction with its coalition plans, Borden had "construed [that warning] as a tender of his resignation . . . . I accepted it."[55] In effect, Rogers was forced out. Western Liberals such as Sifton and Calder could not tolerate his outrageous political tactics, and given the no-holds-barred politics of the era, his removal indicated how far Rogers had crossed the line.

The Gang of Three—Norris in Manitoba, Martin in Saskatchewan and now Stewart in Alberta—was heartened. Two of the new ministers had themselves written endless letters to Borden about resource control. Calder had pestered him. Sifton had composed his first plea when Sir Wilfrid Laurier was prime minister. Both had dodged Borden's demand that they address his three conditions. Surely this was as good as it could get.

Borden romped to victory in the December 17, 1917 election. His Union government won every seat across the West, with the exception of one in Manitoba and one in Alberta. Laurier retained only 82 of the 235 seats in the House of Commons, including 62 in Quebec. The great Conservative gamble had paid off, but Borden was exhausted. He toddled off to Hot Springs, Virginia: because of train delays, his trip took three days, and the Christmas weather in Virginia soon turned as icy as an Ottawa winter. Even on holiday, he was peppered with telegrams. On January 1, 1918, conscription took effect, and on January 7, Borden headed to Washington, returning home on January 10 to the "still terrible problems of War."[56]

His challenges seemed never to end. Borden dealt with problems in proposed railway legislation, compulsory registration for the draft, and the enfranchisement of women. He kept tabs on relief operations for Halifax in the wake of a devastating explosion in the harbour on December 6, 1917, that had flattened the city's downtown, killing almost two thousand people and injuring another nine thousand, including many blinded by shattered glass.

On January 16, provincial labour ministers, union representatives and the cabinet war committee gathered in Ottawa to mull the manpower crunch. To win their goodwill, Borden talked about conditions in Europe, including the manpower shortages in the trenches. Then he asked for their co-operation. Union leaders, in turn, stressed the need on the home front for food and industrial workers. Borden was torn. He set up a cabinet committee to mobilize labour and resources for the war, for food production, and for the "labour supply in essential industries."[57] The nation was deeply divided: many conscripted men were appealing their designation.

The stress was terrible. Finance Minister Thomas White was "ill and dispirited."[58] White had borrowed to finance the early years of the war, because customs and excise taxes, postal rates and other minor taxes could not cover the bills. Between 1913 and 1918 the national debt had soared from $463 million to $2.46 billion.[59] In 1916 White had introduced a Business Profits War Tax on larger firms, and in late July 1917, faced with the "grave conditions" of wartime and vehement public complaints about profiteering, he had imposed personal income taxes. There was now a four per cent tax on single men with incomes over $2,000; the personal exemption for other Canadians was $3,000; and there were variable rates for those with incomes of more than $6,000.[60] The finance minister had stipulated that his successors should review the income tax in "a year or two after the war is over" to see if it were still required.[61] Months later, as inflation soared and the bills piled up, those taxes were clearly more symbolic than even close to sufficient. White wanted to resign. Borden convinced him to wait.

More crises loomed. Fuel was in short supply, and it might be necessary to close factories. Borden set up a War Trade Board, which would work with the American War Trade Board to ensure that essential industries stayed open. On January 19, the owner of the *Ottawa Evening Journal*, Philip Dansken Ross, reminded Borden of the words once applied to U.S. president Abraham Lincoln: "He who is the master of patience is the master of all."[62] Those words, Ross told Borden, could apply to him. The prime minister did not disagree.

By early 1918, however, Borden could glimpse peace on the horizon. He and his ministers had to plan for a postwar world in which the soldiers would come home and new immigrants would bring economic growth. Official Ottawa still accepted the antiquated notion that the unemployed could find prosperity if they were settled on the land. In the last few years before the war, Ottawa had sought farmers for the northern regions of the Prairies. Interior Ministry officials had even suggested the use of colonization firms, despite their decidedly mixed records. Ottawa could also double its land grants from 160 acres to 320 acres, because northern homesteaders might need more acreage to survive in tougher conditions. The Borden government had shelved those suggestions during the war, but the Western premiers knew that federal plans to give away more of their turf were merely on hold.

The Gang of Three had to move fast. When Borden summoned the premiers to an informal conference on February 15 and 16 to discuss wartime supplies, the fate of the returning veterans and postwar immigration policy, the Western premiers seized their chance. *Their* resources were fuelling the war effort. *Their* residents were dragging out the coal. *Their* farmwomen, along with *their* ethnic residents, were ready to sow the crops. *Their* voters had swung behind Borden's Union government, which now included men who had fought for resource control. After three and a half years of war, they reasoned, Borden could no longer ignore the justice or the urgency of their cause.

They could already foresee their postwar challenges. In January, Alberta premier Charles Stewart had warned Borden that veterans were congregating in dangerous clusters on the streets of Calgary and Edmonton. They required "immediate" settlement on the land.[63] (Borden would soon appoint Senator Sir James Lougheed as minister of soldiers' civil re-establishment to direct them onto plots.) On February 7, 1918, when Stewart received his invitation to the conference, he argued that Ottawa should transfer resource control before more veterans applied for land.[64] Alberta also needed resource royalties to handle "the return of Soldiers to civilian life."[65] The premier even had a fallback position: "If, however, it is impossible to go into the question of the transfer of the whole of the Resources [at the upcoming Conference]," he wrote, "some immediate action should be taken with regard to coal and mineral lands."[66]

It was a pre-emptive strike. Stewart was perfectly willing to work *with* Ottawa to settle soldiers and newcomers on the land. As the premier told a Toronto audience two days before the conference: "It is immigration we need more than anything else in Alberta—people to come out and develop our resources."[67] But the premier wanted to control the lands on which they would settle. Saskatchewan attorney general and acting premier William Turgeon echoed Stewart. He did tell Borden, however, that resource talks should occur "only if time [is] sufficient."[68]

Journalists soon learned of the Westerners' renewed campaign. Their reports aroused the fears of the Maritime premiers, who asserted that *their* financial demands should take priority. It was the same old federation song of squabbling partners on the home front. On February 11, New Brunswick premier Walter Foster telegrammed Borden: the upcoming conference should discuss, and "if possible" settle his demand for compensation for the federal lands transferred to Ontario and Quebec in 1912.[69] That same day, Prince Edward Island premier Aubin-Edmond Arsenault also sent a telegram: he had read that the Western provinces would "present claims for additional subsidy" at the conference, and that the "right of other provinces for compensation" for the 1912 boundary extensions was also on the agenda.[70] Was this true? (No.) Remarkably, Arsenault did not mention Western resources, only Western *subsidies*. The West's quest for constitutional equality and the Maritimes' demands for extra cash were now thoroughly jumbled together. The Maritime premiers would be quick to exploit the confusion.

As the Conference opened, it was clear that *The Globe* had gravely underestimated the extent of the challenge. "It is practically certain," it proclaimed, "that the Prairie Provinces will be given their long-demanded control of their natural resources, and this will involve a readjustment of the Provincial subsidies, necessitating agreement with all the Provincial Governments."[71] But the issue would quietly drop from the newspaper's pages over the next few days. It could fairly be asked why *The Globe* believed the problem could be solved in wartime, or how the participants would somehow reach a subsidy deal.

The conference agenda was already packed, with the two-day meeting focusing on the war, including "the imperative necessity" for food conservation—a polite expression for rationing.[72] The three Westerners could not even get their demands on the schedule. Meanwhile, the Maritime premiers claimed they had received no notice of any discussion on resource control. The exasperated Westerners agreed to pull together their case after the conference: "it was understood" that the three Maritime provinces could then react "at a later date."[73] Forty-eight years after the creation of Manitoba, thirteen years after the formation of Alberta and Saskatchewan, these ill-natured quarrels could seemingly go on forever.

Instead, the participants discussed federal plans "for employment and land settlement" for the returning soldiers.[74] Then their fellow Westerner, Immigration Minister James Calder, outlined his scheme to revive immigration, which had effectively ceased during the war: the only way Ottawa could pay for the war and for railroad construction was through increased productivity, and "this in turn could only be achieved by increasing our man-power."[75] Federal ministers did not want the provinces to duplicate their land-settlement efforts. The proceedings dragged late into the evening of Saturday, February 16, until finally, all provinces agreed to play ball. In an era that celebrated efficiency, a strong central government had trumped the Westerners. Once again, the demands of the Gang of Three were inconvenient, and, given the resistance of the Maritimes, virtually inconceivable.

The three petitioners were undaunted. While Borden would not mention the conference in his memoirs, he did note that on Monday, February 18, the three Western premiers gathered in his office. "They did not press for immediate consideration," he later recalled, "but they did ask to discuss resource control before the 1919 Parliamentary session.[76] Ever distracted at home, Borden made a mistake in his terse two-sentence entry: in reality, Attorney General Thomas Johnson had represented Manitoba at the conference; Premier T.C. Norris had been delayed in Winnipeg. The encounter that mattered so much to the West was so unmemorable for the prime minister that he would not even recollect the participants.

A day later, the two premiers and Johnson followed up with a deceptively mild letter to Borden, which all three signed. There were the usual polite disclaimers: they hesitated to bother him; they did not want to embarrass him; they knew that he had so many important problems.[77] *But* the issue of resource control was so important that he should tackle it as soon as Parliament adjourned so that he could take the "necessary legislative action" during the next Parliamentary session.[78] There was steel in their tone. The Westerners would not address Borden's three conditions: after all, his government was "thoroughly familiar" with their claim.[79] When talks were underway, they could "go thoroughly into every phase of the question."[80]

This letter was remarkable for what it did *not* say. It did *not* request another dominion–provincial conference. It implied that the Western provinces could hammer out the transfer of resource control with Ottawa. The three Westerners were trying to do an end-run around their fellow premiers. But they were forestalled. A day later, on February 20, 1918, Borden decided to postpone any talks. Then he added one of those roundabout sentences that would say so little and so aggravate Westerners. "As soon as may be possible after the ensuing session, having regard to other important engagements which must be undertaken by my Colleagues and myself in connection with the approaching Imperial Conference," he wrote, "we shall be prepared to arrange for the *full consideration and conference* which you suggest."[81] Borden sent a copy of this letter to every provincial premier.

The Westerners had lost any hope of keeping the discussion between the prime minister and themselves. They could do no more when the nation was at war. Borden rightly assumed that their patriotism would trump their grievances. More remarkably, on the very day that Borden penned this note, British Columbia Premier Harlan Carey Brewster, who was still in Ottawa, replied to Borden from his hotel. He wanted to hitch his wagon to the Prairie premiers' cause, and he agreed with their desire to handle such issues after Parliament recessed. He had intended to sign their letter to Borden, "and only that I must have been unavailable at the time of its presentation was it presented without my name attached."[82] The absence of his signature was surely no accident. The three Westerners were desperate to deal with Ottawa alone: they had almost certainly dodged Brewster in the Château Laurier Hotel corridors.

Brewster was a determined politician. He asked Borden, once again, for the return of any unused railway lands that British Columbia had transferred to Ottawa in the nineteenth century. The control of those railway lands had been one of the grievances that McBride had stuffed into his legal case for the special commission on subsidies. Brewster now insisted that the Prairie provinces' claims had "the same relative force" as his demands for the railway lands.[83] The connection might have been tenuous, but the three Westerners surely understood that Brewster's efforts to link his case with theirs spelled trouble.

Other premiers added more caveats. Two days after Brewster's letter to Borden, Ontario premier Sir William Hearst joined the fray. He benignly declared that he had no wish to interfere in affairs between Ottawa and the Prairie provinces. But then he pounced on Borden's offer of "full consideration and conference" on the issue. If Ottawa decided to transfer resource control, he wrote, "such action would, I submit, necessarily open the whole question of subsidies."[84] Ottawa should allow *all* provinces "to present their views" and no transfer should occur "without a readjustment of subsidies."[85] Three days after Hearst's warning, Prince Edward Island premier Arsenault asserted that whenever Ottawa decided to discuss the issue, the Maritime provinces should "be notified in time for them to take the necessary action."[86] The premiers were in essence drawing up the battle lines for the November 1918 Conference.

There would be a terrible fallout from that February gathering. British Columbia Premier Brewster had not known a moment of peace since he had taken power in late 1916. The Liberal businessman had probed the misdeeds of his Conservative predecessors. In addition, he had tackled provincial finances and established "a new department of labour, an amusement tax and the passage of bills dealing with a new taxation board, public accounts, workmen's compensation and war veterans."[87] He had introduced a civil service commission, personal income taxes, female suffrage and prohibition. Finally, he had also handled a disruptive smelters' strike in Trail when some of its residents had turned on the industry's largely Italian-born workforce.

Brewster was exhausted before he reached Ottawa—where he had then fought for days to get Borden's attention, and to link his claims with those of the Prairie premiers. He had failed. When he finally boarded his train for Victoria in late February, he was bone-tired and even more harassed than when he had arrived. The February 1918 Conference had been too much. By the time his train reached Calgary, the premier had pneumonia. He was taken to a local hospital, where he died on the evening of March 1, 1918, his body reaching Victoria three days later. His successor, the plainspoken John Oliver, would *never* abandon his predecessor's pursuit of British Columbia's special claims, including the return of the railway lands. At the November 1918 Conference, Oliver's demand for a solution to his claims would throw another spanner into the Prairie premiers' efforts.

Before that gathering, however, Borden and his fellow first ministers had to end the war. No matter what Westerners might urge, there could be no serious discussion of resource control until the guns were quiet.

8

THE REST VERSUS THE WEST:
THE GANG OF THREE LOSES THE PEACE.
MARCH 1918 TO NOVEMBER 1918

—

BY THE SPRING OF 1918, the war was going badly, both abroad and at home. Canadians were saddened, bitter and resentful. Although the Allies had expected a German offensive, they were taken aback when German troops punched through British lines on the first day of spring. Eleven days later, Quebecers rioted against conscription, exchanging gunfire with troops. Four civilians died in the mayhem. Despite those protests, Prime Minister Sir Robert Borden cancelled all exemptions to conscription as the Allies retreated in Europe. Farmers and industrialists objected to the potential loss of their remaining workers. Borden could not escape his critics, whether they arrived in delegations on his doorstep or rose in Parliament, flinging accusations.

The prime minister was fed up with narrow interests. He had little tolerance for regional jockeying for position and was not much interested in the struggles of ordinary Canadians. In 1918, whenever he was in Canada, he was hunkered down, dutiful and occasionally despairing. His diary and his later memoirs, rarely reflective at the best of times, became chronicles of wearying chores. In Europe, carried by the hustle of great doings, he was a good listener, tolerant of quirks, and patient in negotiations. But at home, while determined to fulfill his obligations, he was detached from what he viewed as petty problems when the nation was at war.

Throughout 1918, the three Western premiers struggled to engage him in the issue of resource control. They did not understand, when they refused to moderate their demands, that they were making no progress.

Borden's apparent passivity on the subject did not, as they hoped, indicate his approval for the transfer of resource control along with the continuation of subsidies in lieu of those resources: The prime minister had simply lost interest in their problems. Instead, with a devious display of collegiality and in the name of efficiency and patriotism, he lured them into supporting his postwar agenda for their lands and resources. As secret memoranda make clear, Borden's cabinet was divided on the issue, and he did not have the money or the inclination to buy regional peace. And he knew he could escape blame because Maritime objections would in any case thwart the West's demands.

The Gang of Three assumed their huge wartime contributions would bring peacetime rewards. They missed the fact that Ottawa viewed their demands as excessive and inconvenient. Why would Borden wade into a regional fight? There was nothing in it for him. That crucial misunderstanding between what Westerners wanted and what Borden was willing to contemplate would contribute to the debacle at the November 1918 Conference. The three Prairie leaders never had a chance.

The war at home was tough on all first ministers. After the February 1918 Conference, Borden had returned to his domestic tasks. He was brave and fair, but he could never conceal his dismay in the face of selfish sentiments. In March, he confronted grumpy Canadian manufacturers, and defended his decision to eliminate the duty on farm tractors. "I emphasized the needs of the country," he would recall, "as contrasted with the interests of the Canadian Manufacturers' Association."[1] He faced down the farmers' mass protest: he was conscripting *their* workers, he explained, to ensure that the English Channel ports remained open to *their* exports. Partly to deter labour unrest and the opponents of conscription, he even adopted an anti-loitering law, which obliged all males between the ages of sixteen and sixty to engage in useful occupations.

Belts had to be tightened on the home front. Fish consumption was now a patriotic duty. Beef and pork use "was restricted in restaurants on certain days, and eating places were allowed twenty-one pounds of sugar for every ninety meals."[2] There were Meatless Fridays, and Fuel-less

Sundays. And there was the gripping fear of a telegraph with news of a loved one's death at the Front. But, trudging through his days, Borden was largely out of touch with the annoyances, deprivations, losses, and brave efforts of most Canadians.

The three Western premiers, however, could see the impact of federal actions on the ground, for better and for worse. They were well aware of the food and fuel shortages and the constant hunt for sedition. When Borden introduced votes for all female citizens that spring, he empowered their hard-working rural women while securing the grudging approval of many men.[3] When he adopted Daylight Saving Time to conserve fuel, he irritated their farmers, who complained that it would disturb their routines. In Winnipeg, *The Voice* newspaper sniped in a front-page column that the animals would still wake up and sleep to the schedule of the sun, no matter how Ottawa changed the clocks.[4] When Borden established the Canada Food Board to conserve supplies and control prices, *The Voice* was merciless. Why had the Food Controller not suppressed an eighteen per cent jump in the price of local bread? "Some controller."[5]

The prime minister's personal intervention was sought at every wartime turn. British Columbia premier John Oliver asked Borden to reverse the decision to switch dry dock repairs for Imperial ships from Prince Rupert to Seattle. Borden discovered that the Prince Rupert dockworkers had "insisted on such excessive rates of pay" that Imperial authorities had opted for cheaper service.[6] Prince Edward Island premier Aubin-Edmond Arsenault wanted Borden to widen the gauge of his province's railroad tracks. Their exchange aroused great indignation in the prime minister. "Perhaps you do not fully realize the tremendous strain upon the financial resources of the country," Borden wrote testily, "which is being entailed not only by the war but by the necessity of making provision to pay for supplies of food and munitions required by Great Britain and other allied nations." He added a cheap shot: "If I am not mistaken this was explained pretty fully at the recent [February 1918] conference with the Provincial Premiers."[7]

Few Canadians now viewed the bloodied trenches along the Western Front as a site of glorious sacrifice. (They would revive that improbable notion in peacetime.) Borden was furious at the ineptitude of British military leaders, and during the early months of 1918, he had fought the British War Office decision to absorb the Canadian Corps into the British Army. He had won. But young Canadian men were no longer eager to enlist. Huge numbers were resisting the draft: although 400,000 men would receive conscription notices during the war, the great majority would apply for an exemption. (In the end, the federal government would meet its target of roughly 100,000 men.)

The war permeated daily life. The *Calgary Daily Herald* ran front-page stories of provincial casualties: Private Harold Smith from Pincher Creek was killed in action, "the only child of Mr. and Mrs. James Smith, old pioneers of the district"; Private George Marcellus of Fishburn was gassed after thirty-one months in the trenches.[8] Calgary women raised $1,100 with a tag day, and bought chocolate, tobacco and cigarettes for soldiers overseas, including the members of the 7th Field Ambulance Corps.[9] Workers at Calgary's Maclin Motors said goodbye to their popular salesman Lester Hanrahan, who was leaving for training in the Royal Flying Corps.[10]

Borden's recollections of that springtime in 1918 would perhaps be more revealing than he intended. "I am impressed by the tremendous variety and volume of important questions that continually surged upon me," he would write. "If I had not been keyed up to the utmost limit, it would have been impossible for me to fulfil the immense measure of responsibility which I was constrained to undertake."[11]

The prime minister was surely grateful for the chance to rise above the domestic sphere when he was summoned abroad for the high-level talks of the Imperial War Conference and the Imperial War Cabinet. On May 24, 1918, he left for New York, en route to London. The Dominion leaders were looking ahead: with the American entry into the war in 1917, the Allies would likely triumph—although the fighting would in fact remain fierce for months. Borden wanted to discuss the demobilization of the veterans, the reconstruction of Canada's wartime economy and the prospect of British emigration. He brought along three cabinet ministers:

Interior Minister Arthur Meighen, Immigration Minister James Calder and Privy Council president Newton Rowell.

But Borden had also invited three surprising guests who would play vital roles in postwar immigration and settlement. As he boarded the *Melita* in New York, the members of the Gang of Three were alongside him—the only provincial premiers in Borden's delegation. They were also glad of a change of scene from the bleakness of everyday affairs on the home front. They did not realize that Borden was co-opting them.

There was Manitoba's Tobias Crawford Norris, who had replaced scandal-plagued Sir Rodmond Roblin in 1915 with an administration that took pride in its integrity. The former auctioneer and livery stable owner had resolved that the reforming ideals of the Social Gospel would prevail at home as Canadians fought abroad for a better world. He "placed a high priority on reducing poverty, ignorance, and other social and economic problems."[12]

But Norris now governed a province rife with the discord that would eventually destroy his political career. In 1916, he had cancelled the Laurier–Greenway agreement that had provided for bilingual schools because Anglo-Canadian Manitobans had objected to instruction in languages such as German during wartime. But that decision had also sideswiped French-speaking Manitobans, including the Métis of Louis Riel, who now feared for their heritage. Worse, because his voters were now annoyed by many of Borden's wartime measures—including the end of exemptions from conscription for farm hands—Norris was criticized for his support of the Union government, and his abandonment of federal Liberal leader Sir Wilfrid Laurier. Manitobans were "resentful of [the] sacrifices they were making, suspicious of others who did not seem to be pulling their weight, and angry at governments of all kinds."[13]

The second Prairie delegate aboard the *Melita* was Alberta premier Charles Stewart, who had replaced Arthur Sifton in October 1917 when Sifton joined the Union government. Stewart epitomized the Old West of the homesteaders. In 1905, when he was in his mid-thirties, the devout Anglican had moved his wife and children from an Ontario farm to a hardscrabble Western plot that he kept afloat by working as a bricklayer and stonemason. By 1909, when he entered the Legislature, he was already a comfortable businessman with a dealership in farm implements and a rural real estate firm.

Stewart could never heal the lingering rifts in his caucus, which was now divided between those who supported Laurier and those who favoured the Union cause. His voters were restless. In the 1917 provincial election, labour delegates had combined with farmers to form the agrarian Non-Partisan League, which had run candidates against former premier Sifton's Liberals. The United Farmers of Alberta (UFA), which had not yet transformed itself into a political party, was lobbying fiercely on behalf of its members. Stewart was "an Alberta farmer, a decent family man who believed that government had an important role to play and that good people had a responsibility to be involved."[14] But he was also a political moderate enmeshed in a hyper-partisan environment that would eventually sideswipe him. He, too, needed a break.

The final *Melita* passenger in the Gang of Three was Saskatchewan premier William M. Martin, who had left his seat in Parliament in 1916 to replace the ailing Walter Scott. The Ontario-born lawyer had absorbed his Presbyterian father's ideals of "porridge, progress and predestination."[15] In the run-up to the 1917 provincial election, he had vehemently espoused provincial rights, tariff reductions and the need to acquire resource control. But wartime strife also plagued his province. In 1917 the Mennonites and Doukhobors had fiercely objected to his insistence on compulsory school attendance. Immigrant groups were now embroiled in a bitter debate with the anglophone majority over proposals for English-only school instruction, which Martin would proceed with in late December. Although the premier had shrewdly aligned his government with agricultural interests, farmers still complained about "wartime manpower and requisitioning policies," which had deprived them of workers and supplies. Farmers were also vexed about Ottawa's wartime "grain handling and marketing policies."[16] Martin was self-confident, and more sophisticated than his two Western colleagues—but he was also worn out from the incessant complaints.

The prime minister never publicly explained why he invited those three Liberal premiers to mingle in Britain's beleaguered but still glamorous society. He did owe them—along with their fellow Liberals James Calder and Newton Rowell—for their strong support for his Union government. The war had also enhanced the stature of the Western provinces and their premiers within Confederation because of their Herculean contributions to the war effort. But Borden and Immigration Minister Calder certainly had an ulterior motive. Calder had fought hard for resource control as deputy premier of Saskatchewan before the war—but he now needed the co-operation of the three premiers in his postwar schemes to settle veterans, former British servicemen and other British immigrants on Prairie lands. He also needed them to use their legal powers to expropriate land that private owners, including speculators, had left uncultivated. Such action would free up lands for the newcomers. Calder's appeals would work: he and Soldiers' Civil Re-establishment Minister Sir James Lougheed would secure approval for their settlement plans at the November 1918 Conference.

In the meantime, Borden reaped Western goodwill. The three premiers had relished the very hush-hush way that the prime minister had extended his invitation. In late April, Borden had telegraphed each of them, specifying the code that they should use in their communications with him. The Admiralty required "utmost secrecy." Information about the departure date and the port were "as important as the actual protective measures" that Ottawa would take.[17] The premiers had gleefully replied in code, although Martin's officials had incorrectly deciphered the date he was to appear in Ottawa. The error was corrected. As Manitoba premier Norris had telegraphed Borden, "Secret miscons-true definer reference apology unsparing," or, as federal officials had translated, "Secret message date received and understood."[18] It had been like one of those popular tales in British boys' magazines.

The danger from German submarines was still very real. The *Melita* travelled in a convoy of thirteen vessels, all packed with American troops, including the 2,500 soldiers who were aboard Borden's ship. The prime minister promptly introduced himself to the American officers. He also organized almost daily meetings with his own delegation to discuss the upcoming talks. Only Meighen was missing in action: the interior minister was so seasick that he vowed to live permanently in England and never again venture on the high seas. The ship's doctor prescribed cayenne pepper tucked inside a raisin, which Borden viewed as "an infallible remedy."[19] He did not report Meighen's verdict.

The captain briefed Borden on the fight against German submarines: on his last voyage, U.S. navy officers had deployed depth charges, and they had probably polished off two underwater threats. Borden's convoy practised target shooting. As the ships drew closer to Britain, they zigzagged at sixty-degree angles to avoid interception. When they were in sight of Scotland and Ireland, four British destroyers joined the flotilla; on the final day, two dirigible balloons floated overhead until the *Melita* docked in Liverpool on the evening of June 7. The three Western premiers had not had much opportunity to travel, so Britain was a rare treat. They also saw Borden's self-assurance and diplomacy when they reached London on the afternoon of June 8, 1918, and the prime minister was swept into a press conference and official meetings. The Gang of Three had been evacuated to another wartime front.

The travellers had left behind an economy that was operating close to capacity. Although factories had vastly expanded, they would still be "stretched to their limits before the Armistice."[20] Labour shortages were now common, and inflation was soaring. Many Canadians were worried about the socialist threat. Societal tensions were escalating now that key farm and industrial workers were no longer exempt from conscription. The remaining workers, including many who had been born in Central or Eastern Europe, were getting real bargaining power, and they resented the high prices of everyday goods. Trade unions were expanding, and strikes proliferated.

Through most of 1918, Ottawa was inundated with calls for intervention in those strikes. The situation was particularly difficult in the West because of the animosity between the largely foreign-born workers and their Anglo-Canadian bosses. Since Ottawa controlled the West's lands and resources, it was responsible for overseeing labour disruptions at those mine sites. The situation could get tense—even though Ottawa had effectively nationalized most Western interior coal mines in July 1917 when it had placed "wages, prices, and supplies" under the dual authority of the Coal Operations director for that union district and the Dominion Fuel Controller.[21] During the late winter of 1918, workers at the Moodie Mine in Drumheller, Alberta, had demanded the right to join the United Mine Workers, and unionized workers at a nearby mine had marched in their support. At the request of the Moodie Mine's nervous owners, Ottawa dispatched the Royal North West Mounted Police to keep the peace. Ironically, the Mounties toted along a machine gun.[22]

Those years between 1917 and 1925 would be critical in labour history. "Never before had workers posed such a broadly based and potent challenge" to the Canadian class system.[23] In wartime, Anglo-Canadians saw trouble everywhere. An anonymous informant warned Ottawa that the Moodie mine site was a powder keg, where management and labour despised each other. Fully 607 of the 1,463 miners in the Drumheller camp were "aliens enemies [sic]," and only 264 men were "really of British descent."[24]

Borden eventually put those Moodie workers under the authority of Coal Operations Director W.H. Armstrong, who supervised wages, prices and supplies for the region. Under federal supervision, the war eventually trumped most squabbles—although tensions continued to simmer. During a meeting with federal officials in Ottawa in late April, Western coal-mine operators and union officials agreed to put aside their differences, and increase their output. Western farmers had already agreed to boost their grain production by twenty per cent, and they were "confidently counting on a bumper crop."[25] Miners and farmers would soon be competing for scarce space in eastbound freight cars.

Ottawa's economic interventions accumulated throughout 1918. The federal government was adopting "government regulation, government control, and, in a vital sector of the economy, a healthy dose of government ownership."[26]

———

Borden, far from these troubles, revelled in his life abroad. His British and Dominion colleagues provided distinguished, if occasionally pompous, company. When the prime minister told the British Parliament about his anti-loitering law, a member of a prominent ducal family whispered a question to Interior Minister Meighen, who was alongside him in the galleries: "Am I to understand from your Prime Minister that there are no gentlemen of leisure in Canada?"[27] When Meighen answered in the affirmative, the dignitary "threw himself back in his seat with an expression which indicated both astonishment and disdain."[28]

But Borden also preferred to discuss the larger issues of life or death—rather than domestic tariffs. The Dominion leaders in the Imperial War Cabinet were waging war "in no less than ten theatres, thousands of miles apart and scattered over three continents."[29] In that company, Borden came alive. A mere week after his arrival, he was sending lengthy cables to Ottawa, denouncing the disorganization, lack of foresight and sheer incompetence of the British Army's leadership. He was collegial: before he addressed the war cabinet, he consulted with the three cabinet colleagues who were with him in London. He cared about the soldiers, and he showed it. He attended memorial services, inspected the troops, and visited Canadian soldiers in British convalescent hospitals. In late June, he journeyed to France with British prime minister David Lloyd George, and experienced the reality of a nighttime air raid. "Incessant firing amid searchlights for 30 minutes," he would write. "Two terrific explosions from bombs that seemed to be bursting very close by."[30]

His schedule was dizzying yet productive. During the Imperial War Conference meetings in July, he tackled the health and welfare of the frontline troops, meat supplies, mineral resources, sea traffic, ocean freight and the difficulty of bringing hundreds of thousands of battle-hardened soldiers home to civilian life. At Immigration Minister Calder's urging, Borden had put postwar immigration on the agenda. The Dominions would eventually create a central authority to oversee emigration from Great Britain, and they would extend credit to British veterans who wanted to farm abroad.[31]

The news from the Western Front was heartening. In late July, the French Army, with invaluable American assistance, rolled back the Germans from Soissons. By early August, with the taking of the Marne salient, any remaining threat to Paris vanished. The Germans were in full retreat. Borden visited the British fleet in Edinburgh, and inspected the newest model of submarine. On August 8, the Canadians launched a successful attack, flanked by Australians and French, and, a day later, those forces had seized fourteen thousand German prisoners and huge stocks of armaments. Borden attended another gathering of the Imperial War Cabinet, and paid more visits to army hospitals and camps. On August 15, he found time for a "flying visit" to souvenir shops.[32] Two days later, he settled into his comfortable quarters onboard the *Mauretania*, en route back to New York.

It had been a hectic trip, but, in his diary and subsequent memoirs, Borden sounded invigorated. Meighen and Calder had returned home in mid-July; Rowell had left in late July. Meanwhile, another minister in Borden's cabinet—Marine Minister Charles Ballantyne—had joined Overseas Military Forces Minister Sir Albert Edward Kemp in Borden's inner circle. Borden mentioned all of them in his memoirs.

But he did *not* mention the three Western premiers. They stayed in the same hotel as Borden; they received invitations to Parliamentary luncheons. They had admission tickets to the British House of Commons and the House of Lords. Like Borden, they tramped through Canadian camps and hospitals, the Ministry of Munitions, armament factories and shipyards. They toured the Grand Fleet and aviation facilities. They even went to France.[33] But they appear nowhere in his pages. In the context of the intriguing world of foreign affairs, they simply did not figure in his consciousness.

The three Western premiers had used their time in Europe to get the measure of the challenges ahead, including their role in the postwar settlement of Canadian veterans, British soldiers and other British immigrants. They knew what Borden expected from them, even if he did not pay much heed to what they expected from him. The war was drawing to a close, and leaders were scrambling to manage the transition to peace.

Borden returned to Ottawa on Saturday, August 24, and went directly into a cabinet meeting. In the next few weeks, he dealt with a strike at west coast steamship companies, Canada's collaboration with the United States on war production, the treatment of draft dodgers and the removal of fish tariffs. He was doing his duty, and he frowned on those who had lost sight of the bigger picture. In a revealing luncheon speech at the Canadian National Exhibition in Toronto on Labour Day, he called for workplace harmony. "In all the annals of history there never was a war like this," he told his audience. "Subtract from the national effort what you will by controversy, by division, by discord; *by so much have you weakened the national purpose and the national endeavour.*"[34]

Borden was linking labour peace with national wartime success. But those words also hinted at how he viewed the Western premiers and their recurring talk of resource control. The Prairie premiers never caught the significance of what Borden did *not* say during wartime. His memoirs would offer only an enigmatic reference to their cause, which cropped up during a discussion of land settlement with the Manitoba-based Meighen that summer. "He thinks we are already too much in [the] provincial arena and should withdraw rather than advance."[35] The transfer of resource control was low on Borden's list of priorities.

Behind the scenes, the omens were dire for the Western premiers and for British Columbia premier Oliver, who had taken up his dying predecessor's campaign for the restoration of the railway lands. Throughout the late summer, after his return to Canada, Immigration Minister Calder had worked with a special cabinet committee to hammer out detailed homesteading proposals. In September members of that committee agreed that Ottawa should start negotiations with the provinces. The governments should co-operate with each other: Ottawa would find the settlers and the provinces would settle them on the land. Both governments would extend credit to the newcomers, and Ottawa would hold a conference to decide how to divide those costs. But, crucially, those federal ministers could *not* agree on "the advisability of transfer-ring to the four western provinces such of their natural resources, includ-ing farm lands, as are at present owned by the Dominion."[36]

Calder inserted that stunning admission into a memorandum to Borden. After decades of negotiations, with an imminent victory that would owe so much to Western resources, the cabinet *still* could not agree to transfer resource control. Many ministers preferred to retain jurisdiction over the lands so that Ottawa could distribute them, likely because federal control was simply convenient. Meighen was likely a rare dissident, as Borden's cryptic note indicated: at the very least, the interior minister did not want to intrude further into Western affairs. But, in future years, he would never concede enough to get a deal.

The Prairie premiers did not know about this cabinet deadlock. In Britain, the three had resolved to secure control before Ottawa finalized its settlement schemes. At home, they took action. On September 27, 1918, Alberta premier Stewart telegraphed a reminder to Borden: when Stewart had left London, he had understood that the prime minister "would call the Western Premiers together at an early date to discuss natural resources."[37] Although Borden had not yet convened a meeting, Calder was already distributing copies of his land settlement scheme to provincial capitals.[38] Stewart called for "as early a date [for talks] as would be convenient for you."[39] The West still wanted to deal only with Ottawa.

Typically, Borden had more pressing international concerns. In mid-October, he met with British and American officials in Washington and New York to discuss the peace talks and war bond sales. He even found time for lunch with that ruthless financier Henry Clay Frick at his New York City home. Borden was agog at the "wonderful collection of paint-ings and bronzes which have cost Frick 20 millions and which, with his house (wonderfully spacious and beautiful) he is leaving to New York City."[40] The prime minister was so much happier when he was away from home. Back in Ottawa, he listed his obligatory chores, including prepara-tions for the upcoming federal–provincial conference, which he relegated to a phrase in his memoirs.

As a dutiful host, Borden officially invited the premiers to the

November 1918 Conference with a telegram on October 26. The gathering would consider "the problem of soldiers' settlement, the general problem of land settlement and the request of the Prairie Provinces for the transfer to them of their natural resources."[41] His invitation added a line that would create endless trouble, especially with British Columbia: "Other subjects for discussion may be proposed."[42]

The problem with land settlement was already very real. On October 28, the Great War Veterans' Association of Canada warned that the available Crown land was "insufficient to meet, and unsuitable for the estimated requirements" of the returning soldiers.[43] The association's acting secretary-treasurer, Robert Stewart, who brashly publicized his letter before it reached Borden's office, demanded the expropriation of idle agricultural lands, or for the assurance that those lands—"if not brought under cultivation within a definite period (say ten years)"—would revert to the Crown.[44] The issues of resource control and land for veterans were now dangerously entangled: the West wanted immediate control while the veterans needed land immediately.

Perhaps predictably, the Prince Edward Island premier pounced as soon as he slit open his invitation. Aubin-Edmond Arsenault wanted more information on the Western provinces' case for resource control. Borden forwarded this request, knowing full well that the spoilers were gathering. On November 1, Saskatchewan premier Martin replied that it would take two to three weeks to copy the Prairie provinces' extensive summary of how Ottawa had used their land. That summary would be "of very little interest": it was so detailed and its conclusions were so brief that it would be "useless to furnish them with the report."[45] Not a good answer. The Western premiers were still struggling to keep the other provinces out of the loop: they were unwilling to provide full lists of what had been lost, and what might be gained. But the other premiers were not about to endorse a blank cheque.

Borden was not co-operating with the Westerners. On November 2, he distributed copies of that 1913 proposal from the original Gang of Three: Saskatchewan's Walter Scott, Manitoba's Rodmond Roblin and

Alberta's Arthur Sifton, who was now conveniently neutralized as a minister in Borden's government—at least in public. This was the convoluted letter that asked for resource control along with the continuation of federal subsidies in lieu of those resources. It had raised hackles in other provincial capitals for five years. Unaware of Martin's brusque dismissal of Arsenault's request, Borden added that, if the Western premiers had more documentation, they would "endeavour to supply copies to the other Premiers before the Conference meets."[46] The Gang of Three, however, would go into the conference with only that one-page proposal. Western governments had endorsed it, and the current premiers saw no need to expand upon it.

All nine premiers accepted Borden's October 26 invitation. However, the prime minister's mind was already, as so often, elsewhere: on October 28, British prime minister David Lloyd George had asked Borden to return to England. Borden surely saw this as a reprieve: his Liberal and Conservative cabinet colleagues agreed to continue their Union government so that the prime minister would face no immediate political crisis. The cabinet also decided that Trade Minister Sir George Foster and Customs Minister Arthur Sifton would accompany Borden to Britain; Justice Minister Charles Doherty would follow at a later date.[47] Borden would soon be free to talk about peace abroad.

First, he had to tidy up loose ends at home. Canadians were anxious. Would there be jobs for the returning soldiers? How would Canada cope with its growing labour militancy? Would prices ever stop rising? Borden's cabinet colleagues were worried about the optics of his absence. His last trip to Europe had provoked "very grave and widespread dissatisfaction" with his government.[48] Voters might not look kindly on another prolonged absence while they endured ongoing labour disruptions, food and fuel shortages, and the lingering rancour of conscription. Worse, as Privy Council president Newton Rowell predicted, there was "likely to be considerable unrest" as munitions factories closed, and men were thrown out of work.[49]

Rowell, who vividly recalled his summer trip to Europe, concluded that the problem during Borden's last trip had been lack of publicity. The prime minister should embed a journalist in his delegation, and take the writer into his confidence. The Canadian Press wire service could circulate his reports across the nation. Rowell proposed John Dafoe, the editor of the *Manitoba Free Press* and a long-time Liberal who had staunchly supported the Union government. Borden agreed, and Dafoe consented. In today's world, Rowell would be hailed as a public relations genius.

The prime minister also had to reassure uneasy Maritimers that Ottawa would not give away the shop to the West at the November 1918 Conference. On the surface, it might seem odd that Borden would appoint a highly partisan former Liberal MP from New Brunswick as Conference chairman, but there was sly federal method behind that selection. Businessman Frank Carvell had joined Borden's Union government in 1917 as public works minister. Politicians played hardball in New Brunswick, and Borden had dropped a prominent Tory minister to find room for Carvell at his cabinet table. Now the high-profile politician worked alongside men that he had once pilloried; they simmered with mutual dislike. But Carvell also represented a strong voice for Maritime interests. Accordingly, as the prime minister packed for Europe, he wrote a letter of subtle reassurance to his prickly minister, who was fuming about the possible transfer of Western resources.

The confidential note was succinct. Borden passed along titillating war news to his public works minister, who was more interested in paving New Brunswick roads. There were "apparently well-founded" rumours that U.S. president Woodrow Wilson would attend the Peace Conference, and Borden approved.[50] Revolutionary ideas were now popular in Germany, so Allied armies might have "to maintain law and order by an army of occupation" until the establishment of a stable government.[51]

But it was Borden's observations on the upcoming Conference that mollified Carvell. The prime minister agreed that cabinet should consider the conditions for any transfer "with the least possible delay."[52] Accordingly, "the committee"—which was almost certainly Calder's special committee—should meet "in the immediate future."[53] (That committee had already failed to reach a consensus.) Borden clamped another condition on any transfer: the Prairie provinces would have to co-operate "in providing suitable and adequate arrangements" for veterans who wanted to settle on the land.[54]

Then Borden added two bland sentences, which conveyed his tacit understanding that the Western premiers' campaign was doomed. "It [the natural resources question] is of course bound up with the co-relative question of compensation to the Maritime Provinces," he wrote. "There will probably be an all round demand by the Provinces for the increase of subsidy which we cannot possibly entertain under present conditions."[55]

What did this mean? What was Borden really saying? Carvell was pushing for more cabinet meetings on the issue before the conference so that federal ministers could air their grievances. Borden encouraged his New Brunswick attack dog. As well, after his trip to Europe with the Prairie premiers, Borden knew that they would not amend their demands, and that they would reject any federal conditions on the transfer. He knew that the Maritime premiers wanted a special subsidy increase because their population growth was stagnant. He certainly recalled Ontario premier William Hearst's declaration that if Western demands were met, Ottawa would have to raise *everyone's* general subsidy.

The prime minister would never support the transfer of resource control when the other premiers vehemently opposed it. The Western premiers would still go along with Ottawa's scheme to distribute their lands to veterans and immigrants. Chairman Carvell could relax: the Western premiers would *not* reap huge benefits while the Maritime premiers went home empty-handed. Ottawa could not afford to pay for peace. Once again, those old jealousies—that blight of the federation since the heyday of Sir John A. Macdonald—could work like a smokescreen when Ottawa did *not* want to do something.

Borden was free to depart. On November 8, with his internal chores settled, the prime minister left Ottawa for New York. Two days later, at midnight, he and his large party, including Trade Minister Foster and Customs Minister Sifton, boarded the *Mauretania* en route to England. Special Press Representative Dafoe was among them, with the "duty of ensuring accurate and instructive reports" to the Canadian Press. (Dafoe would return home at the beginning of March 1919.) Decades later, Borden would assert that Dafoe had "admirably discharged" his duty, an assessment that would have pleased his champion, public relations whiz Rowell.[56] His reports would be a propaganda coup. Now no one, not even the Western premiers, could doubt that the prime minister was a central player abroad.

The *Mauretania* was at sea when Borden learned from the ship's porter that Germany had signed an armistice, and hostilities had ceased at 11 a.m. on Monday, November 11. It was a bittersweet triumph. More than 600,000 Canadian men and women had enlisted; nearly 60,000 had died; another 141,000 were wounded, and that figure did not include those who had suffered mental breakdowns in the trenches. That day, in his diary, the prime minister was unusually thoughtful. "The world has drifted far from its old anchorage and no man can with certainty prophesy what the outcome will be," he wrote. "I have said that another such war would destroy our civilization. It is a grave question whether this war may not have destroyed much that we regard as necessarily incident thereto."[57] He would not live to see his prophecy almost fulfilled in the Second World War.

There would be a startling omission in Borden's memoirs, however, and it underlined his domestic isolation. Spanish influenza had first appeared in Canada in September 1918 when returning veterans stepped ashore in Quebec City. The airborne virus brought aches and chills, and lethal pneumonia. Many businesses lost their remaining workers, and families lost their breadwinners. Orphans wandered the streets or huddled alone in their homes. City governments eliminated routine services, and provinces enacted quarantines—insisting that their residents wear masks. The impact would be almost as terrible as the war: by the spring of 1919, fifty thousand Canadians, who were vulnerable after the deprivations of war, would be dead. That was almost equal to the number of Canadian soldiers who had died during the war itself.

Borden scarcely noticed. On October 19, 1918, the day after his delightful lunch with Frick, he had arrived back in Ottawa, and jotted an observation in his diary: "Influenza epidemic very widespread and many fatal cases. Cars everywhere on streets bearing red cross emblem. Mayor said to have done excellent work."[58] Provincial health ministers could handle the outbreak. As the *Mauretania* plowed through the waves, the prime minister did not know that the Western Provinces had asked Ottawa to postpone Thanksgiving services in the churches because the afflicted might infect the healthy.[59] Once again, Borden had left problems at home unacknowledged.

Official Ottawa was ready for the November 1918 Conference. Behind the scenes, in the November gloom, bureaucrats had assembled reports on soldier settlement and agricultural training, and on the West's natural resources. Interior Ministry officials had already dismissed the West's demands. The Western premiers may have been unwilling to spell out Ottawa's use of their resources for the Rest of Canada, but federal bureaucrats filled that gap with a thirty-nine-page compendium, along with meticulous charts.

Each report was a condescending display of administrative competence. There were tallies of the acreage that the Dominion Lands Branch had distributed since 1870, along with the amount of money that Ottawa had pocketed from those sales. Officials noted that they had *already* reserved "all vacant and available parcels of Dominion Lands within fifteen miles of a railway" for returning veterans.[60] Because so many soldiers wanted to farm, however, those officials urged Ottawa to give money to the provinces so that they could purchase or expropriate land "now in private hands and uncultivated."[61] (Ottawa was also reclaiming land from failing homesteaders.)

There were also lists of the alienated railway lands in British Columbia, including those set aside for homesteads, town sites, irrigation and mining. There were huge lists of Prairie school lands, attached to an older memo from the head of the school lands branch, Frank Checkley. "It would be difficult to administer them more economically than at present," Checkley had sniffed on February 25, 1916, shortly before Borden had rejected the Gang of Three's wartime request for school-lands control.[62]

Ottawa's holdings were vast. Before 1907 it had sold coal-mining plots; now the Mining Lands and Yukon Branch leased the rights to those plots in return for royalties. Ottawa had interests in petroleum and natural gas, quartz mining and dredging. Officials produced an apparently scrupulous accounting of forestry operations, irrigation administration, timber and grazing lands, and hay production. A six-page memorandum on waterpower even warned against the Western premiers' grasping ways: they might hike the prices for hydropower even though "cheap power is one of this country's greatest assets in the post-bellum industrial rivalry of nations for world trade."[63]

Another memo pulled the classic trick of invoking a menace and then supplying the preferred solution. Suppose Ottawa did not quickly settle the returning soldiers and British veterans on Western lands: they might "crowd into our cities and create conditions of unemployment that will be most difficult to cope with."[64] The implication was fascinating: without Ottawa's efficient administration of Western lands and resources, there could be riots in the streets. Given such rhetoric, the Westerners had no hope of cutting a deal with Ottawa, securing their resources, and distributing the lands to the soldiers themselves. They were deluding themselves.

As the premiers headed to Ottawa, Borden arrived safely across the Atlantic. The *Mauretania* reached the mouth of the Mersey River, twelve miles from Liverpool, on the evening of Saturday, November 16. Officials from the Colonial Office, the Admiralty, and the Canadian High Commission clambered aboard from a motor launch to welcome him. At noon the next day, the delegation stepped ashore in Liverpool, and immediately boarded a special train to London where they were given "a very impressive reception."[65] Prime Minister Lloyd George had left his country home to greet Borden on the platform. A high-ranking member of the king's household welcomed the group on behalf of George V, and Borden delivered a dignified response. He inspected and congratulated the three Nova Scotia battalions in his guard of honour. "A great crowd . . . gave the Canadian Prime Minister a very warm welcome."[66]

Borden swung into action. He and Lloyd George went immediately to the War Cabinet Offices to discuss the upcoming peace talks. Then, along with Trade Minister Foster and Customs Minister Sifton, he huddled until midnight with the first Canadian commander of Canada's overseas forces, Sir Arthur Currie, and two other officers. A day later, on the morning of Monday, November 18, he met with South African Jan Smuts, who had helped to establish Britain's Royal Air Force. In the afternoon, he discussed General Currie's report with his delegation before he and Foster hurried off to a meeting of the Imperial War Cabinet.

By Tuesday, November 19, Borden was in the Royal Gallery in the Houses of Parliament to hear the king's address, and "everyone agreed that a cheer at the end . . . would have made an excellent finale."[67] That evening, after a British Privy Council meeting, he had an audience with the king, who explained that he had delayed his speech until Borden had reached Britain. The prime minister preened. The two men agreed that German Kaiser Wilhelm II should not be put on trial, but left in his "present condition of contempt and humiliation."[68] Borden did note that the king was exhausted from "the great nervous strain" of war.[69] He, however, was in his element, at the centre of Great Events, far away from Ottawa, where the first ministers were gathering in the temporary Senate Chamber in the Victoria Memorial Museum.

Few Canadians were paying any attention to their premiers' conclave. Wartime news dominated Canadian headlines that day. As *The Globe* trumpeted: "Canada Raises $676,000,000 On Victory Bonds"; "Ships To Surrender At Sea: German Navy Ends Career"; "President Wilson To Attend Peace Conference In Europe."[70] Tardy news of Borden's arrival in London was relegated to a snippet on page sixteen.

The premiers had brought along their stale grievances for a fresh hearing. British Columbia still wanted additional subsidies for its special circumstances and the return of its railway lands. Maritimers wanted to revive their flagging influence and secure special subsidies. Quebec and Ontario wanted higher general subsidies because their needs had escalated in wartime. Arrayed in a lonely clump, the three Prairie provinces clung to their 1913 demands. While Borden made peace elsewhere, his ministers had their marching orders for this postwar landscape.

After the first ministers had arranged their papers on the wooden conference table, the austere Toronto trust-company executive Finance Minister Sir Thomas White welcomed them. He then turned the session over to Chairman Carvell. Ominously, the first item on the agenda was a consideration "at length" of the Prairie provinces' terse demand for resource control.[71] Interior Minister Meighen immediately put the cat among the provincial pigeons with a brief summary of Ottawa's approach. The federal government would transfer the resources *if* certain conditions were met. Although Meighen did not elaborate on most of those conditions, he was almost certainly referring to Borden's long-time caveats: the three premiers must provide free lands for homesteading, maintain an open door for immigrants and lower their financial expectations.

The interior minister did add one crucial condition that tipped the balance against the Westerners. As Manitoba premier Norris would reveal more than two years later—when he was *still* angry about the conference—Meighen said that Ottawa would continue the West's subsidies in lieu of resources, "provided that [this] was mutually satisfactory to the other Provinces."[72] Meighen added that Ottawa could not afford a general subsidy increase. This was simply mischievous. The interior minister knew full well that his caveat—the Westerners would get special treatment if the other provinces consented—would upset the other premiers, particularly the Maritime contingent.

The room erupted, as Meighen almost certainly intended. Decades later, Prince Edward Island premier Aubin-Edmond Arsenault would explain what happened when Meighen sat down. An unnamed delegate asked Carvell if it was Ottawa's decision to transfer the lands to the West. Carvell—that fierce advocate of Maritime rights—"simply answered, 'Yes.'"[73]

The federal trap snapped shut. Such largesse was all too much for veteran Quebec premier Lomer Gouin, who had held office since 1905 and who had somehow steered an even line between his support for the war and his opposition to conscription during those terrible anti-draft

protests in Quebec. Gouin was a Montreal corporate lawyer and businessman, taciturn and reserved, and he had largely stayed out of the
West's battles for resource control. Now he moved swiftly. To Arsenault's
amazement, Gouin flourished his own resolution: the other six provinces
would agree to transfer resource control, but the subsidies in lieu of
those resources should be withheld from the West, "and the sum total
of these subsidies [should] be divided among the other Provinces and that
no Province [should] receive less than $325,000 annually."[74] As Arsenault
would gleefully recount, the Western provinces strongly objected, "and
no progress could be made."[75]

The discussions became so intense that Carvell adjourned the proceedings. The nine premiers then held a "Special Conference of Representatives
of Provinces"—which agreed that the premiers would devise a resolution
"embodying the view of the Conference" on the West's request.[76] The premiers would submit this resolution to Ottawa at the next session. This
decision was catastrophic for the Prairie premiers. They had already lost
control of their agenda: they had wanted to discuss their demands with
Ottawa alone. Now, they were arguing with the Rest of Canada, away from
any federal cabinet ministers who might have toned down the squabbling.

As the Western premiers fretted, Borden focused on his London obligations. On Wednesday, November 20, he strolled into the Imperial War
Cabinet, chatting with Lloyd George about how absurd it was that the
two New Zealand leaders, who uneasily shared power, wanted to attend
the peace talks together. (In the end, they would both attend, but only
one would officially represent New Zealand.) At the cabinet table,
Borden called for a report on freedom of the seas and briefed his peers
on U.S. president Wilson's approach to the fate of the German colonies.
Wilson wanted an impartial assessment of the views of the governed as
well as those who sought to govern them. Borden suggested that the
Dominions defend the British Empire against Wilson's democratizing
urges and, more important, promote Britain's claims to Germany's lost
empire. The prime ministers then argued about the Kaiser's fate.

———

Across the Atlantic, the second day of the November 1918 Conference opened with no mention of that promised resolution on the West's demands. Instead, Immigration Minister Calder outlined the "best methods" for federal–provincial co-operation on land settlement—based upon Ottawa's Soldier Settlement Act of 1917.[77] Ottawa would lend money to the provinces to expropriate undeveloped land; new farmers could buy that land with down payments of just ten per cent; and soldiers would also be eligible for deferred-interest loans to purchase stock and equipment. Ottawa and each province should create a central organization to ensure that the veterans received agricultural training. The morning session adjourned after just seventy-five minutes. Ottawa's plans were rolling out, on time and on target.

Two hours and fifteen minutes later, after what was no doubt a tense luncheon, the West's problems worsened. First, representatives of six provinces—Nova Scotia, New Brunswick, Prince Edward Island, Quebec, Ontario and British Columbia—solemnly tabled their response to the West's requests. (Even British Columbia had abandoned its erstwhile Prairie allies.) Their two-paragraph statement was grim. The six provinces had heard Meighen's airy declaration that Ottawa was willing to transfer "the ungranted or waste lands and other natural resources" to the Western provinces "*under certain conditions and restrictions.*"[78] But Ottawa surely had no right to act unilaterally. The six provinces wanted to enter their position in the official record. If Ottawa shifted those lands and resources to the West, and "maintained in whole or in part" their subsidies, "a proportionate allowance" should be granted to *them*.[79] No matter what the Gang of Three proclaimed, the Rest of Canada had a financial stake in any transfer.

Opposition to the Westerners deepened. *Another* Gang of Three now materialized at the conference, and the Borden government was deeply indebted to its Maritime members. First in seniority was Nova Scotia premier George Murray, a progressive Liberal lawyer from Cape Breton. Murray was a canny powerhouse, who had been premier since 1896. His voters respected his "honesty, integrity, and modesty . . . his approach was open, his character unblemished, and his demeanour friendly."[80]

Unlike Borden, the premier *was* interested in ordinary people: he would stop ditch-diggers to ask about their work. In 1910, he had lost a leg above the knee because of a blood clot, and now walked with a cane and an artificial leg. Ironically, his government depended heavily on coal royalties for cash—revenue sources that his Western peers could not tap—but wartime labour unrest at those coal mines had weighed heavily on him. The premier had troubles at home: to make ends meet, he had introduced taxes on everything from personal incomes to public utilities. He had also deeply divided his party when he supported Borden's Union government despite his friendship with Sir Wilfrid Laurier.

Alongside Murray was New Brunswick premier Walter Foster, whose Liberals had held office for only eighteen months. Born in the small seaside town of St. Martins, Foster had become a bank clerk in Saint John at sixteen. A decade later, he had joined a prosperous dry goods firm, married the owner's daughter, and become managing director of the company. Although New Brunswick politics had been a cesspool during wartime, the boyish Foster had struggled to remain aloof: "He continued to be the mild-mannered 'boy Premier' with the wooden personality."[81] But Foster also had enemies: like Murray, he had divided his party and his caucus when he supported the Union government against Laurier.

The final member of this trio was the dignified Arsenault, who could trace his ancestry back to a native of Normandy who had arrived in Acadia in 1671. His prosperous family had moved to the Island in 1728 when it was still a French colony. Arsenault was born to succeed. He had studied the classics in New Brunswick, and then taken law in Charlottetown and London, where he had qualified as an English barrister. He had led his Conservative government since 1917, after his predecessor had become a judge. A fierce proponent of his province's interests, it was not surprising that he had hounded Borden to replace the gauge of his railroad tracks during wartime. The conflict had unsettled the Islanders. Throughout 1918, the premier had often consoled farmers whose sons were conscripted when they were needed at home. In Charlottetown, Government House and the local sanitarium were now military convalescent homes. Like his fellow premiers, Arsenault was pinching pennies and he worked long hours. "It was no picnic," he would recall.[82] His patience was exhausted.

The West had powerful opponents. That two-paragraph statement from the Rest of Canada had concluded with a final shot: in addition to the demand for a general subsidy increase, the six governments had reserved the right to make special claims. Now the Maritime provinces asserted their right to have their special claims *"adjusted at the same time"* as Ottawa dealt with the West.[83] As Borden had assured Carvell, the Maritimers would link their cause with the West's demands. This new Gang of Three's pithy statement put that linkage on the record. They were seething. Decades after the November 1918 Conference, PEI premier Arsenault would still be enraged at the Western premiers' effrontery in demanding their resources *and* the continuation of their subsidies. "No arguments made any impression on them," he would mutter, "and a report to that effect was made to the main conference and that ended it."[84]

The Western premiers were equally furious, outflanked once again in a battle that their predecessors had waged for decades. All three were tall men. Alberta premier Stewart towered over his contemporaries, as did the silver-tongued Manitoba premier Norris. Saskatchewan premier Martin was almost as imposing. The trio formed a striking contingent, certain that their wartime contributions far outweighed those of the other six provinces. But their cause was lost, and they knew it. Martin simply replied that the Westerners were sticking with their 1913 position. For the moment, at least, there would be no official reply. The issue was no closer to resolution.

The first ministers returned gratefully to land settlement. Interior Minister Meighen now outlined his scheme for soldiers: veterans would need a down payment of only ten per cent, and Ottawa would extend development loans of up to $2,500. The premiers then pored over a letter from the National Council of Women, which called for a federal health department and a child welfare bureau. (Ottawa would set up a health department in 1919, after the ravages of Spanish Influenza.) New Brunswick's public works minister asked his federal counterpart Carvell to talk about federal aid for highway construction. The meeting adjourned at 5:45 p.m., as dark clouds loomed over the waiting limousines.

Remarkably, on the third day, on Thursday, November 21, the rifts among the provincial factions deepened. At 10:45 a.m., British Columbia premier Oliver handed an icy memorandum to Carvell that denounced Ottawa. "Honest John," a sturdy self-made man and the son of English farm labourers and miners, had no patience for Ottawa's obstructionist ways. A farmer from the Fraser Valley, he seemed a "rustic sage, a proponent of simpler values in an increasingly complex age."[85] But he was also a stubborn politician who could give as good as he got. At the November 1918 Conference, furious at the ill treatment of his province, he was an angry sage. His fierce little memorandum added spicy details to the tale of the first two days of Conference proceedings that the official record had ignored.

Oliver first pointed out that Borden had promised to let the participants put other topics on the agenda. Accordingly, when the first ministers had discussed the Prairie premiers' claims, he had asked that they consider his claim for the return of the unused railway lands. Those federally controlled lands dotted the strip that ran twenty miles on either side of the Canadian Pacific Railway's line—roughly 9.6 million acres—along with a block of more than 3.2 million acres in the Peace River area. After all, his claims were "of an even stronger character" than the West's.[86]

Interior Minister Meighen had rejected his request, because it "was not specifically upon the list of subjects prepared by the Dominion for consideration by the Conference."[87] Oliver had fought back. Eventually, Meighen's position "was modified," and the premiers had agreed to consider his claims during their Special Conference on Tuesday afternoon.[88] Oliver was *already* angry.

Away from the federal contingent, Oliver declared, his fellow premiers had seemed open-minded. His claim "was discussed as fully and as freely" as those of the three Prairie provinces. [89] He had been optimistic. "As far as we could judge from the tenor of the discussion . . . the claim of British Columbia was favourably considered," he related. "Certainly so far as we are aware, no objections were made to the validity."[90] He had detected only one problem: the nine premiers could not decide if Ottawa should increase Maritime subsidies if it granted the Prairie provinces' wishes.

Oliver had shrugged. If Prairie claims created an inequality, it would be Ottawa's job to fix it. So what? *His* claims were valid, and Ottawa should address them when it tackled those of the Prairie provinces. But, he related, when the first ministers had assembled on the morning of the second day, Interior Minister Meighen had *still* insisted that he could not deal with British Columbia's claim because it was not on the agenda. "We were very much discouraged," Oliver related.[91] If Meighen's attitude persisted, "we shall certainly feel that we have been very unfairly treated."[92]

Oliver's letter, which Carvell read to the first ministers, sparked more uproar. Meighen "took prompt exception" to this direct attack with a masterfully convoluted retaliation.[93] "What I said was that a subject not upon the Agenda of the Conference should not become a matter for decision by Conference Resolution," he harrumphed.[94] That is, the premiers alone could not put Oliver's request on the agenda. Instead, any addition required "the consent of all, *inclusive of the Federal Government* who had called the Conference."[95] Although the minutes did not elaborate, Meighen thwarted Oliver's demand by withholding Ottawa's consent. Long after he had calmed down—and it took a very long time—the British Columbia premier would confide that Meighen was "a clever fellow all right, but he and I don't pull."[96] British Columbia's hopes were as dead as those of the Prairie premiers.

For the remainder of the third day, the first ministers mulled demobilization, industrial and social reconstruction, the opening of employment exchanges, the federal government's estimated $100 million shortfall, and programs to help the returning soldiers put their lives back together.[97] In Borden's absence, four ministers—Finance Minister White, Labour Minister Gordon Robertson, Militia Minister Sydney Mewburn, and Soldiers' Civil Re-establishment Minister Lougheed—delivered the briefings.

The Westerners remained livid. On Friday, November 22, on the final day of talks, the morning session had been underway for only twenty-five minutes when the three Prairie premiers presented a letter to Carvell, who read it aloud. The premiers thanked Ottawa for its consideration of their request. They admitted the obvious: their fellow premiers could not agree on their demands. But they would never allow this impasse to happen again. Despite their formal prose, their anger at their fellow first ministers' meddling was incandescent.

They would not go along with the other premiers' insistence on linking their issues with Prairie claims. That would "virtually establish an admission on our part that the other provinces have a right to share" in the West's resources. They also rejected assertions that the continuation of their subsidies in lieu of resources would be unfair to the Rest of Canada. It would become their rallying cry. "Not one of these provinces has had its natural resources used as ours have been by the Dominion for the general benefit of Canada," their statement declared, "and we are of [the] opinion that the use of our resources in such a manner is in no way compensated for by the subsidies we are receiving on that account."[98]

There was more. The three premiers had heard the other provinces ask for higher subsidies, and they had listened to Ottawa's bleak financial report. Therefore, they were "disposed at present to postpone, until a more favourable condition arises," any discussion of a general subsidy increase.[99] If Ottawa did look favourably on a special claim from another province, they would not object. But if the federal government did increase its general subsidies, "we would thereby have a claim to be dealt with in exactly the same manner."[100] This was a tortuous, but subtle, warning. Suppose Ottawa transferred control of their resources to them, and maintained their subsidies: if, at the same time, Ottawa increased the general subsidy to appease the other six provinces, the Western provinces would also be eligible for a share of that money.

Finally, they turned to Meighen. The interior minister had not spelled out Ottawa's reservations. In the absence of explicit conditions, they would stick with their demands. They should be placed "entirely upon the same footing as all the other provinces."[101] It was a plea for constitutional equality. They added the pious wish that Ottawa would carry this request "into practical effect at an early date."[102]

With that five-paragraph statement, the Westerners did significant collateral damage: they told the other six premiers to get lost. Westerners owned their resources, and the other provinces had no "beneficial interest" in them.[103] Ottawa's plan had worked: the premiers had squabbled among themselves over resource control. Ottawa had escaped most of the blame. It was a set-up. Borden had tacitly admitted as much in his brief letter to Carvell, when he had linked Maritime and Western demands. That might have been a brilliant political ploy, but it would damage national unity.

A little more than two hours later, after talks on housing, technical education, and roads and highways, the November 1918 Conference reached its weary end. Nova Scotia premier Murray congratulated his fellow delegates on their achievements. Carvell thanked them for their acceptance of Ottawa's proposals on soldier-settlement and immigrant settlement. (Borden had known for months that the Westerners would co-operate with those plans.) Everyone accepted, however grudgingly, that Ottawa faced an estimated $100-million shortfall, so the federal income tax would likely continue, at least temporarily. There was no mention, however, of natural resources. Most premiers left Ottawa as soon as the proceedings adjourned at 1:15 p.m. on Friday, November 22.

The three Prairie premiers lingered in the capital, postponing their departure until the next day. They wanted to press their case "still further" with federal cabinet ministers.[104] But as always, they could only spin their wheels. An anonymous Westerner had an optimistic take on the stalemate: now, he told *The Globe*, "the Dominion Government knows exactly what we want and how we feel about it. They know what the eastern Premiers want as well."[105] But, of course, Ottawa had known what the West wanted long before the delegates had gathered under the museum's arched windows.

Borden remained in England, and no one could blame him for the failure of the negotiations with the West. That Friday, November 22, as the Conference adjourned, he happily wrote to White about the nation's prospects for a share of the postwar reconstruction contracts and when Canadian troops might come home. He even updated White on the Siberian Expedition, in which Allied troops, including Canadian forces, would intervene against the Bolsheviks in the Russian civil war. Borden was not curious about resource control, even though the legacy of the November 1918 Conference would bedevil Canada for years. After all, he was establishing Canada's new international status. He was meeting with British ministers on the upcoming peace talks, and discussing raw materials and international trade at the new Canadian Mission in Whitehall Gardens. He was otherwise occupied. What more could anyone expect?

Three months after the November 1918 Conference, Finance Minister White confided to Prince Edward Island premier Arsenault that the war had cost too much, and that soldiers' pensions would cost even more. "I think it is extremely improbable that any increase in provincial subsidies will be considered."[106] There could not be peace at any price on the home front. Borden would continue to dash between heavyweight talks and glittering gatherings. The prime minister would not leave for home until May 19, 1919. By then, the West would be in turmoil.

Louis Riel.

Manitoba premier
John Norquay.

North-West Territories
premier Frederick Haultain.

Minister of the Interior
Clifford Sifton.

Galician immigrants, circa 1911.

Alberta premier
Arthur Sifton.

Manitoba premier
Rodmond Roblin.

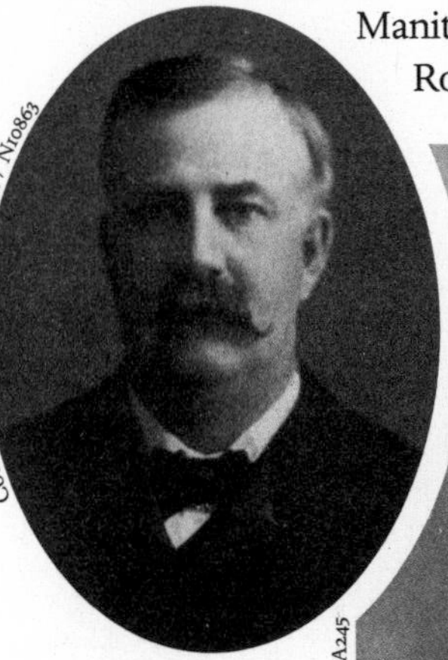

Courtesy of Archives of Manitoba / N10863

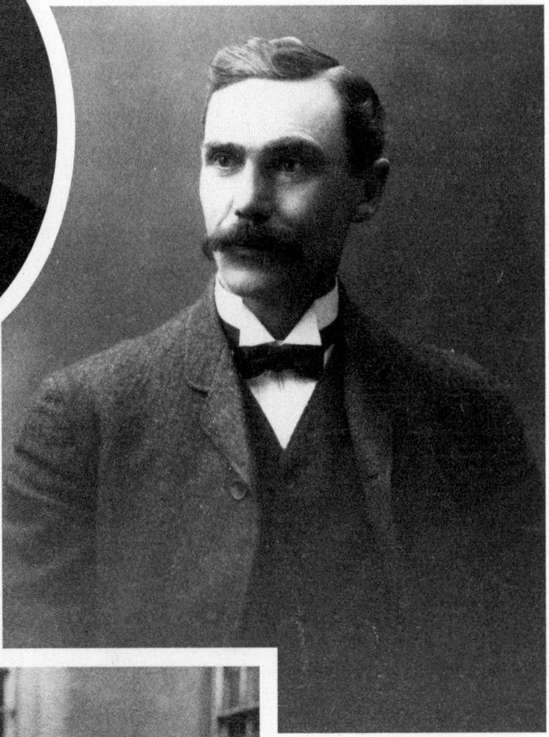

Courtesy of Saskatchewan Archives Board / R-A245

Saskatchewan
premier
Walter Scott.

Library and Archives Canada / C-002082

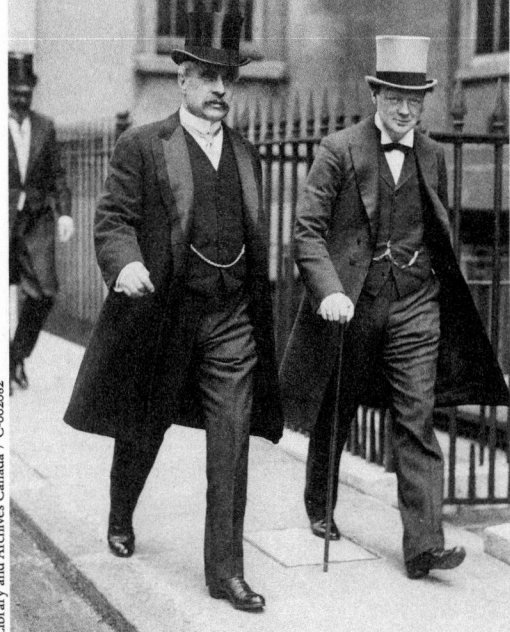

Prime Minister Robert
Borden and First Lord
of the Admiralty
Winston Churchill, 1912.

Manitoba premier
Tobias Crawford Norris.

Saskatchewan premier
William Martin.

Alberta premier
Charles Stewart.

Prime Minister
Arthur Meighen.

Saskatchewan premier
Charles Dunning.

Alberta premier
John Brownlee.

Prime Minister Mackenzie King
and premiers of Quebec and
Ontario, 1927.

Saskatchewan premier
James Gardiner.

Saskatchewan premier
J.T.M. Anderson.

"THE PRINCIPLES HAVE BEEN COMPLETELY LOST TO SIGHT": MEIGHEN SPLITS THE GANG OF THREE. 1919 TO 1921

—

THE GREAT OLD MAN DIED only three months after the Armistice. Sir Wilfrid Laurier had survived the angst of war, including the brutal debates over conscription and the catcalls of "traitor." In peacetime, his spirits had soared with the hope of putting his party back together again. He was organizing a grand congress to woo back the dissidents when he collapsed from a stroke in front of his devastated wife Zoe. "This is the end," he told her during a moment of lucidity.[1] His body could no longer handle the stress. The next day, on February 17, 1919, the Liberal leader died. He was seventy-seven.

It was the end of an era, and Canadians recognized its passing. They put aside their quarrels with Laurier, their tangled loyalties, and their harsh words. Official Ottawa closed down. More than 50,000 people— half of Ottawa's population—filed past his body in the Victoria Memorial Museum, where the regions had clashed so disastrously only three months earlier.[2] The *Calgary Daily Herald* saluted him as a "great Canadian, a statesman of international reputation, a man whose kindly disposition and great abilities won for him the sincere devotion of his friends and the high admiration of the keenest of his opponents."[3] Among his notable achievements, the newspaper added, was the granting of autonomy to Alberta and Saskatchewan. Five days after his death, the former prime minister was given a splendid state funeral in Ottawa's Notre-Dame Cathedral. More than 100,000 people walked behind his body for two kilometres to Notre Dame Cemetery, as government cameramen filmed the procession. Footage was rushed into Ottawa theatres that day.[4]

The postwar age saluted its past, when newcomers had spilled across the West, dreaming of prosperity on their own Promised Land. Prime Minister Sir Robert Borden was at the Paris peace talks when he learned of Laurier's death. Privately, he confided that it was a "great shock." Publicly, he praised Laurier's magnetic personality, his "unfailing grace of diction in both languages," and his charm.[5] The death of his long-time rival might have eased the political pressure on his Union government, but it had also reminded the nation of its lost dreams.

Those early post-war years were turbulent. Despite Borden's best plans for reconstruction, the economy slipped into a downturn. There was no money to buy civil peace, and labour disruptions were now frequent. Ottawa clung to the notion that salvation could be found on the land, and it was advertising its post-war prospects in the United Kingdom and the United States. Potential farmers were "specially desirable," and Ottawa would welcome former Allied soldiers.[6] Because the Union government had conveniently retained control over Western lands, it could still dangle the offer of a free if often remote homestead, and after the wartime lull, immigration resumed.

At home, Canadians were not so optimistic. Why were their lives not better? Why had so many young men died? What had made so many people turn on each other at home? How could it be that so many disabled and disenchanted veterans were on their streets? What had so stunted their economic prospects in this industrializing age? Why were so many fatherless families to be crammed into inner-city tenements? Despite the failure of the November 1918 Conference, the natural resources issue remained on the Western agenda. Ottawa was still mulling that application from the Shell Transport Company for exclusive oil and gas exploration rights over 328,000 square miles of Western land, including most of central and northern Alberta and a large portion of the Northwest Territories. That area was roughly 75,000 square miles larger than the entire province. By now, Westerners had heard rumours of Shell's application, and they were horrified.

The November 1918 Conference had left a bitter legacy. Premiers had drawn lines in the sand that no tides could wash away. The regions were divided against each other, and the Prairie provinces were isolated. The absent Borden would never find the patience or the money to broker harmony at home, and his successor, Arthur Meighen, had irritated too many premiers at that conference with his sharp tongue and decisive rulings. Meighen would make two efforts to broker Western peace; he would even commission a ghostwriter to prod Westerners toward compromise. But the participants at the November 1918 Conference remained too disgruntled. Their days in office were numbered: postwar dissatisfaction would eventually oust every politician at the November 1918 Conference table—with two surprising exceptions. Quebec premier Lomer Gouin would step down gracefully in 1920; British Columbia premier "Honest John" Oliver would somehow survive the postwar wrath.

After their defeat at the November 1918 Conference, the Western premiers resorted to damage control: they had to restrain Ottawa's resource handouts while they negotiated a deal. Only two days after Laurier's death, before his funeral procession coiled through the chill Ottawa streets, the Alberta Assembly passed a unanimous resolution that clearly referred to the Shell Transport bid. It was a pre-emptive strike. Ottawa was apparently about to grant "extraordinary and extensive concessions of petroleum territory . . . on terms oppressive to the people of this province."[7] Since negotiations on resource control were "pending," Ottawa should stop doing deals. The Legislature was exaggerating: no negotiations were scheduled—but in principle they were always "pending."

The Westerners did not know that even Official Ottawa had qualms about Shell's brash bid. One federal report had questioned whether it was "in the public interest" to grant exclusive rights for an indefinite period "over so large an area."[8] After all, other energy companies had asked for exclusive rights over smaller areas, but Ottawa had rejected those bids.[9] The Shell proposal even intruded on Borden's European sojourn. On March 30, 1919, he somehow found time to see Shell Transport director Sir Reginald MacLeod in Paris. Neither of them knew that, only two days before their meeting, Interior Minister Meighen had already rejected Shell's application.

When MacLeod learned about Meighen's decision, he was outraged. He suspected that the Standard Oil Company had lobbied against Shell, and he asked Borden for another meeting "when next passing through London."[10] MacLeod's suspicion was correct: W.J. Hanna, the president of Imperial Oil Ltd., had indignantly wired Borden at Claridge's Hotel in London on February 11, arguing that approval of Shell's request would be "highly prejudicial" to his firm.[11] Hanna had clout: in mid-1917, he had accepted the thankless role (for which he had refused any salary) of Dominion Food Controller, investigating supplies and calling on Canadians to change their diets. When Hanna died in late March 1919, one pastor at his funeral charged that the job of "laying down sound economic rules for a carping and ungrateful people [had] sapped his strength."[12] Still, Hanna's intervention probably had less effect than MacLeod suspected: Shell had defeated itself with its greed. But that bad-tempered episode did dramatize the value of the West's resource heritage.

Borden gave the matter little attention. He had just declined an offer to become Britain's ambassador to Washington. His memoirs made no mention of Sir Reginald MacLeod or the Shell bid. The prime minister likely dismissed the kafuffle with the same resignation that he showed on April 26 when he contemplated proposed modifications to an international treaty. "My colleague Sir T. [Thomas] White has often told me that political life consists of one d–d thing after another. I well believe that this is so; but one must do one's best."[13] Shell's proposal died.

While Ottawa fended off Shell Transport, its offer of virtually free land was attracting war-weary settlers. Before the war, American Clyde Campbell had worked as a pharmacist and a Harvard-trained metallurgist. In 1918, while he was supervising ammunition-manufacturing sites, including forty-four sites in Canada, for the U.S. War Department, he contracted Spanish influenza. By the spring of 1919, he was exhausted, and Ottawa's offer of free land "in the far reaches of northern Canada" was so alluring.[14] Desperate for a respite, anxious to be his own boss, he broke with his past, and claimed a homestead in the Peace River district of northwestern Alberta. Campbell had virtually no idea how to farm the northern Prairies, but he had fallen in love with the notion that this was the Last Best West.[15] His letters to his family in Ohio, which Edmonton scholar R.G. Moyles has edited into a book for the Historical Society of Alberta, captured his romance with the fickle West.

While Borden talked peace in Europe, and the Western premiers struggled to protect their resource heritage, Campbell painstakingly began to clear his land for grain and livestock. From the start, it would be a precarious existence. His expenses would be high, his profits negligible. He would survive on an allowance from his parents in Toledo. During those first months, as his wife and twelve-year-old daughter prepared to join him in early August, his letters were achingly optimistic. But there was already an inkling that his luck would *always* be around the next corner. "This is a happy land for me," Campbell affirmed soon after his arrival, "and I don't want to hear about strikes and discords."[16]

Unfortunately, strikes and discords dominated the headlines. By the time Borden finally left Europe on May 19, 1919, Canada was in an uproar. Labour militancy had escalated throughout the spring. On May 15, two weeks after Winnipeg building and metal trades workers had gone on strike, more than 30,000 local workers had walked off the job in a General Strike. Factories and stores closed. Trains stopped. Public-sector employees "such as policemen, firemen, postal workers, telephone operators and employees of waterworks and other utilities" joined their private-sector brethren.[17] Across the nation, from Nova Scotia to Vancouver Island, workers staged sympathy strikes. Interior Minister Meighen and Labour Minister Senator Gideon Robertson headed to Winnipeg, where they met with the Citizens' Committee that opposed the strike. But, concerned about the Communist threat, they unwisely took sides, and refused to meet with the workers' Central Strike Committee. (Meighen, however, was "on hand" to watch strike leaders urge Post Office workers to ignore a federal cabinet order to return to work.)[18] Ottawa threw its weight behind the employers, broadened the Criminal Code definition of *sedition*, and actually amended the Immigration Act to allow for the deportation of British-born immigrants.[19]

As Borden disembarked at Halifax on May 25, 1919, Canadians were profoundly angry and divided. They could not cope with the lethal combination of high unemployment and inflation, and they were also concerned about the loss of civility. Borden was shocked by the "abnormal" state of mind. He detected "a distinctive lack of the usual balance; the agitator, sometimes sincere, sometimes merely malevolent, self-seeking and designing, found quick response to insidious propaganda."[20] The prime minister had exchanged the glamour of Versailles for domestic turmoil, and he did not understand that dire working conditions had provoked desperate responses. The Winnipeg strike ended in mid-June with the arrest of the leaders and a disastrous encounter between the protesters and the Royal Northwest Mounted Police in which one striker was killed.[21] The prime minister could no longer delude himself: he could not resolve the problems of all frustrated workers by settling them on farms.

Meanwhile, resource firms were leaning on Western politicians, who in turn were leaning on Borden. On June 7, 1919, Customs Minister Arthur Sifton forwarded a complaint from an Edmonton coal executive. H.A. Lovett, the president of North American Collieries Ltd., who was understandably frustrated by having to deal with two jurisdictions. What could be done about this? The former Alberta premier had an ingenious if impractical suggestion: why not shift jurisdiction over minerals to the three Prairie provinces? To ensure that the Rest of Canada did not squawk, the West could promise to use any resource revenues for new schools, to keep the peace and to administer the resources. Borden should act "at once."[22] Coming from Sifton, one of the original members of the Gang of Three, this proposal represented a huge concession: that Western governments would accept less-than-equal status, earmarking their revenues to meet pivotal expenses to appease the Rest of Canada, which claimed a stake in their resource wealth.

That proposal likely exasperated Borden. Nine days after Sifton's letter, the prime minister wrote to coal executive Lovett with bad news.

Such schemes were never going to work. How could the provincial governments control the minerals when Ottawa retained ownership of the land? Surely it was better to leave one government in control. Any other approach would lead to "difficulty and confusion" as well as injustice. [23] Anyway, Borden added, in the "don't-bother-me" tone he often took with petitioners at home, his time was limited because of "the imperative and extremely urgent matters which continually press upon my attention from day to day since my return." [24]

The prime minister's skill at domestic diplomacy may have been wanting on the resource issue, but in this period he really did make an effort to confront pressing problems, along with a new opponent. On August 7, 1919, almost six months after Laurier's death, William Lyon Mackenzie King won the leadership of the Liberal Party by thirty-eight votes. It was a tough contest that went through three full rounds of ballotting. King was relatively inexperienced: he had sat in Parliament from 1908 to 1911, including a two-year stint as labour minister in Laurier's last government. But he won because his main competitor, Laurier's long-time finance minister, William Fielding, had broken with Laurier over conscription, while King had not openly challenged the old lion. The party remained splintered. Convention delegates also adopted a nebulous resolution: the West should get control of its resources "on terms that were fair and equitable . . . to all other Provinces." [25] With that resolution, which would return to haunt them, the Liberal Prairie premiers could have no illusions about the Rest of Liberal Canada's views. King soon won a Prince Edward Island seat, and set about mastering his volatile world.

Borden was wary. His coalition, which had stuck together in wartime, was now restive. In early June, before the Winnipeg General Strike had ended, Agriculture Minister Thomas Crerar had quit the cabinet to protest high tariffs, less than two years after he had joined the Union government. That wily operator King was now wooing other lapsed Liberals back to the fold. In contrast, Borden's government wore its tumultuous history like a string of tin cans.

Borden opted to make peace. In September 1919, he promised his caucus that, during the life of the current Parliament, he would adjust tariffs, control spending, and ease veterans into civilian life.[26] It was an inadvertently candid list: ten months after the Armistice, the prime minister was still grappling with soldier settlement, labour unrest and a budget crunch. He also vowed to ask Parliament to transfer resource control to the West, "under fair terms and conditions," while settling with the other provinces on such terms *as may be practicable.*[27] Borden recognized, however fleetingly, that he had to do *something* after the debacle of the November 1918 Conference. He simply did not want to tackle that exhausting domestic spat at that moment.

But there was no escape. That same month, Nova Scotia finally forwarded a five-month-old House of Assembly resolution. The timing was deliberate, and the message was curt. Nova Scotia was not giving up *its* claims. It wanted compensation for the West's school lands, because Prairie governments received interest on money from the sale of those lands. It demanded compensation for the West's subsidies "purporting to be" in lieu of resources. It also wanted compensation for the boundary extensions of 1912. And, it spluttered, it wanted compensation for its unfairly paltry share of federal subsidies *since Confederation.*[28]

This was a calculated power play. That cunning Liberal premier George Murray, who had supported Borden's Union government despite his friendship with Laurier, now wanted payback. He had waited until such immediate postwar problems as demobilization had subsided. Then he had swooped in. The resolution did not mention the Prairie provinces' claims. No matter what Ottawa did with the West, Nova Scotia wanted consolation and compensation for past wrongs. Clearly, Ottawa could not settle with the West if it did not appease the East. The regions were at war.

Voters were as impatient as their politicians. The first participant in the November 1918 Conference to feel their wrath was Prince Edward Island premier Aubin-Edmond Arsenault. The Liberals trounced his Conservative government in late July 1919, amid charges that he planned to raise education taxes. Two years later, Arsenault would contentedly accept an appointment to the Prince Edward Island Supreme Court. Three months after Arsenault's defeat, the United Farmers of Ontario (UFO), which now had more than fifty thousand members after five years of advocacy work, defeated Borden's long-time ally, Conservative premier William Hearst. The veteran premier was shocked. As he told Borden, his supporters had insisted "we would sweep the Country. . . . The spirit of unrest however, was deeper than we thought."[29]

His belated diagnosis was correct. Rural Ontario voters were still smouldering over the conscription of farmers' sons. The UFO had also tapped the labour unrest, working informally with the Independent Labour Party during the election. It now formed a coalition government with Labour MLAs, and selected an activist farmer, Ernest Drury, as premier. He set to work with the declaration "I can think better when plowing."[30] Such bewildering change—such reversal of the pre-war social order—was a warning to every politician.

Borden was ill throughout the fall. He escaped to the United States several times, to recuperate and to discuss the postwar world. On December 9, 1919, still exhausted, he consulted his doctor in Montreal. "He examined me for more than an hour," Borden would recall, "after which he told me that, in his opinion, I could not remain in public life."[31] Borden returned to Ottawa and resigned. With the persuasion of his cabinet colleagues he agreed to stay—on condition that he be allowed a lengthy holiday. He appointed Trade Minister Sir George Foster as acting prime minister, and headed off to the Caribbean and Britain. He did not lose sleep over Western resource control. A month later, in January 1920, at a Winnipeg Conference, Thomas Crerar became the first leader of the new Progressive Party of Canada, which espoused such pro-farming policies as lower tariffs.

On his Peace River farm in northwestern Alberta, Clyde Campbell was already worried about the high costs and low returns of farming. He had been devastated when his horse died with its head in his arms before Christmas. "It is a tremendous loss where there is no income," he wrote to his parents.[32] His wife, Myrle, was more upbeat. She wore her silk dress for the first time in Alberta when the Campbells slid across the snow by bobsled to celebrate New Year's Eve at a neighbour's home. They warbled church songs and popular ditties with almost sixty homesteaders, and returned home at 3:30 a.m., laden with apples, nuts and candy. This first unforgiving winter charmed Myrle. "This is no lonesome land out here," she wrote to Clyde's parents.[33] She loved the sunsets and the sunrises, and the way that the water under the snow crackled when the temperature dropped.

Clyde, however, sounded worn out. He was hauling ice to stack on the north side of the house as drinking water, and it was "like dragging out tombstones and just about as heavy."[34] He was also daunted by the mid-January cold, which "at fifty below is strangling—like a faint touch of chlorine gas."[35] His news was scattered. The cold had stopped the trains, which carried the mail, for ten days in mid-January. The Mounties had arrested a neighbour for housebreaking, instructing the offender to stay put until his trial. There was no risk of escape: "When you are here in the winter, you are here, and that's certain."[36] Clyde had decided to grow flax in the spring because the prices were high. If his crops were good, he would not have to do odd jobs for other homesteaders for cash in the fall. Despite the hardships, he tried to remain sanguine. "Believe me, folks, the safe place to be in the next few years is the FARM," he wrote, capitalizing his location, "and don't forget it for a moment. . . . Just give me five years and I will show you something."[37] The world outside his door was in ferment.

Borden was in Britain in February 1920 when Alberta premier Charles Stewart dashed off a quick note to Acting Prime Minister Foster. In the spring of 1919, he had set up an Alberta Coal-Mining Industry Commission to scrutinize the industry that now produced almost six million tons of coal each year.[38] When the commission had reported in December 1919, Stewart was unable to implement the most important recommendations "on account of our not having control of our Natural Resources."[39] He asked Foster to tackle the issue when the Parliamentary session opened on February 26, 1920. (MPs would then meet for the first time in the new House of Commons Chamber, rebuilt after the 1916 fire.) Foster was non-committal, stalling for time.

Borden was still travelling. In early March, the prime minister, then on vacation in South Carolina, learned that "our members were uneasy, especially with regard to by-elections."[40] Voter discontent was palpable, but Mackenzie King was winning cautious applause. The premiers were restless, too, and backbench MPs feared for their political lives. Sir Robert Borden had ushered Canada through the war, acquitting himself admirably abroad—but an absentee prime minister with an eroding Union government was now a political liability.

Borden retreated to Asheville, a resort town nestled in the Appalachian Mountains in North Carolina. By mid-April, he was "making steady progress in regaining strength" in that "very delightful and restful place."[41] His MPs expected he would be back to work in early May. Perhaps predictably, on the same day that Borden expressed that he felt stronger, Nova Scotia added more fuel to the regional fights. In an astonishing thirty-two-page speech in the House of Assembly, Public Works Commissioner James Tory outlined "The Claims of Nova Scotia Respecting Western Lands."[42] The title alone would have infuriated the Prairie premiers.

It was the same old story, just with new math. Tory reviewed Ottawa's acquisition of Rupert's Land and the Northwest Territories, along with the school-lands trust and the subsidies in lieu of resources. By his arithmetic, when *all* costs were considered, Ottawa had purchased Rupert's Land for $1.5 million, administered Western lands at a net loss of $40 million, and handed out almost $21 million to the West in extra subsidies. Maritime taxpayers had "received absolutely no compensation" for this largesse.[43] Even worse, according to Tory, Ottawa was neglecting Maritime transportation, immigration, agriculture and the local fisheries. Nova Scotia demanded $75 million as compensation for past injustices: It should get an extra $562,000 per year if Ottawa continued its subsidies in lieu of resources to the West.

Maritime demands trumped Western claims: by Maritime reckoning, Ottawa must compensate the Maritimes for decades of injustice *before* it could transfer resource control and continue its extra subsidies. Tory admitted that Nova Scotia had made "no progress" since that legislative resolution of April 1919.[44] But the times were finally right—"now that the settlement of the war obligations is pretty well at an end."[45] If anything, the November 1918 Conference had bolstered the Maritime sense of grievance: it now wanted more money, virtually immediately, even if Ottawa did *not* transfer resource control in the West.

A few days after this fusillade, Alberta premier Stewart returned to the attack. On April 19, 1920, he forwarded a unanimous resolution from his Legislature to Acting Prime Minister Foster. The natural resources issue should "be settled without delay."[46] Ottawa should simply ask Parliament to approve "at its present session whatever authority may be requisite or necessary for the transfer."[47] The issue was solely between Ottawa and the Prairie governments, no matter what the Rest of Canada claimed.

As always, Borden could not abide such regional rivalries, and especially now, when the economy was faltering and peace remained far from certain. Britain had just assured France that it would consider the use of force if Germany did not follow through with its promise to disarm. The British government was also imposing an excess profits tax of sixty per cent, while France had just placed a ban on the import of luxury goods. In Ottawa, veterans' leaders were wearing overalls to their offices to protest the high cost of clothing. In the West, seeding was lamentably late because the snow from that harsh winter had lingered. The anticipated joys of the postwar era remained elusive.

Tensions within Borden's caucus were knife-sharp. On April 30, the six Western members of a special committee on natural resources actually scoffed at Borden's vague promise to transfer control "on such terms as will be equitable to the other provinces." The Westerners were blunt: such dreams were futile. An agreement "is no nearer being reached than it was eight years ago, nor does there appear to be any likelihood of a settlement . . . in the future, near or remote."[48] They called for an immediate settlement. Ottawa should simply ignore the other six provinces because "the question is solely one between the Western Provinces and the Dominion."[49] But in the National Archives there is a letter attached, without explanation, to that report. It is the ultimatum from the other six provinces, which they had delivered at the November 1918

Conference: if the Western provinces received their resources *and* continued subsidies, they wanted more cash.

Eighteen months after the November 1918 Conference, no side had budged. If anything, the positions were harder and tougher.

Not everyone liked Clyde Campbell's neighbourhood. In mid-April of 1920, a business colleague from Milwaukee and his wife arrived in Grande Prairie, and Clyde fetched them from the railway station to his homestead, roughly forty miles to the west. Sam and Helen had planned to homestead as partners with the Campbells, and Clyde was counting on Sam's financial contributions. But the newcomers loathed the place. "They hate it so much they want to go back on the next train," Clyde wrote grimly. "If the roads were not impassable they would have me take them back right now."[50] They still managed to leave within days. Helen's only positive observations were about the homesteaders. "She said she would never have believed that people of such good breeding could live in such a primitive country," Clyde reported.[51]

It was a huge setback, but the Campbells looked for joy wherever they could. Their cabin appeared cramped from the outside, but inside it was "roomy and as cozy as can be."[52] They loved the kaleidoscopic colours of the sky above the mountains. They prized the birds, including the grouse, which they would not kill because the "brave little fellows" had survived the toughest winter in a generation.[53] Clyde and his neighbour were going to work together to build a chicken house and a cattle barn on each other's property. Despite the difficulties, it was the season of new beginnings.

Borden and his wife, Laura, finally disembarked at Ottawa's Central Station at 12:45 p.m. on Wednesday, May 12, 1920, to the cheers of his ministers and Union MPs. His government had taken over that ailing problem child, the Grand Trunk Railway, which had never recovered from its prewar construction of a transcontinental line. Now the price tag of dreams—the cost to reimburse investors—was coming due. His ministers had just decided *not* to raise MPs' salaries, because any generosity to themselves would spark renewed demands to extend federal pensions for disabled veterans to *all* veterans. Borden was six weeks away from his sixty-sixth birthday, and no one, including the prime minister himself, was "now any nearer to knowing whether he [could] carry on or not."[54]

His caucus eyed him carefully. He was tanned and thinner. Some saw a spring in his step. Others detected more grey in his hair. Some ministers speculated that he might want to become Canada's minister plenipotentiary to Washington. But there was a difference in status between the post of British ambassador and a mere minister plenipotentiary, and those ministers assumed that Borden would not want the job unless London raised the status of that position. As *The Globe* ruefully observed: "the uncertainty and indecision that have prevailed for so long still prevail."[55]

Less than week later, Borden found himself embroiled in the budget debate. Finance Minister Sir Henry Drayton slapped a one per cent sales tax on all consumer goods, except coal and food, along with excise taxes ranging from ten to fifty per cent on everything from better quality textiles to luxury goods. He also hiked personal income taxes for anyone earning more than $5,000 per year. To deal with peacetime unrest, spending on the Militia Ministry soared by $4 million, or almost thirty-three per cent. The headlines were dire: "Consumer Is Badly Hit By Drayton's First Budget. . . . No Tariff Reductions To Help Masses."[56] The *Edmonton Journal* put the best possible spin on this huge tax grab: "New Taxes Fall On Shoulders Of Those Best Able To Meet Payment."[57] To the dismay of Westerners, however, Drayton delayed tariff revision.

Such scattershot bad news certainly did not constitute a pre-election budget. In the midst of complaints that Ottawa's policy was "being given to the country piecemeal," federal ministers promised to delay any trip to the polls until at least 1922.[58] They spoke too soon, however: Borden was flagging from the long sessions and the domestic sniping. "I soon discovered that I had reached the end of my strength," he would recall.[59] On July 1, after eight and a half years as prime minister, he gathered his caucus, and once again resigned. When his caucus allowed him to choose his successor, Borden ascertained that his ministers would prefer former finance minister Thomas White, while his backbenchers favoured Interior Minister Meighen. White refused the honour: he, too, was worn out from the war.

So, on the evening of July 7, Borden and Meighen visited the governor general at Rideau Hall, where Meighen accepted the Duke of Devonshire's formal invitation to form a government, which he did under the clumsy name of the National Liberal and Conservative Party, in tribute to its remaining Liberal members. The new prime minister inherited tumult. "The political pot was boiling madly everywhere in Canada in 1920," wrote Meighen's biographer Roger Graham. "The war had had a cataclysmic effect on politics, disturbing traditional loyalties and dissolving the old two-party system."[60]

Everything seemed amiss. The economy was mired in recession. Western workers were still seething, and Western farmers resented the continued tariffs on farm equipment. Quebecers remained angry over conscription, which Meighen had pushed through Parliament. Immigrants could not forget that many naturalized Canadians had lost their vote in wartime, and Meighen had also pushed that bill through Parliament. (Most regained their vote in 1920.) Few politicians could have mollified those groups, and Meighen in particular did not have the temperament. He was a logical man, firm in his convictions, staunch in the face of opposition. In July 1920, he had just turned forty-four, and he had sat in Parliament for twelve years. Despite his experience, however, he could not heal the federation's rifts.

Decades after that summertime handover, and long after former Prince Edward Island premier Arsenault had retired from the bench, he would muse that Meighen "had probably the keenest intellect" of any MP of his generation.[61] The meticulous Meighen could explain the complicated legislation that had amalgamated the railways without even glancing at his notes. Arsenault thought that he was "far abler" than Mackenzie King, but he lacked the latter's political acumen. Meighen also tried to do everything himself, when "it would have been preferable for him to allow some of his lieutenants to carry on under fire."[62] He would never deflect blame: he seemed to attract it.

The new prime minister was a transplanted Westerner. Although born in Ontario, and raised on his family's dairy farm, he had moved to Winnipeg in 1898. Four years later, he had joined a small law firm in Portage la Prairie. The Prairie premiers drew little comfort from his ascension. They remembered only too well the disastrous November 1918 Conference when Meighen had deftly set the premiers to quarrelling among themselves. Anyway, Meighen's immediate focus was not on them, but on the lingering disaffection over conscription in Quebec, where Mackenzie King's Liberals held wide appeal. In the West and Ontario, Thomas Crerar and his farmer-based Progressive Party were building on the success of the United Farmers of Ontario, eroding other Conservative bastions. The West's natural resources were not high on the new prime minister's list of priorities. In his intricate biography of Meighen's pivotal years from the summer of 1920 to 1927, historian Graham did not even deal with resource control.

The only first minister who slipped away on a high note in 1920 was Quebec premier Lomer Gouin, who had so vehemently opposed the continuation of Western subsidies in lieu of resources if Ottawa transferred resource control. Gouin had easily won re-election in 1919, largely because his Conservative opposition was still reeling from its federal counterpart's support for conscription. A year later, in early July 1920, mere days after Borden reached his own career crossroads, Gouin stepped down himself. His anointed successor, Quebec City lawyer Louis-Alexandre Taschereau, took over. Taschereau was an ardent advocate of private investment—including American investment—in Quebec's resources and hydropower. He would eventually prove far less inclined than Gouin to oppose the West's bid for continued subsidies. This quiet transition was an exception to the general turmoil.

That summer, despite his back-breaking toil and his list of chores, Clyde Campbell was relieved to be on the land, away from the threats of the modern age. He had read about the Bolshevik scares. He suspected that the Spanish flu epidemic, which had felled him once, might resurface. "We are happy out here," he insisted, "and feel lucky to be away from all the worry of the nations."[63] When his parents fretted that he might be depressed, he countered that it was only natural for a farmer to want to get ahead. "This is paradise," he insisted. "Here all is serene and peaceful. There is no strife or discontent."[64] His greatest worry was that he would not be able to cultivate enough acreage to satisfy the government inspectors—and earn title to his land. So many veterans were looking for land, and there was so little good land, that whenever a homesteader failed, the inspector would evict the failed farmer, and settle a soldier in his place. "All this is worrying me quite a little."[65] So many people were chasing his dream.

Meighen's political future did not look promising. But the prime minister was not a defeatist. In the fall of 1920, he embarked on a one-month Western tour in the company of Immigration Minister James Calder, the former deputy premier of Saskatchewan who had pestered Borden about resource control in those halcyon pre-war years. It was perhaps at Calder's urging that Meighen mentioned the possibility of an eventual transfer of resource control during a speech in Medicine Hat. It was almost a throwaway line, but word travelled fast. Prominent Calgary businessman A.E. Cross, who was one of the founders of the Calgary Stampede, immediately penned a rare Western objection to any transfer—because the provincial government "would be almost obliged to make extravagant expenditures."[66] Of course, Cross was a Conservative, and Alberta premier Stewart was a Liberal.

More important, the Gang of Three stirred. On November 30, 1920, Manitoba premier T.C. Norris called for discussions on the issue. But he added a new twist to his old tale: he wanted Ottawa to account for the money that it had pocketed, and for the money that it *would have* received if Manitoba's lands and resources had been sold at their actual value. The veteran premier had suffered a stinging rebuke at the polls in late June, when his Liberals had been reduced to a minority. Because his survival now depended upon a disparate clump of rural and independent legislators, Norris needed great deeds to prove his worth. Alberta premier Stewart seconded Manitoba's request. Meighen obligingly invited

Saskatchewan premier William Martin. The gathering was set for Wednesday, December 15 at 3 p.m. in the prime minister's East Block Office. The Gang of Three from the November 1918 Conference was back in business. Temporarily.

In the few weeks before that meeting, Meighen set to work like a diligent tactician. Ottawa's right to control the West's lands and resources was spelled out in the three constitutional acts that had created the Prairie provinces. If Meighen were to secure agreements with the Gang of Three, Parliament and the three provincial Legislatures would have to endorse those deals. The British Parliament would then have to approve those amendments because Canada did not have its own amending formula. The other six provinces had vehemently opposed the West's proposals at the November 1918 Conference. If anything, their opposition had grown stronger since that confrontation. Any unilateral deals with the West would be dead if those six premiers—and perhaps more significantly, MPs from the Rest of Canada—objected to them: the agreements would never get through Parliament.

Meighen wanted a deal. But he wanted the West to see things his way, and compromise. On December 2, he wrote to Ottawa-based writer R.E. Gosnell, who had been a B.C. government public servant and was a prominent Conservative. Western journalists were demanding resource control, Meighen noted, and they were "repudiating the idea that any other provinces have concern or interest in the question."[67] Those scribes were ignoring how much Ottawa did for the West. So Meighen spelled it out. There was the subsidy in lieu of resources. Ottawa spent $900,000 each year on forests and railway construction. It had created huge parks, and had paid for surveys and irrigation projects. It was developing waterpower. It fostered homesteading and agriculture. Such benevolence had ensured that many immigrants had settled in the West.

Meighen had a remarkably duplicitous request. He wanted Gosnell to draft a letter "showing how the Maritime Provinces are interested, particularly from the stand point of subsidy allowances" in Western resources.[68] The Prairie premiers should not pretend that the Maritimes would benignly accept any unilateral deal on the West's terms. The letter should also mention that, at the 1919 Liberal leadership convention, Alberta premier Stewart and Saskatchewan premier Martin had accepted that the West should get control of its resources "on terms that were fair and equitable" to all provinces.[69] So those two premiers had recognized the interest of other provinces in the West's lands. Gosnell should ensure that the letter was published, "say in the Winnipeg Free Press," under the signature of "a prominent and highly regarded Eastern Province man, say, from the Maritime Provinces."[70] Meighen even suggested that Gosnell ask Interior Minister Senator James Lougheed for permission to interview key bureaucrats. It was a set-up: if the West would not compromise, the Maritimes or even Westerners themselves would take the blame; if the Westerners compromised, they might get a deal, and Meighen would be a hero.

The prime minister was taking no chances. He also asked Northern Ontario MP Francis Keefer, who was an expert on natural resources policy, to burrow through Ottawa's files. Keefer dutifully summarized the competing views over the decades, and enclosed an earlier report that he had prepared on the brink of the November 1918 Conference. His approach would become achingly familiar. Long after the natural resources were transferred, federal politicians would still invoke its condescending sentiments, to the fury of Westerners.

Ottawa, Keefer wrote, might be a trustee for the Western provinces, "but she is also a Trustee for Canada as a whole."[71] The federal government should figure out which Western resources were provincial, and which were interprovincial or even international. Parks were *national* assets. Ottawa was a superb administrator of the school-lands trust. Timber sales and forestry should remain under federal control. Non-fuel minerals such as nickel could go the provinces, but Ottawa should have the power to prohibit specific exports. Fuels such as coal should remain under Ottawa's jurisdiction "on account of its scarcity and demand." Finally, Ottawa should control "the use of any waters that are interprovincial or international and not purely domestic."[72]

Keefer used the word "trust" frequently. When the MP referred to interprovincial waters, he observed with delicious naïveté: "The Dominion can be trusted by each Province to do what is best for both with no particular selfish interest to serve."[73] He articulated another sentiment that would linger in the hearts of many federal and provincial politicians from the Rest of Canada: "Would it be nationally sound . . . to transfer fuel minerals [oil and natural gas] upon which the whole country depends not only to one Province, but practically to one small district of the Province?"[74] It was an invocation of Sir John A. Macdonald's vision of a strong central government. And—of course—it was for the West's own good.

In late November, even though he was desperate for cash, Clyde Campbell could not find a job. He applied for a position with the government land agent, and received no reply. Another government office sent a list of factories that might be hiring, but the closest was in Vancouver. He and Myrle and their daughter, Isabel, had been hauling hay over the muskeg in a huge horse-drawn sled, when it had tipped over, injuring Myrle's leg. Clyde had not found time to build a chicken shed or a cow barn or an icehouse. He wanted to buy a cream separator but "the banks are not lending a dime to anybody."[75] In early December, the hardware store in Grande Prairie sold him the separator, which cost a whopping $100, along with other supplies such as milk pails, without a penny down. The first payment would not come due until July 1921. There was "one bright spot in town": Imperial Oil had leased 33,000 acres of land from the government in mid-November, and it had just discovered oil. "People were lined up at the [federal] land office filing leases."[76]

The prime minister did not leave Norris in the dark. Eight days before the gathering, he pointed out that any deal would have to receive Parliamentary approval. "From the standpoint of its fairness," the deal also had to secure the support of MPs from *all* regions.[77] Meighen flatly rejected Norris's demand that Ottawa account for the money that it *would have* received if the lands and resources had been sold at their actual value. It would be impossible, he scoffed, to estimate the theoretical value on the open market. Ever the lawyer, Meighen would not deal with such "might-have-beens." Anyway, those homesteaders who had responded to the offer had boosted provincial economies. Norris could be assured that Ottawa was now diligently calculating its Western subsidies, its expenditures and its revenues. But it would be virtually impossible to calculate "how much of federal expenditure was due to the retention of such resources and how much was not."[78]

Then Meighen added a twist, using the same aloof tone that had so aggravated the Western premiers at the November 1918 Conference. In 1914, he asserted, Borden had asked the Prairie provinces to endorse Ottawa's homesteading and immigration policies. If they had agreed, Borden "was quite prepared to transfer the resources *if the Provinces were prepared to relinquish the subsidy.*"[79] Meighen struck. "*The question therefore at issue is one affecting subsidies rather than one affecting resources.*"[80] In Meighen's eyes, the talks were not about constitutional equality. Norris had to negotiate: he could not have the resources *and* the full subsidies in lieu of those resources. Meighen was rarely subtle. That might have been a virtue at times, but it could also be a tragic political flaw.

Norris was furious. He wrote a twenty-one-page letter to Meighen, which the prime minister received on December 14, and which the premier also delivered in person at the next day's gathering. It was a scathing review of Manitoba's past dealings with Ottawa, and it dwelt upon the stalemate at the fateful November 1918 Conference. Then, Ottawa had maintained that it was not prepared to increase its *general* subsidies so nothing could be settled. But, he spluttered, the Prairie provinces had made "no suggestion of *increased* subsidies" in their proposal.[81] Norris blamed the failure of that conference on "the lack of any accurate knowledge or acquaintance with the historical and constitutional basis" of the relationship between Ottawa and the West.[82] In effect, federal politicians and officials were ignoramuses.

The premier now snatched that December 13, 1913 proposal—which the Gang of Three had espoused for seven years—off the negotiating table. The Prairie premiers had demanded resource control along with the continuation of their subsidies in lieu of resources to compensate for

Ottawa's past usage. But Ottawa and the other six provinces had seized upon that letter as an excuse to confuse two separate issues: Ottawa's per capita subsidies to *all* provinces and the transfer of resource control. "The principles have been completely lost to sight," Norris fumed.[83]

The Rest of Canada had meddled in an issue that concerned only Ottawa and the West. This would not happen again: Norris demanded that the December 15 meeting deal *only* with the transfer of natural resources, along with compensation for the resources that Ottawa had already given away. The transfer of resource control would be "nothing less than the completion of Confederation itself."[84] Meighen should stop fussing over Ottawa's past expenditures. "Nothing whatever is to be gained at this stage by discussing details until the principle . . . is unreservedly conceded."[85] That was a clumsy approach to Ottawa's top legal beagle.

The meeting went badly. Norris would not negotiate: he wanted Meighen to concede the principle. But Alberta premier Stewart and Saskatchewan premier Martin were not so stuck on principles. There would be no official records of that meeting in the late-afternoon winter gloom. But nine days later, on Christmas Eve in 1920, Meighen formally answered Norris's screed with a recap of the proceedings. The four first ministers had discussed the issue "very thoroughly," but Norris had ignored his offer to negotiate. Instead, the Manitoba premier had insisted that Ottawa "concede what is described as a principle."[86] Thereafter, supposedly, a "mere system of accounting" would settle the issue of resource control.

That stubborn stand on principle had ruptured the Gang of Three. As Meighen gleefully observed, "There has not been acquiescence in this position by the Governments of the Provinces of Alberta and Saskatchewan."[87] Those two governments were eager to make a deal. Indeed, as it would later emerge, Alberta was almost desperate to sign a pact.

Meighen repeated his bottom line. Ottawa had picked up a hefty tab for such expenses as Western immigration, railways, irrigation, and the Mounted Police. The mere acceptance of a principle was useless if Ottawa and the Western provinces could not agree on how much Ottawa had spent. Any attempt to do the accounting *afterward* would simply ensure continued bickering, or "in a word, every difficulty would arise that now confronts us."[88] Meighen added his final barb. "I think you will agree," he observed archly, "that after [the] discussion of the 15th this was the view of the majority of those present."[89] Norris had lost his two comrades, and the prime minister was rubbing it in.

They were at a standoff. Meighen knew that he faced an election relatively soon, and so much had changed since the Union government had romped to victory in 1917. On January 21, 1921, Secretary of State Arthur Sifton died. It was a huge loss for Meighen: Sifton had been one of the high-ranking Liberals to join Borden's Union government in 1917. Although the former Alberta premier had been in poor health for several years—he had needed a car to cover the few hundred yards from his apartment in the Château Laurier Hotel to Parliament Hill—his colleagues had prized his concise contributions at the cabinet table.[90]

Meighen's MPs in the National Liberal and Conservative Party were losing heart. Most ministers had not wanted Meighen to succeed Borden, and few MPs were now ardent loyalists. Nor had voters warmed to him. Ottawa needed cash to deal with regional problems, and although he would likely have transferred resource control if the Gang of Three had compromised, Meighen had valid financial reasons for dodging any immediate settlement of the issue.

Norris had provided a good pretext for inaction. Federal wartime obligations were close to resolution: the Soldier Settlement Board had put almost half of its approved applicants on the land by December 1920. Theoretically, it should have been possible to transfer the resources, but fortunately for Meighen, Norris was inflexible. He would not bargain, even though Meighen had to placate the Rest of Canada with a compromise. He had even fallen out with his provincial allies. Meanwhile, with every day that passed, resource-rich Alberta was losing more revenues from royalties and leases than its fellow Western governments. Almost two decades after that dismal encounter, historian Chester Martin, who was a fierce Western partisan, would reveal that Alberta "was prepared to accept half the existing schedule of subsidies" in 1920.[91] Mackenzie King would later exploit that willingness.

The three Prairie provinces met again with Ottawa in May 1921. Chester Martin, ever the Norris loyalist, would later assert that the Manitoba premier stuck to his principles. "The fiscal aspects of the problem," Martin would declare loftily, "were subordinated to the central issue of constitutional right."[92] The Manitoba brief, however, did hint at compromise. As soon as Meighen conceded the principle, the province would not "stand uncompromisingly upon the rigid letter of the law . . . equitable adjustments could be arranged by common consent, or in the last resort, by arbitration."[93] But Meighen would *not* concede the principle before the two sides agreed on who owed what to whom. Those talks also failed. After ten years of on-and-off negotiations, the Western provinces had made little progress. Meighen had cracked their common front. The Gang of Three had ruptured. But the prime minister was not a sufficiently skilled politician to exploit their differences and do a deal that the Rest of Canada would accept.

There would be few winners anywhere during those early years of the 1920s. Voters were demanding new social services that provincial governments could not afford. Federal subsidies now constituted an ever-smaller share of provincial revenues, but Ottawa could not afford to increase them. The economy was stalled. Western farmers were beset; many veterans and new immigrants were already worried about their future on the land. Grain prices were falling, and the shortage of railway cars was acute. Many new homesteaders were settled on unsuitable lands.

Ironically, for grain farmers the problems had actually increased in peacetime. During the war, under the authority of the War Measures Act, the federal government had established a Board of Grain Supervisors, which had monopolistic powers over the purchase and sale of grain. With the arrival of peace, Ottawa had dissolved the board. In the fall of 1920, when the price of wheat had plummeted, farmers had strongly protested. When prices declined even further during the summer of 1921, "their demand for the resurrection of the board became more shrill and insistent."[94]

Meighen's political standing in the West was precarious.

———

On his Peace River farm, Clyde Campbell had lost his rose-coloured glasses. In March 1921, he was unable to sell his potatoes in Grande Prairie because the market was glutted. His hens were not laying eggs because he had no wheat to feed them. Local farmers were earning only ten to fifteen cents for each bushel of oats when it had cost twelve cents to thresh the crop. Poignantly, the only people getting rich in his community were the oil drillers. Sometimes, depending on the humidity, Campbell could smell crude oil. Local Aboriginals had told him that there was a flowing natural gas fissure to the south of his property, but he was reluctant to invest in a lease because he had "seen so many perfectly sure things sicken and die."[95] When he finally tried to interest an agent in the oil on his property, the agent explained that the price of oil was too low to make a profit. Campbell concluded that the big companies would be fine while "small cautious investors" were left on the sidelines.[96]

He was fed up with the anti-Bolshevik propaganda sweeping the nation, but equally skeptical about Bolshevism. It was almost spring, but the mercury in his outside thermometer was "knocking the bottom out of the glass."[97] The cost of his supplies was prohibitive, and his three horses were eating him out of house and home. Campbell spoke for many Prairie farmers when he lamented: "The hand that feeds the world is being bitten."[98]

Two months later, he was still disconsolate. There were no jobs on the roads around Grande Prairie, "and nothing else stirring." Butter prices had plummeted. "Why, it looks as if universal and all-conclusive stagnation has hit the world." Clyde could not even feign optimism. "Prosperity is on its way but I sure wish a job would show up right now, when I need it most."[99] A week later, a husky broke into his hen house, ate three mother hens and killed most of the chicks. He could not find wage-paying employment. "The outlook is decidedly rotten all right," he observed. "The governments, Federal and Provincial, will have their hands full this fall and winter trying to keep people from starving to death."[100]

Westerners were fearful and at the end of their rope. On June 27, 1921, the voters overwhelmingly defeated Meighen's party in a by-election to replace Arthur Sifton in Medicine Hat. The irony of that Alberta defeat was surely not lost on the prime minister: it was there in the fall of 1920 that he had talked about the transfer of resource control. The contest had been a two-way fight between the government candidate and a Progressive farmer who had the backing of the United Farmers of Alberta. Meighen had ignored advice to call the by-election immediately after Sifton's death in January—because it would have been difficult for farmers to get to the polls during the winter.[101] The Progressive candidate had now won by almost ten thousand votes—one of the largest margins of victory in Canadian politics to date. After years of complaints, the "aroused and well-organized farmers" had sent a fierce message to Ottawa.[102] Meighen was at the Imperial Conference in London when he learned of the "stunning blow."[103]

Three weeks later, Alberta premier Stewart was also swept away in a dramatic election upset. The United Farmers of Alberta, which had run candidates in less than seventy-five per cent of the seats, suddenly found itself with a comfortable majority—and the UFA didn't yet have a political leader. (UFA president Henry Wise Wood declined the chance to serve as premier.) Some members seriously suggested that Stewart should stay on the job; the departing premier declined. In the end, the UFA settled upon a prosperous farmer, Herbert Greenfield, who did not even have a seat in the Legislature. Within a dozen years, this non-partisan lobby group had transformed itself into a political powerhouse because farmers had revolted against the established parties and their own difficult lives. Although Greenfield vowed to rescue drought-plagued farmers in the province's southeast, he would not be able to stop "the worst farm abandonment in Canadian history."[104] But, albeit briefly, his stolid presence would inspire hope.

On his Peace River farm, in mid-August, Clyde Campbell was grateful that the drought was relatively mild in his region, although his hay was scant and short. Then the rains came, and he was torn: should he join a threshing team to earn scarce cash or should he stay at home to bring in the hay that would see his cows through the winter? Debt-collectors were at his door, and Clyde figured that he would soon lose his heifer to cover some bills. That autumn, he dug down thirty-one feet through heavy blue clay, but his new well would only produce a barrel of water a day. Another man was threatening to take away his horse and wagon if he did not pay more bills. Even though his parents sent cash regularly, Campbell was always behind, and always waiting for the railroad to snake west from Grande Prairie so that he could easily sell his grain. His letters had lost their "gee whiz" quality, as he described his weary days, and his worrisome nights. He had decided, however, that he "was actually accomplishing something of worth, that I am a definite value to society."[105] The homestead would stand as a monument, he concluded, to his family's efforts.

Although he professed to hate politics—and could not vote until he was naturalized—he was now the secretary of his United Farmers of Alberta local. He had great hopes for the party, "there being but few lawyers in the House."[106] He also detected class oppression in his plight. Prior to the UFA victory, "the big interests, the railroads, the manufacturer, the various ranks of labor and professional life had the government, both provincial and federal, by the scruff of the neck. . . . the farmer was left as the puppet to be kicked and moved at will."[107] The UFA provincial government would provide good roads and civic improvements. "We are now getting ready for the Dominion elections," he added ominously, "and if it is within our power we mean to have a Farmer's federal government."[108]

The writing was on the ballot for Meighen. He had inherited a messy coalition, and he simply could not keep it together. In the House of Commons, Mackenzie King and his feisty Liberals, including some former Union MPs, were on the offensive. Former agriculture minister Thomas Crerar, who had joined the Union government with such fanfare in 1917, was leading a small group of Progressive MPs that fiercely espoused farmers' issues. Meighen's cabinet strength was fading. Sifton had died. James Calder, who had fought so strongly for Saskatchewan resource control, was heading to the Senate.

Meighen's cabinet and caucus wrestled with the timing of the election call. His close adviser, Senator Billy Sharpe, reported that the majority of the cabinet, most Tory MPs, "and most Conservatives in western Canada [were] in favour of deferring an election until 1922."[109] Meighen worried, however, that his party would lose an upcoming string of by-elections, which would be death by a thousand cuts. His majority might be perilously diminished. So he opted "to appeal to the people before it got worse as it seemed likely to do."[110]

The prime minister headed out for three months on the campaign trail. He defended protectionism on the Prairies against what he viewed as Western radicalism. He attacked Mackenzie King's slippery stances. He warned against the limitations of the Progressives' farmer-based approach. He stood up in Quebec for the decision to impose conscription. He promised to bring back the wheat board, but it would be without monopolistic powers—so the grain-exchange markets would determine the price. Resource control was barely an issue. Meighen campaigned for the status quo in a nation that had largely rejected it.

On December 6, 1921, his government was decimated. That awkward amalgam, his National Liberal and Conservative Party, elected only 50 members, compared to 65 Progressives and 117 Liberals.[111] Meighen even lost his own riding of Portage la Prairie. Meanwhile, all sixty-five Quebec ridings elected Liberals. Seventy government candidates lost their deposits, and the Conservatives did not win a single seat in the three Prairie provinces. "The public had rejected Meighen's bid for a mandate with no uncertain voice," his biographer Roger Graham noted grimly.[112] Meighen resigned immediately as prime minister. Mackenzie King and his new government took office on December 29, 1921. In late January, Meighen won a by-election in Eastern Ontario, and headed back to Parliament—but his confidence was badly shaken.

Mackenzie King would now become the new face at the bargaining table on Western resource control. He would embark on a long and painful learning curve.

In the Peace River district in early 1922, Clyde Campbell was braced for his "hardest year" on his very hard farm because he had to pay off loans and buy fence wire along with a harness and a wagon.[113] "If I can pull through this year things will be o.k.," he assured his skeptical family.[114] His sunny hopes of only three years ago had faded, however. His well was almost dry. His stomach troubled him. He was afraid that the coming summer would be even drier than the last one. He wired his parents for more money to stop the bank from selling his cattle for debts, but before the money could arrive, he had surrendered the cows. So he used the money to buy more cattle at a better price. If his crops succeeded, he planned to buy hogs in the fall, and feed skimmed milk from the cows to the pigs. "I am just frantic to have the farm producing *something*."[115]

Campbell would celebrate the New Year of 1924 with a wistful look at his years on the land. He had made many mistakes when he first arrived, "not knowing much about the country or the weather." But, at least, he had not slipped "back into savagery."[116] Another homesteader had killed seventeen horses, left his chickens to starve for a week, and "tied up milk cows to trees for days in a row without a bite to eat or water to drink."[117] While that homesteader had failed, Campbell had persisted. Now he would soon receive the title to his land. "It's a grand and glorious feeling, believe me," he would write. "We are the most thankful people in the world."[118]

He never did strike it rich, but in 1924, he wrote a novel along the lines of Sinclair Lewis's dyspeptic *Babbitt* as a diversion from his farm chores. Ever hopeful, he mailed the manuscript to publishers. He even penned a second book, but had no luck. In late 1924, he started to suffer severe headaches that would afflict him for five out of every six days. He was still earning no money from the farm: the railroad had not arrived, and he was hauling his goods by road to Grande Prairie, where few merchants would buy anything anyway. "It takes a wagon load of profitless hogs to buy enough groceries to fill a shoe box," he wrote. "We all love this country, its breadth, its fertility. But one cannot subsist on snow-capped mountains."[119] His condition worsened until he was bedridden. Eventually, a perceptive doctor diagnosed kidney disease.

Campbell would sell his beloved farm in 1927, and he and his family would return to Toledo. His unpublished novels would disappear. He would die in 1930, at the age of forty-four. By then, Mackenzie King would have worked his wonders on the Western premiers.

10

"WE ARE VERY DESIROUS OF HAVING THIS LONG STANDING QUESTION SETTLED": MACKENZIE KING'S ON-THE-JOB TRAINING. 1921 TO 1925

—

IN GRAINY PHOTOGRAPHS FROM THE 1920S, William Lyon Mackenzie King never looks as imposing as his name. He is undeniably dapper, although his vest sometimes strains across his pudgy stomach. He is short, his face full. By the mid-1920s, he is balding at the back of his head, but he has combed his hair carefully across the front. His gaze is usually pleasant, almost avuncular, but there is a hint of appraisal and subtle calculation in his eyes. He would be prime minister throughout most of this decade, but he seems unremarkable when compared with his elegant predecessors. He shared that notion. When he watched the "moving pictures" of the Parliament Hill ceremonies for Canada's sixtieth anniversary, he compared himself with his guest of honour, star aviator Charles Lindbergh, seeing in himself only "a little fat round man, no expression of a lofty character"[1] while Lindbergh was "like a young god."[2] Actually, as King's biographer Blair Neatby aptly notes, King looked "like a conservative and reliable businessman."[3]

In the beginning, this apparently unremarkable politician was beset. He had inherited the social and economic chaos that Sir Robert Borden did not have the strength to tackle and Arthur Meighen did not have the savvy to resolve. This long-time Liberal activist was still mastering the art of politics. With a scant majority in the House of Commons, he was dependent on the goodwill of Progressive MPs with their agrarian platform to stay in power. He had a gruelling agenda and a difficult cabinet that included former Quebec premier Lomer Gouin and his former

274

leadership rival William Fielding. He was also afraid of Conservative leader Meighen's razor-sharp criticism. In short, he did not know if he was up to his job.

But, as King resolved before the portrait of his beloved mother, he would do his best. Within weeks of taking power on December 29, 1921, he tackled the transfer of resource control. When it became clear that the Gang of Three had lost its unified negotiating position, he often dealt with each Prairie province separately, trying to exploit their weakened negotiating positions. During those first years, his efforts were sporadic. He was concentrating on wooing Western voters—not Western premiers—and the farmers who had voted for the Progressives were more interested in tariff cuts than resource control. But his tactics in the resource fight dismayed and disconcerted premiers from multiple regions.

King did not understand the country. He did not accept that *every* region, particularly the Maritimes, claimed a stake in *any* transfer—that he could not simply do a limited one-on-one deal with the Prairie premiers. It was only in the mid-1920s—after years of on-the-job learning, after the economy bounced back and after a great electoral shock—that King finally learned the *real* lessons of the November 1918 Conference.

The transfer of Western resources had become *the* focus of regional jealousies and financial ploys. There were so many resentments, so many demands and so many of those "If you do that for them, you have to do this for me" ultimatums on the table that only a master strategist could understand them. It was only then that this prime minister with lofty aspirations and a *very* political soul would work magic with his squabbling federation partners.

He may have appeared uninspiring on Parliament Hill, at least to himself. But, by then, after a dizzying learning curve, King was weaving his web.

After a private morning prayer and a perusal of his Bible, Mackenzie King took his oath of office at mid-afternoon on December 29, 1921. The new prime minister was a forty-seven-year-old political economist who had been a senior civil servant and a labour consultant. He viewed himself as a social reformer. An intensely private man, he had few friends. In his diaries, he critiqued his faults and rationalized his ploys, leaning heavily on signs from his ancestors and the Almighty for guidance. Wherever King's advice came from, his contemporaries underrated him to their peril.

He was drawn to the West, which he regarded with "a complex mixture of genuine sympathy, self-deception and political expediency."[4] He romanticized the pioneers and clung to the Christian dream of a moral life on the land. He also wanted to entice the Progressive MPs, who held thirty-nine of the forty-three Prairie seats, into the Liberal fold. So he reached out to Westerners. In last-minute negotiations on the chilly winter day before his ministers took their oaths, he had lured former Alberta premier Charles Stewart into his cabinet as interior minister, even though Stewart did not yet have a seat in the House of Commons. King paid attention to such gestures and symbols. At the swearing-in ceremony, he presented each minister with a Bible. Then he signed their official appointments with the last pen that Sir Wilfrid Laurier had used.[5] He was proud, nervous and eager to get started.

So it was not surprising that, after only seven weeks on the job, with the cheery insouciance of a newcomer, King wrote to the three Western premiers—Alberta premier Herbert Greenfield, Saskatchewan premier William Martin and Manitoba premier Tobias Crawford Norris—with an offer on resource control that they would surely not refuse. King and his colleagues, and that certainly included Interior Minister Stewart, were "very desirous of having this long standing question settled."[6]

Times had changed. Ottawa's reasons for continued control were "not necessarily sound . . . when the three Provinces have reached maturity."[7] (It was characteristic of the cautious King that he added the words "not necessarily.") Previous negotiations had floundered because the other six provinces had insisted upon compensation if the West's demands were met. But surely those other provinces were mistaken, King remarked disingenuously, when they asserted that the West wanted its resources along with continued subsidies in lieu of those resources. Perhaps the "earlier claims" of the Prairie provinces had provided "some warrant" for that belief. Everything could be resolved, however, if the Prairie provinces would take over resource control, and "surrender the subsidy."[8]

It was a classic King ploy, spinning away from the past, smoothing the path to the future that *he* envisioned, and, with any luck, charming the Progressive MPs along the way. But the letter was also a clear indication that he did not understand the historic competitiveness over Western resource control: he assumed that the other six provinces would drop any objections if the subsidies ceased. King even declared that there was little to discuss, as the money that Ottawa had pocketed from the resources and the money that it had spent on administration were "probably fully balanced."[9] If the provinces agreed, the deals could be done quickly, and Parliament could ratify them. If the three governments objected to this quick-and-easy method, an independent tribunal could rustle through the accounts, and award compensation, which could include sums that *the Provinces might owe the Dominion*.

It was a bold, almost brazen bid to end more than fifty years of squabbling. The Gang of Three, so unified at the November 1918 Conference, had split in late 1920 over what constituted an acceptable settlement. The members now accepted that the terms of any future settlement would likely be different for each of them—because Ottawa had used their resources in different ways. Ottawa had created Manitoba in 1870, and had exploited Manitoba's resources for thirty-five years before the creation of Saskatchewan and Alberta. Ottawa had also used a different formula to calculate Manitoba's subsidies until 1912, when Borden had taken back control of the province's swamplands and brought its subsidy formula into line with those of its new neighbours. Such differences would become troublesome.

King's blithe proposal attracted no immediate takers: Ottawa had tapped the West's resources for too long to settle the problem with the scrawl of a pen. But the prime minister had piqued the Prairie premiers' interest.

In early March, on the day that Parliament opened, King was anxious. As a strong wind howled outside his window, he read and reread his speech. The words blurred before his eyes. He tried to write letters but he could not concentrate. He wandered from his rooms in the Roxborough Hotel to his House of Commons office, where he found a large new table with no drawers for papers or pencils. All was not right. After the ceremonies in the Senate, he walked into the House of Commons where, he later wrote, "I found it difficult to speak, to get my voice & think on my feet."[10] By evening, he was exhausted: "I pray God I may have strength to go on in a right spirit."[11] He had 117 seats in the 235-seat House of Commons and had to watch his every step (he could generally rely on the support of the three Labour MPs and one Independent MP). Self-doubt gnawed at him. But he had work to do.

Perhaps unsurprisingly, given his precarious minority position, the first premier to reply to King's invitation was Manitoba premier T..C. Norris. Although he "very keenly appreciated" the prime minister's interest, the once affable auctioneer was almost as peevish with King as he had been with Meighen.[12] Manitoba could not simply ignore the past and pretend that the ledger of Ottawa's revenues and expenditures evened out nicely. Norris demanded compensation for the *true* value of every scrap of land and resources that Ottawa had handed out since 1870 "for the purposes of the Dominion."[13] The federal government had given away so much land, while pocketing so much customs revenue because of the new arrivals, that Manitoba had been "literally impoverished" by Ottawa's giveaways.[14] As a parting shot, Norris reminded King that any potential deal would have to be endorsed by the Manitoba Legislature, so it had better be good. It was a fierce opening gambit.

Alberta premier Herbert Greenfield was more gracious because his government needed the money. He first dismissed the demand that Norris had made when Meighen was in power: Ottawa had *clearly* accepted the principle of provincial resource control because Meighen had offered to transfer the resources—so the parties merely had to agree upon the terms. But Ottawa could not simply assert that its revenues and its costs evened out. Greenfield wanted an independent arbitrator, and the accounting "had to be wide enough" to include the value of what Ottawa had given away "*for the benefit of the Dominion as a whole.*"[15] If Ottawa agreed, he would not ask for continued subsidies.[16] That sounded more promising.

In Saskatchewan, Premier William Martin was reeling. His Liberals had easily survived an election in June 1921, partly because he had promised to keep his distance from the federal Liberals. He could curb his caucus, but he could not contain himself. On the brink of the December 1921 federal election, Martin had endorsed a Liberal candidate, and trashed the Progressive Party for its narrow focus on farmers. When the Progressives took fifteen of the province's sixteen federal seats, Martin was in the glue. (Ironically, the Liberal that he had endorsed won.) The United Farmers of Alberta were in power to his west; the United Farmers of Manitoba were threatening Norris's minority government to his east. The farmers in his province now distrusted him, and on April 4, 1922, he stepped down. Three months later, he would accept an appointment to the Saskatchewan Court of Appeal, where he would work with Chief Justice Sir Frederick Haultain. Politically, it was a very small West.

So it was Martin's successor, Charles Dunning, with his strong connections with the farming community, who answered King's letter. Born in England in 1885, Dunning had arrived on the Prairies as a penniless teenager. He had survived as a transient farm labourer until he secured a homestead in southeastern Saskatchewan. Struggling to improve his life, Dunning had joined farm organizations, and he had honed his business skills as general manager of the farmer-owned Saskatchewan Co-operative Elevator Company. By 1916, he was in the Legislature; by 1922, he had been a very capable treasurer for six years. As commodity prices plummeted, farmers knew that he was a sympathetic ally—and that was a formidable political asset.

Dunning reached out to King only six days after he became premier in early April of 1922. He was pleased that King had contacted *only* the three Prairie premiers who had previously been "in the position of, in reality, negotiating with the other Provinces."[17] But Saskatchewan could not ignore its past: Ottawa had handed out millions of acres to compensate the Canadian Pacific Railway for construction in *other* provinces, including more than 4.5 million acres in the three Prairie provinces for tracks that were actually built in British Columbia. Like Greenfield and Norris, he would welcome independent arbitration. But no tribunal could just examine Ottawa's revenues and expenses. Saskatchewan wanted compensation for those resources "alienated for the general advantage of Canada."[18]

It was a start.

The West had so many problems—including labour unrest at those treacherous coal mines—that they often overwhelmed the discussion of resource control. In Alberta alone, there would be 3,300 accidents every year between 1920 and 1924.[19] The stories were haunting. Ukrainian immigrant Nikola Wirsta had left his family's farm in Galicia in 1910 because young men from his village had written about their good salaries and better lives in Canada. He had worked on the railroads in Saskatchewan and done construction work in Winnipeg. He had threshed grain and had become a miner. But during wartime, he had almost died in a cave-in, and in 1920, he had fought a terrible fire in the mines at Cadomin, just east of Jasper National Park. "We worked sixteen hours a day in that inferno," he would recall, "with death at our backs every minute."[20] The miners had finally given up, abandoned their firefighting, and opened a tunnel into the mine from the other side of the hill.

It was not surprising that in 1922, fed up with such harrowing work, the Cadomin miners went on strike. Premier Greenfield deplored the turbulence, sympathized with the mine owners, and fretted ineffectually. After several wretched months without pay, Wirsta went back into the mines—probably as a strikebreaker—where he "crouched and crawled on all fours digging coal."[21] In 1925 he would be seriously injured in a mining accident. Unable to walk for two months, he ended his mining career. "My legs have been weak ever since," he would muse in 1942, when he owned a beer parlour and a hotel. "I have lived through many harrowing experiences in Canada. . . . What intrigues me now is how a human being can endure so much."[22]

The Prairies struggled during the early years of this new postwar age. Many coal mines were losing money, despite their cheap labour and their tolerance for unsafe conditions. Wheat prices were pitifully low, and federal–provincial efforts to create a new Wheat Board were stalled.

Farmers across the West could scarcely make their fraying ends meet. Provincial governments were still saddled with pre-war obligations, including the interest on those defaulted railway bonds that earlier governments had so carelessly guaranteed. Alberta's total spending on railroads drained an astonishing $6.7 million from the Treasury in 1922.[23] Bootleggers were running wild, flaunting Prohibition, which many Westerners now ignored. In September 1922, two bootleggers would shoot and kill an Alberta police officer—an event that would briefly unnerve the scofflaws. But although the two killers were eventually hanged, the illicit trade would remain brisk. It seemed as if that endless Prairie sky was falling.

Eager to please the Progressives, Mackenzie King hosted the three Prairie premiers in Ottawa in mid-April. Because the three recalcitrant leaders were no longer bargaining as a team with a common position, King exploited their differences. Saskatchewan premier Dunning and Alberta premier Greenfield wanted compensation for the resources that Ottawa had used before 1905. King wrestled with this notion: Saskatchewan and Alberta had a "certain moral claim" because Ottawa had handed over lands to the Canadian Pacific Railway "just before" provincehood; but Ottawa had also spent lavishly in the West before 1905.[24] He took Dunning and his agriculture minister to dinner at his home in the "The Roxborough," an elegant eight-storey apartment building that he cherished for its "quiet and comfortable atmosphere [and] notes of beauty and refinement."[25] The next morning, he rejected their request. Dunning would not budge from his demand for pre-1905 compensation. Greenfield "held out but not so strongly."[26]

On that same day, April 21, 1922, King and Manitoba premier Norris concluded a hasty pact. King agreed to examine Ottawa's use of Manitoba's resources since those long-ago days of the Riel Resistance in 1870. If Manitoba and Ottawa could not agree on specific questions, those issues would go to arbitration, and both governments would have to ratify the deal. Norris was amenable because he faced an election, and he needed good news. The premier left for the West that evening, bragging about the advantages of his partisan ties with King. That would be a costly mistake:

Manitoba farmers were seething about Ottawa's high freight rates and tariffs.

Alberta premier Greenfield desperately wanted a deal himself. As Norris boarded his train to Winnipeg, King took Greenfield and his coolly competent attorney-general John Brownlee to dinner at The Roxborough. The prime minister warmed to Greenfield, who seemed "a good sensible fellow, [a] broad-minded honest farmer, no pretensions."[27] Host and guests all recognized that they were novice leaders, as was Saskatchewan premier Dunning. King drew comfort from that notion, and two days after that dinner, he decided, "I am beginning to get a better grasp of questions coming up & not to feel the same anxious concern."[28]

Greenfield and King seemed close to an agreement. Greenfield offered to settle if Ottawa paid for the Alberta lands that it had given away to the railroads for tracks in *other* provinces between 1900 and 1905, and King regarded this offer as "a very fair basis of settlement." But he was afraid that his Maritime cabinet ministers who were "much narrower" than the Quebecers would oppose it.[29]

His fears were justified. Three days after his last talk with Greenfield, King ran into a buzzsaw: seventy-three-year-old finance minister William Fielding from southwestern Nova Scotia. As an advocate of Maritime rights, Fielding objected vociferously, and King still did not understand enough about the nation to know why. "[We] could get nowhere with Fielding who is like a dog in the manger, when it comes to making any allowance on an equitable basis," King complained. "You would think Alberta was out to rob N.S. . . . We have lost a good chance to do a good piece of work."[30] The Maritime habit of linking its grievances with the West's resources baffled and annoyed him. The deal with Greenfield was off.

Two days later, King visited his summer property at Kingsmere in the Gatineau Hills, twelve miles from Ottawa. He loved that land. It was his refuge from Ottawa's social gatherings and the constant political importuning. It had been a busy week, including his meetings with the premiers, a speech in Montreal, hasty cabinet gatherings, "everything at once." The day was bright and beautiful, and he decided to buy the land adjoining his lot. He would offer $1,000 or maybe even $1,200, which was "quite a sum." But, now that he was prime minister, he could see "greater certainty in the future."[31] He could relax, and even splurge—comforting himself with the thought that he had survived. In the privacy of his diary, he did not reflect on the federation's ills. He had little idea that regional identities almost subsumed any national identity—and he did not realize how much the regions defined themselves by their grievances.

King's deal with Manitoba was as tentative as Premier Norris's grasp on power. In early July, when Norris finally understood how angry his farmers were with Ottawa, he struggled to "put distance between himself and the King government."[32] Now he emphasized how *unpopular* he was in Ottawa. But it was too late: the farmers deserted him. On July 18, 1922, the last member of the wartime Gang of Three, the politician who had fought so stoutly for resource control at the November 1918 Conference and who had split the Gang of Three because of his stubborn stand-off about principles with Meighen in late 1920, was ousted. The former auctioneer could no longer sell himself.

He could not compete with the grassroots forces and the agrarian platform of the United Farmers of Manitoba. The UFM had even won the unlikely support of the business community—because it deplored the Liberals' spendthrift ways. There was only one problem with this astonishing victory: the party was leaderless. The stunned UFM caucus interviewed three candidates, including Thomas Crerar, who had just stepped down as head of the National Progressive Party, and UFM secretary Robert Hoey. Both these men refused. The remaining candidate, the resolutely non-partisan agronomist John Bracken, reluctantly accepted the job. Bracken was the principal of the Manitoba Agricultural College and had no political experience. But he was ambitious. He also "did not want to see the UFM fail or their unique opportunity squandered."[33]

The new premier set out to balance the books, and to figure out his agenda. He would prove to be a competent leader, partly because he embodied a "union of moral simplicity and applied science" that appealed to rural Manitoba.[34] But, as historian W.L. Morton commented dourly, "the abrupt limits of his interests [were] almost as narrow as those of the people he had undertaken to lead."[35] In the beginning, at least, he would be no match for Mackenzie King.

When Alberta premier Greenfield asked for more talks in the fall of 1922, the desultory bargaining resumed. At a mid-November meeting in his Ottawa office, the prime minister laid out three options: Ottawa could return the resources along with a payment equivalent to three years of subsidies; Ottawa could return the resources, and then pay extra cash if its resource revenues outweighed its administrative costs; or Ottawa might even pay for lands it had exploited for reasons that had nothing to do with a province's needs.

The discussions hit a brick wall. The four governments agreed that it would be impossible to get a settlement "with all three combined, and that the provinces [should] be dealt with separately."[36] Bracken demanded the return of the resources along with compensation for the market value of *all* resources. King refused. He repeated the offer that Norris had so happily accepted: Ottawa would calculate its revenues and its costs, and send any disputes to arbitration. While Bracken mulled that offer, King briefed his key ministers. To his astonishment, this time, his own solicitor general, Daniel Duncan McKenzie from Cape Breton, "was very much the 'stumbling block' all the way through. . . . He is very cranky and full of objections."[37] A day later, Sir Lomer Gouin, who had vehemently opposed the West's request for continued subsidies at the November 1918 Conference, also resisted the deal. King acidly pointed out that cabinet had agreed to the very same deal with Norris in the spring, and Gouin was now humiliating him. "The truth is Sir Lomer had quite forgotten what was previously agreed to," he remarked.[38]

But King had not yet grasped that the battle over the West's resources epitomized the corrosive regional jealousies that had plagued Canada for decades. The next day, in full cabinet, when Solicitor General McKenzie again protested, the prime minister erupted. He was tired of ministers who did not pay attention, missed meetings, tried to reverse decisions, and discussed topics *"with which they were not familiar."*[39] He threatened to quit. If ministers did not agree with this offer to Manitoba, they could resign. "It was the farthest I have gone at any time," the prime minister wrote with obvious satisfaction.[40] But he did not realize that Maritime minister McKenzie was perfectly familiar with the issue and was objecting on *his* region's behalf.

There were, however, no deals to be done. Bracken concluded that the two sides were too far apart for arbitration, so King should probably send the dispute to the Judicial Committee of the British Privy Council, Canada's highest court. But King wouldn't do so. Then Saskatchewan premier Dunning observed that it would be better for his province, and probably for Manitoba as well, if Ottawa kept the resources—because the administrative costs were now so high: "It wd. soon pay us to give them the resources."[41] King was stunned.

Dunning shrugged and went home to Regina. His farming community faced huge problems, and his feisty Progressive Opposition viewed agriculture as a higher calling. Perhaps his most formidable critic was Independent MLA Harris Turner, who had lost his sight during the war. Turner had espoused the agrarian cause in the Legislature since 1917. In 1924 he would become Opposition House Leader as well as the co-founder of the *Western Producer* newspaper. Such idealists were formidable threats: Dunning had no time to bicker with King about resource control.

He left in Ottawa a shaken prime minister. King was desperate to conserve cash. His government was borrowing $400 million in 1922, which was "the tragic part of the present business of government."[42] His ministers had gone to the New York markets to raise money that spring, and the venerable Fielding was concerned about the interest payments on Canada's mounting debt. When King realized that the resources were costing more than they contributed, he expediently decided that his bottom line matched the greater good. He owed it to future generations "to close this matter off quickly if at all possible."[43]

But he could not even do a deal with Alberta premier Greenfield—the one who really wanted to settle. King had repeated his offer of resource control plus the equivalent of three years of subsidies, and Greenfield went home to consult his cabinet. On the second day of January in 1923, as temperatures plummeted across northern Alberta, he rejected King's offer as "quite inadequate."[44] But he had a counter-offer: he would settle for compensation for the 6.4 million acres of land that Ottawa had given away to subsidize the construction of railways outside Alberta just before 1905. Ever the negotiator, even when it was in his interests to settle, King retaliated. He was *so* sorry that Greenfield had rejected his terms. His offer was "definitely terminated."[45] The discussion would go back to where it was before their November 1922 meeting.

Greenfield did not stand a chance. The prime minister waited. Two weeks later, Greenfield asked for more talks after the Legislature adjourned. King replied that he would be pleased.[46] And then he left Greenfield to stew. The Western premiers had lost their strength of unity in the Gang of Three, and King was picking them off, one on one, like a hard-line businessman facing a desperate trade union.

Like most Canadians that winter, the prime minister was feeling his way through this postwar world of slow economic growth yet dizzying social change. Prohibition was not working. In mid-February 1923, the captain of a British schooner en route to New Brunswick claimed that armed pirates on a mysterious trawler had intercepted his vessel alongside Long Island, overwhelmed the crew and absconded with more than four thousand cases of Scotch, worth roughly $300,000 at bootleg prices. United States customs officials, who had already forced the vessel out of their waters, were skeptical of this tale. New York City alone had thousands of speakeasies and illegal stills.[47] In Canada, provinces such as Quebec and British Columbia had already backed away from their wartime embrace of Prohibition. Others were questioning that policy, if only because they were losing "millions of dollars of untaxed revenues [that] were pumping through the bootlegging and rum-running networks."[48]

The mid-February weather was dreadful. Ontario residents were hunkering down against a fierce blizzard, praying that the cold would deter the spread of a mild influenza outbreak. In the West, another storm brought "intense suffering": the transportation system in Moose Jaw even shut down because of the snowdrifts.[49]

Meanwhile, the mysteries of the ancient enthralled the beleaguered moderns. In Britain, in an impassioned letter to *The London Times*, Sir Rider Haggard pleaded for respect: scientists should photograph, examine and model in wax the newly discovered mummies of the Pharaohs, including Tutankhamen, and then replace them in the Great Pyramid, which should be sealed with concrete. Perhaps Egypt was so tantalizing because the modern world was unsettling. In Cumberland, British Columbia, after a disastrous explosion that killed more than thirty people, miners at a mass meeting decided not to return to work until management improved underground safety and expelled "Orientals" from their underground jobs. *The Globe* interspersed such news amid nuggets of the familiar. It carried "The [syndicated] Adventures of Raggedy Ann and Raggedy Andy," and devoted extensive coverage to the Toronto conference of North American Presbyterians.

The prime minister, however, was becoming more confident. "Until 1923, Mackenzie King had been more of a conciliator than a party leader," his biographer Neatby observed. "[Now] his authority within the party was growing."[50] King was steering his fragile majority through talks about tariffs, the construction of the Hudson Bay Railway, monetary reform and combines investigations. His cabinet was debating ocean and freight rates. On so many issues, he "was threatened by open revolt if he took action and by the revolt of others if he did nothing."[51] As Neatby added admiringly, "It was no small achievement to be still in office after two years."[52]

The Canada that he governed was incredibly disparate. In Calgary, during those early months of 1923, the Unemployed Central Council was planning a tag day to help its distressed members. Albertans were debating the behaviour of "Today's Young Girls," who daringly went to bowling alleys and smeared "shoe-black" on their eyelids.[53] The *Calgary Daily Herald* was selling up-to-the-minute maps, perhaps because the first Packard Single-Six touring car in the West was rolling into town. In Brandon, Manitoba, local farmers were boycotting purchases of new farm equipment until the prices dropped. The United Farmers of Ontario were warning Premier E.C. Drury that he should not forsake his party's class-based roots. Meanwhile, a Legislative committee was considering the elimination of four rural seats because so many people were leaving the country for the city.[54] In Nova Scotia, Cape Breton steelworkers were on the verge of a strike. Treasure hunters were digging in vain for Captain Kidd's loot on Oak Island, off the South Shore. More ominously, in Halifax, a Conservative MLA was demanding a referendum on secession from Canada, and the creation of an independent self-governing British Dominion—because Ottawa's high freight rates violated the spirit of Confederation.[55]

King really did not comprehend Atlantic Canada. For decades, Maritime governments had complained about low subsidies, their diminishing political influence, and Ottawa's offhand neglect. But Maritimers now nursed a new and dangerous grievance. Regional manufacturers had survived Ottawa's high tariff walls because the Moncton-based Intercolonial Railway had provided low-cost transportation since the late 1870s. Manufacturers could compete in Western and Central Canada because their freight rates were twenty to thirty per cent lower than those in Ontario.[56] Capital investment in Maritime manufacturing had actually *quadrupled* between 1900 and 1920.

When the war ended, however, the federal government had combined the nation's struggling railroads into one entity. It had switched the flagging Intercolonial's head office to Toronto, replaced many senior executives and clamped the revered Maritime railroad under the jurisdiction of its Board of Railway Commissioners. When Central Canadian manufacturers and Prairie farmers had demanded lower freight rates like those on the Intercolonial, Ottawa had simply hiked Maritime rates. Worse, in 1920, the railway commissioners had raised *national* freight rates by forty per cent. They had also eliminated special freight rates for products such as Caribbean sugar, which were unloaded at Eastern ports. Between 1916 and September 1920, Maritime freight rates had risen between 140 and 216 per cent.

Atlantic Canada seethed with resentment. Merchants were devastated. The protest against rising freight rates united labour and business

groups, along with farmers and fishermen, against the rest of the country.[57] In the 1921 federal election, the Liberals had exploited this anger:
King had captured twenty-five of the region's thirty-one ridings in a protest vote against Conservative rate hikes.

Despite that clear mandate, the prime minister had cooled his heels.
He did not "get" the link between resource control and those Maritime
protests against freight rates. In December 1922, Prince Edward Island
premier John Bell, who had ousted the Conservatives in 1919, staked out
his region's "undivided share or interest" in the West—because the Prairie
provinces were acquired as "an asset of the partnership."[58] King had no
right, said Bell, to transfer the lands and resources unilaterally. Even if
Ottawa gave away *only* the resources and no cash, "we could not concur
in the proposition" *unless King also settled Maritime claims.*[59] This was no
flowery address, but an ultimatum from a politician in trouble. Seven
months later, Prince Edward Island voters would defeat Bell and swing
behind the Conservatives.

Nova Scotia politics was also changing. In early 1923, Premier George
Murray stepped down after almost twenty-seven years in power. He was
ill, and he was tired. If the veteran Murray could not get King's attention, his successor Ernest Armstrong would have even less luck. But still
he tried. On April 4, the new premier bravely forwarded an Assembly
request that Ottawa cut its high freight rates "at the earliest possible
moment."[60] King blandly thanked Armstrong, and sent the resolution to
his acting railways minister, George Graham, who was a rural Ontario
MP. One week later, King appointed a Nova Scotia protectionist, Edward
Mortimer Macdonald, as his minister without portfolio. That purely
symbolic act did nothing about freight rates, but it did aggravate the
Western Progressives who wanted *lower* tariffs.[61]

It would take a very tough lesson to make King finally listen.

By the spring of 1923, the prime minister was more at home with his
Ottawa responsibilities. In early May, he addressed Liberal MPs and senators at a dinner in his honour. Before the address, he had lived "in fear
and trembling lest [he] might not be equal to the occasion."[62] But his

audience had cheered. Afterward, standing alone on the House of Commons stairs, he sensed a message of encouragement from his parents in the Great Beyond, and took heart in the face of his changing world. In mid-May, as a cold blast from the Arctic swept unseasonably icy rain across Ottawa, Ontario MP Agnes Macphail captured headlines with her contention that men should no longer give away brides at weddings. On Parliament Hill, senators were denouncing the principle of public ownership of railways as "nothing but a return to serfdom . . . going back to the middle ages, to the time of the guilds."[63] There were dark hints that the privately owned Canadian Pacific Railway was behind the senators' campaign against the Canadian National Railways.

In the House of Commons, Finance Minister Fielding, weak from a cold, his voice quavering, tabled a budget that reduced the duty on sugar, the tariff on British imports that arrived through Canadian ports, and the duty on cigarettes. Remarkably, Fielding showed a $34 million surplus on ordinary spending, although railway and merchant marine obligations still added $49 million to the debt. *The Globe* was ecstatic: Canada's "eldest and most widely revered statesman" had delivered a "cheering message to the common people . . . the man who toils and the woman who keeps the home."[64] King was relieved. But he also noted the contrast between Maritime praise for the budget and Western complaints that the tariff cuts were not deep enough. "There is the tragedy,—our East & West [divided] on economic & to a certain extent on racial & religious lines, it is a serious national situation."[65] He was learning.

But he was still taking a hard line with the West on resource control. In mid-May, Manitoba premier Bracken forwarded a Legislative resolution, which asked King for arbitration on the dispute.[66] (That resolution tactfully ignored Bracken's earlier insistence that the two sides were too far apart for arbitration.) Six weeks later, Alberta premier Greenfield also asked for more talks. When the three Prairie premiers could not find a mutually agreeable date in early August to meet King, they agreed to wait until the prime minister returned from the Imperial Conference in London.

Unlike Borden, who was so much happier abroad than at home, King did not want to leave Canada. The prime minister was fretting about domestic problems—"I keep feeling I am out of touch with the country"— and he was rattled. "I am filled with a kind of terror & a terrible sense of my own inability adequately to cope with matters."[67] But he was determined to oppose British plans for a common Imperial defence and foreign policy. His intercession in London was successful, and King returned home in early December.[68] And then . . . well, it was Christmas, and that was no time to talk about resources.

Two of three Prairie premiers, however, had New Year's resolutions. On January 1, Bracken abruptly notified King that he would be in Ottawa within four days, on Friday, January 4. He asked for a "short conference" so that he could go home on Friday evening because he had a proposal on school lands—"which we think will appeal to you."[69] The startled prime minister told Bracken—"on board Canadian National No. 2 East Bound"—that he would somehow find time for him, and the Premier should call his secretary on arrival.

Two days later, Alberta premier Greenfield met with King and seven cabinet ministers. (King grumbled that six of those ministers had not even bothered to peruse the briefing notes from Interior Minister Stewart.) Greenfield offered to settle for the equivalent of six years of subsidy payments, but King stuck to his original offer of three years, or roughly $1.8 million. To the prime minister's astonishment, former Alberta Premier Stewart, who was supposedly on *his* side, "handicapped us" by observing that an official reckoning would probably show that Ottawa owed $5 million to $6 million.[70] What did King expect? Perhaps he had forgotten Stewart's presence amid the Gang of Three at the November 1918 Conference. The next day, on January 4, 1924, Greenfield again met with King—who would not budge from his offer of three years.

King then dashed into his meeting with Bracken. The cash-strapped Manitoba premier presented his scheme: Ottawa should transfer the school lands and the money in the school lands trust to Manitoba before the two governments concluded any deal on natural resources. King listened, and Bracken went home. The premier had already lost. As King confided in his diary, any discussion of the school lands trust would spark controversy over the size of the Catholic share. He did not want another religious war: it would be "better to settle everything together."[71] Three and a half weeks later, he formally rejected the proposal: the resources should "be dealt with as one."[72]

Bracken was snippy. His government wanted to show progress on *something* "after nearly five years of almost continuous negotiation."[73] Actually, it was *more* than five years since the November 1918 Conference. King once again refused his request, and Bracken then demanded extra money because Ottawa had placed the school lands fund in a low-interest savings account. Otherwise, he warned, he would go to court. King sent that threat to his minister of justice, and forgot about it. Bracken was going nowhere—except down in King's estimation.

King and his government had somehow survived into even more interesting times. Provincial governments were growing in importance: they now delivered social services such as allowances for struggling widows that the Fathers of Confederation could never have foreseen. Protestant ministers were fanning out across English Canada, bringing the Social Gospel message of reform, and ministering to the inner-city and rural poor. In Quebec, the Roman Catholic Church, which controlled social and educational institutions, was struggling with an urban influx of poor workers who had abandoned their subsistence farms. The economy was finally stirring. The boom "came faster to some areas than to others; in some parts of the country, it never came at all."[74] Wheat prices were slowly rising, but although agriculture would remain important, it would never regain its pre-war economic strength.[75] Canada was becoming more urban than rural, and the new jobs were in the cities. The dreams of the postwar Prairie homesteaders were becoming ever more elusive.

The old order was splintering.[76] Industries such as coal now languished, while automobile manufacturing flourished. Middle-class Canadians were on an investment and consumption binge. Central Canada and the Pacific Coast were thriving, and the West was edging its way back to prosperity— although Prairie governments were struggling. The Maritimes and marginal agricultural areas would "not have a good decade."[77] Such inequalities would be a recipe for more regional competition and strife.

King's talks with the three Prairie provinces were almost at an impasse. Saskatchewan premier Dunning had too much on his plate. Bracken was sulking about the rejection of his request for the school lands trust fund: he needed money for education and transportation. Only Greenfield remained in the game. At the end of March 1924, the Alberta premier took another run at King. His Legislature would probably not approve King's offer, but the equivalent of five years of subsidies without any accounting would be a "fair compromise."[78] King was polite . . . and implacable. Parliament would not approve more than the equivalent of three years. "We feel we are going just as far as existing circumstances will justify."[79]

Greenfield was likely confounded when King's government only nine days later presented a budget stuffed with goodies, including $24 million in tax cuts. Ottawa was in the black in its operational spending for the first time since 1913. King had also secured cabinet approval for major tariff reductions on farming equipment—because two key protectionists were sidelined. Finance Minister Fielding had suffered a debilitating stroke before Christmas; Sir Lomer Gouin had resigned in a fit of pique in early January and, to his consternation, King had accepted his departure. The tariff cuts elated Progressive MPs. King was "very proud, very happy—and very grateful to Providence who has guided me despite all my errors & shortcomings & failures." Although he was concerned about a protectionist backlash, "[the budget] may bring us back into power by getting the Progressive forces and ours united."[80] The prime minister readily mixed religion with political calculation: after all, God was surely on his side.

Greenfield was desperate. Many Albertans, particularly the farmers, were heavily in debt. His government had added almost $20 million to its own debt since 1922; it now owed more than $83 million. In 1924, provincial revenues were $14.7 million, but expenditures were $20.9 million. Greenfield was juggling bills for interest on the debt, advances to railways, guarantees to provide seed grain, and the completion of public works projects.[81] A few days after the federal budget, the Alberta premier offered to settle for the difference between Ottawa's revenues and its costs since 1905. In early May, Greenfield and Attorney-General Brownlee met with King in Ottawa.

This time, the prime minister was cocky. He produced his calculations, and offered $1.8 million once again. Greenfield went home to mull the figures: his auditor regarded Ottawa's tally of its administrative costs as suspiciously high. Greenfield parried with a request for $3.2 million, or almost double Ottawa's offer. King asked his ministers to examine Greenfield's arithmetic. The Alberta premier could not win. On November 19, 1924, Greenfield and Brownlee went back to Ottawa and agreed on a deal for roughly $1.8 million. King had set the terms. Forced into a corner, Alberta accepted the equivalent to three years of subsidies. Now the two parties had to decide which services, such as forestry maintenance, were provincial, and which should remain federal. By late 1924, the premier anticipated "no serious difficulty."[82]

He counted his cash too soon. On the other side of the country, on June 25, 1925, Nova Scotia voters trounced Armstrong's Liberals: almost sixty-one per cent of the voters swung behind Conservative leader Edgar Rhodes, who had proudly run on a Maritime Rights platform. This was akin to a revolution: the Liberals had been in power in Nova Scotia for forty-three years. Mackenzie King was no fool, and five days later, he delayed the transfer of resource control. Given the "national importance and magnitude" of the transaction, he told Greenfield, each federal ministry would have to weigh the deal's implications *for other provinces* and for the Dominion itself. Although King politely promised to speed up the process, "it is not possible to say just how soon our expectations may be realized."[83]

Greenfield surely despaired, if only because his own future was on the line. As an election approached, the members of his United Farmers of Alberta caucus were worried that Greenfield was not up to the job. After all, in 1923, in a startling turnaround, Ontario voters had ousted Premier E.C. Drury and his United Farmers of Ontario, putting the business-like Conservative Howard Ferguson in power. What if a similar anti-agrarian sentiment erupted in Alberta? The UFA's opponents might depict Greenfield as "well-meaning, perhaps, well-intentioned, possibly, but despite all, not nearly competent enough to be trusted with the serious business of governing Alberta."[84] The caucus members turned to Brownlee, who initially spurned them. In the fall of 1925, they enlisted the moral support of that esteemed UFA president, Henry Wise Wood. Brownlee would only accept the job if Greenfield personally assured him that he would hold no grudges. On November 23, 1925, Brownlee would become the premier of Alberta's financially sinking ship.

By then, King had his own troubles. Throughout the summer, he and his cabinet had anxiously weighed the timing of an election call. The economy might be even stronger in 1926, but his Liberal MPs might be exhausted after Parliament adjourned. There was discontent among voters, and the mood was uncertain. True, in early June, Liberal premier Dunning had won a new term in Saskatchewan, but in late June, Nova Scotia's Liberals had fallen, and Premier Rhodes now promised to stop the flight of young people to other regions in search of better jobs. In mid-August, the New Brunswick voters also ousted the Liberals; the new Conservative premier, John Baxter, was a strong advocate of Maritime Rights, including freight-rate reductions. King scrambled to bolster his regional appeal. He tried to convince Dunning to leave his premier's perch in Saskatchewan and join his cabinet. But Dunning stayed put.

In King's cabinet, everyone had a view, and no one could agree. The discussions were agonizing. King retreated to Kingsmere, and cleared the underbrush. Many cabinet members, he decided, were "like barnacles rather than fighters."[85] He resolved to throw the dice, and called an election for October 29. "To wait is to be surrounded by intrigue for months, to have life made a sort of hell by almost impossible situations."[86] He even persuaded former New Brunswick premier Walter Foster to become secretary of state—and to run in Saint John. (Foster would lose, but King would eventually appoint him to the Senate.) On the campaign trail, he boasted that "taxation had been reduced, the budget had been balanced, and the national debt slightly reduced."[87] But he was running against two strong parties—the Progressives and Arthur Meighen's Conservatives—in a fundamentally divided nation.

The results were confounding. The Liberals won 101 seats in the 245-seat House of Commons; eight cabinet ministers lost their ridings; the prime minister was also defeated in his own riding to the north of Toronto. The Liberals slipped from twenty-three seats in the Maritimes to six. But they soared from three seats in the Prairie provinces to twenty, including fifteen in Saskatchewan. The Conservatives now had 116 seats—seven short of a majority and fifteen more than the Liberals. Meanwhile, the Progressives had only twenty-four seats, down from sixty-three. Their heyday had passed. There was one independent Liberal, along with three Labour MPs.

Now what? King decided to remain in office until he could test the will of Parliament, which would open on January 8, 1926. Then he set out to remake his image by remaking the regions. He finally secured Saskatchewan premier Dunning's agreement to come to Ottawa as minister of railways and canals. He drafted a Speech from the Throne that would appeal to the Progressives, while Meighen rebuffed their advances. He wooed reformers by quietly promising to introduce old-age pensions. On January 9, 1926, *before* he even had a seat in Parliament, King and Alberta premier John Brownlee finalized that deal to transfer resource control—and on Friday morning, January 15, 1926 at 12:56 a.m., King's Liberals squeaked through a crucial vote of confidence with a majority of three. In early February, before Dunning officially left Regina for Ottawa, the Legislature formally asked him to negotiate the transfer of resource control "without further delay, on terms fair to the Province."[88] So far, so smooth.

On February 15, King won a by-election in the riding of Prince Albert, Saskatchewan. He had stayed in his office until late that evening, waiting for the returns, and when he learned of his victory, he headed into the Liberal lobby of the House of Commons where MPs carried him on their shoulders. He spoke to them "of Quebec & Saskatchewan being as the supporting structure, of a great bridge making for unity in the

Dominion."[89] It was his new vision of regional harmony. He walked home, kissed his late mother's statue, said a prayer before her portrait, and then randomly opened his Bible. The message seemed clear, and he underlined the text: "For unto whomsoever much is given, of him shall much be required; and to whom men have committed much, of him they will ask the more."[90] King prayed for the grace and power to succeed.

But it would take more than the Almighty to keep him in office. He would need to charm and bully and bribe MPs and his fellow first ministers. He was, after all, reliant on the tentative support of the Progressive Party to save his fragile minority government.

The West was awakening from its slump. In Alberta, the Scottish Immigration Aid Society was ordering one million feet of lumber to build houses and barns for the one hundred immigrant families that it was shepherding into the province. An Alberta legislator demanded better roads for the existing settlers, and warned that half of all newcomers would desert their lands because the agricultural conditions differed so greatly from their homelands. Most Albertans ignored him in all the excitement of the modern. Australian explorer Captain George Wilkins was planning to fly his Fokker plane from Alaska over the Ice Pole to Norway, and Western Canadians were breathlessly following his every move. (Wilkins would eventually make it in April 1928.)

Western firms were clamouring to harness the region's immense waterpower.[91] On the Saskatchewan–Manitoba border, the Hudson Bay Mining & Smelting Company was chipping out copper, gold, zinc, silver and cadmium. Miners had plunged into the densely forested northern shore of Lake Athabaska.[92] A lifestyle was changing: in Alberta, "cars, roads, and the telephone broke down rural isolation. . . . The end of pioneering in most regions and growing farm mechanization gave farmers more time for social activities."[93] Families savoured their new freedom on the bumpy roads.

Western newspapers were now bedecked with sensationalist tidings. The Maharajah of Indore was stalling British efforts to haul him before an inquiry into the attempted kidnapping of a dancing girl. American immigration authorities would not allow an adulterous British countess to come ashore in New York, but they did admit her partner, the Earl of Craven. In Nova Scotia, a recluse who had spent forty-eight years in the forests after the woman he loved had rejected him, died. The Age of the naughty Flapper reached the West. The *Calgary Daily Herald* warned women that "Excess Fat is not in fashion," while a Calgary haberdashery advertised the "Student Prince" hat with its "racy air," named after a popular operetta that had debuted on Broadway in 1924.[94] There was even an unsettling glimpse of the future on those front pages, as an Italian general who had taken refuge in Nice warned that Fascist leader Benito Mussolini was a murderer.[95]

In Ottawa, King had finally learned the lesson of the November 1918 Conference. His Alberta deal could not win Parliamentary approval as a solitary achievement. He could not pick off the Prairie governments, one by one, like a federal sharpshooter. He needed the goodwill of every region to mend the inequalities of the past. He needed a strategy to modernize the federation, and would have to move incrementally, tackling each region simultaneously, if he wanted peace and popularity. He was determined to settle the natural resources issue, to hammer out the deals that Borden would not make, and Meighen could not make. Alternately daring and devious, he set out to save the nation. He would move slowly, of course.

"THIS MAKES A VIRTUAL RECONSTRUCTION OF CONFEDERATION": THE DEAL, THE DEPRESSION AND THE GREAT DISCOVERY. 1926 TO 1947

—

IN LATE JANUARY OF 1930, Mackenzie King desperately needed a favour. "My dear Rogers," the prime minister purred in a note to Norman Rogers, who had been his trusted private secretary for three years before leaving to teach political science at Queen's University. The prime minister was in a rush. Surely Rogers could quickly turn out a pamphlet on *The Dominion Government and the Provinces*, which would boast about "what we have done since coming into office in the way of removing all sources of friction" in the federation.[1] King cajoled: he knew Rogers was busy, but he understood the file. He pleaded: Rogers would know "how all-important" the pamphlet would be during the upcoming Parliamentary session when the resource agreements between Ottawa and the provinces would be presented to the House of Commons.[2]

King had worked miracles. Resentments about resource control had multiplied like prairie gophers since Sir John A. Macdonald and Sir George-Étienne Cartier had brought Canada to life, and ushered Rupert's Land and the Territories into their National Dream. King had finally understood the strength of those grudges during the 1925 election, which had left him with fewer seats than Arthur Meighen's Conservatives. Since then, if only to save his own political skin, he had tackled those regional resentments. He had finally found the cash to buy off the complainers. He could finally stop handing out Western land because the days of the pioneer farmer were almost over—and the administration of the forestry and mineral resources was costing more than the royalties. He had turned a pivotal federal–provincial conference into a love fest. He had soothed the East, which claimed a stake in the West's resources, and had mollified the West, which resented the East's claims and its own second-class status in the federation. He had dreamed an implausible dream of regional harmony, and it was about to come true. He listed his deals, and then he crowed: *"This makes a virtual reconstruction of Confederation."*[3] Essentially, he had herded provincial cats.

Every premier at the November 1918 Conference would have been proud of King's work. He had inched ahead, scheme after scheme, appointing Royal Commissions to sanction his plans and playing hard-ball to isolate resistance. He had been blessed by being able to tap the prosperity of the Roaring Twenties to make deals with the provinces who had remained aggrieved after the November 1918 Conference. He had wheedled and stalled and charmed and blackmailed his fellow first ministers, all the while certain that he was fulfilling his God-given des-tiny. Despite his thousands of pages of diaries, which could rationalize almost anything for *his* greater good, King would never truly know him-self. But, by 1930, he recognized that he was an *unheralded* Father of Confederation. He had moved so slowly and taken such exquisite care with each accord that few Canadians realized what he had done.

Mackenzie King had always regarded compromise as an evangelical virtue. In early 1926 he tackled his minority government with a formi-dable combination of piety and politics. One day, he resolved, "Above all else I want to be a good man."[4] A day later, he could pray, pore over the picture of his departed mother, read his Bible, sing a hymn, greet well-wishers and then settle in for a dead-serious talk with two cabinet ministers. He might tartly observe that a fellow Liberal "always stresses a policy to fight for"—and it was not a compliment. He might dismiss down-and-dirty strategizing in his diary ("This is the sordid side of poli-tics which I loathe") but he did not miss a trick.[5] King could always bring low cunning and high principles to his quest for victory.

The prime minister had clinched his first resource deal on Saturday,

January 9, 1926, when he finally signed the agreement that he had made with Alberta in late November 1924. Then, he had held all the cards. Now, shaken to his political core, facing a vote of non-confidence in the House of Commons, he had quickly finalized the legal text of the deal with Premier John Brownlee in his Ottawa office. It seemed a blessing for both politicians. Although Brownlee would have preferred better terms, he now had what his predecessors had long sought. King would have preferred to avoid a cash payment; nevertheless he had bolstered his dicey political status.

But that first deal on resource control—after decades of acrimony, so much strife between the Gang of Three and the Rest of Canada, and so many harsh words between premiers and prime ministers—could not escape the devil in the details. Mere weeks after Brownlee had reached home, Justice Minister Ernest Lapointe raised the issue that dared not speak its name in the early decades of the twentieth century: religious education. Lapointe insisted that the deal could imperil Alberta's funding for Roman Catholic schools, because the province would control the school lands trust fund. The legal text, he said, should stipulate that Alberta would administer its separate schools in accordance with the Alberta Act of 1905, which guaranteed that, where numbers warranted, there would be "no discrimination" in public funding for Catholic minorities.[6]

It was an ill-advised amendment. There was *nothing* in the resources bill that would have removed Alberta's obligations to Roman Catholic minorities. But Lapointe was worried that fiery Quebec nationalist Henri Bourassa, who now sat in the House of Commons as an Independent MP, would denounce the Liberals if the agreement did not explicitly reinforce Alberta's obligations.

King was devastated. He remembered the uproar in the 1890s when Prime Minister Wilfrid Laurier had secured guarantees for French-language schools in Manitoba. He could already imagine the tumult in the West over this federal meddling. "There is the danger of another 'school question' arising in this connection," he confided. "Should such prove to be the case it would be the most embarrassing feature thus far."[7] To King's relief, Brownlee did not initially object to the addition. "For a moment," King wrote, "I feared a really serious situation might arise, and another religious question be on our hands."[8] Everything seemed fine; even Henri Bourassa was "most agreeable."[9]

The harmony did not last. On March 22, 1926, Quebec premier Louis-Alexandre Taschereau kindly warned King that the Alberta deal had already provoked Maritime complaints—which the prime minister should have anticipated after the November 1918 Conference. Former Quebec premier Lomer Gouin had taken a "very strong stand" against any transfer of resource control at that Conference, Taschereau noted, and the Maritime provinces had not abandoned their demand for higher subsidies.[10] Now Prince Edward Island premier James Stewart was grumbling about Alberta's windfall in letters to him and to Ontario premier Howard Ferguson. Stewart even claimed that the other two Prairie provinces would soon make similar deals with Ottawa. "The time has arrived," Stewart had declared, when the other six provinces should demand higher subsidies at an interprovincial conference.[11] Ferguson had agreed. Liberal Taschereau forwarded his letters from those two Tory premiers to King, and asked for advice.

Once again, that plague of Confederation—one-upmanship over control of the West's resources—threatened King's schemes. The prime minister was grateful for the warning. He told his Liberal ally that the premiers were of course free to get together and ask for more money, but that the odds against them were high: the federal government still carried debts from the world war. "The Dominion is not giving up an asset to Alberta, but is being relieved of a liability."[12] This was a crucial assertion. During the last ten years, King insisted, Ottawa had lost roughly $6 million because its costs for Alberta resource administration and subsidies had exceeded its revenues. King advised Taschereau to stall, using the excuse that deals with Manitoba and Saskatchewan were nowhere close to settlement. Taschereau countered that he could not refuse Stewart's request, although he would mention that the prime minister was busy. King could already hear the same old song from the November 1918 Conference: if the Western premiers got their resources along with a cash settlement, everyone should get more cash. No good deed could go unpunished. It was tiresome.

While Mackenzie King pleaded poverty to the scrappy premiers, many Canadians were belatedly revelling in the prosperity of the Roaring Twenties. The "haves" were on a roll. In an unusual full-page advertisement on the front page of *The Globe* in early April, Willys-Overland Fine Motor Cars boasted that it had sold more than $150 million worth of vehicles in 1925, including its Six "66" with "an engine you'll never wear out."[13] In the West, resource development dangled enticing prospects. The Calgary Bond and Security Company was offering one-dollar shares in the Highland Oil Company, which was erecting a derrick in the Turner Valley field. Prospective investors were advised to snap up this bargain "Before It Is Too Late."[14]

In April, as chill weather swept over southern Ontario, fashionable women in Hamilton celebrated Easter in lush furs and new hats. One "smart set" man wore a suit that was the colour of mother-of-pearl, along with lemon-coloured gloves and a flowered vest. The "have-nots" were not so fortunate. Household maids wore "short little, cold little suits," so their noses reddened. "Some of them looked rather pathetic," *The Globe* unkindly reported.[15] Over the Easter weekend, former Methodist minister Albert Edward Smith, who was now an avowed Communist, won the presidency of the Ontario section of the Canadian Labour Party. The have-nots were jostling against the haves.

King knew he had to remake his world. He could no longer shrug off Maritime complaints when Maritimers had ousted three provincial Liberal governments within the last three years. The prime minister would later be frank about his peacemaking priorities: "We began with the Maritimes when they were in a state of insurgency."[16] On April 7, 1926, with the flourish of a proud magician, King unveiled a three-man commission, under the chairmanship of British industrialist Sir Andrew Duncan, to report on Maritime claims. It was "most undesirable," King declared, that Maritimers should believe that their interests were "being knowingly prejudiced."[17] The Duncan Commission would scrutinize freight rates on the former Intercolonial Railway, which the government-owned Canadian National Railways had swallowed in 1919. It would also look at how such federal policies as customs duties affected the three provinces. It was a heady moment: such scrutiny, the *Halifax Herald* reported, would "ensure a return of contentment and prosperity."[18]

Oddly enough, it was not until the day *after* his announcement that the prime minister fully grasped the depths of Maritime rage and envy. King had viewed the grievances as almost intractable theoretical policy issues, which he had to address if he wanted to stay in power. He had not understood the emotion. On the afternoon of April 8, however, he met with a young man from Nova Scotia, who was not identified in his diary but was probably a journalist. "Sinclair of N.S. astonished me by speaking of the strength of the secessionist movement in the Maritimes," King wrote. "It is a sort of council of despair."[19] Maritimers lacked markets, and "their people are leaving for the U.S. largely. There is need for radical change in freight rates & tariff policy."[20] Remarkably, it had taken more than four years in power before this savvy politician got the message.

The prime minister could take nothing for granted. No sooner had he put something in place for the Maritimes than Alberta premier Brownlee demanded the removal of the provisions for religious schools from the resources legislation. The cautious lawyer was spooked: the *Sentinel*, the newspaper of the Grand Orange Lodge of Canada, had charged that Ottawa was forcing Alberta to support Catholic schools, even though education was an area of provincial jurisdiction. (The issue was considered so sensitive that King's go-betweens communicated in code.) The Alberta Act of 1905 remained valid, Brownlee argued, so minority schools were protected. Why didn't King leave well enough alone? The premier was exasperated with an addition that had "nothing to do with [the] Natural Resources Bill."[21]

King was now trapped between Bourassa and Brownlee, who was facing an election in late June. Eventually, Brownlee suggested a ploy to buy time: the two sides should agree to re-examine the transfer of minerals and parks "as a means of delaying."[22] That scheme squelched the Protestant complaints. Meanwhile, Ottawa asked the Supreme Court of

Canada to rule on the constitutional status of separate-school protection in the Alberta Act. If that protection remained valid when the trust fund was transferred, there would be no need for the new clause. Brownlee romped to a healthy majority.

Another petitioner now popped up on King's doorstep. British Columbia premier John Oliver was the only political survivor among the first ministers at the November 1918 Conference. "Honest John" had clung to power by adopting the tactics of his irrepressible predecessor, Sir Richard McBride: he had run against Ottawa. The shrewd Liberal warhorse had warned King in January that if Alberta received control over its resources, he would demand the immediate return of the unused railway lands, including the 3.23 million acres in the Peace River tract, which Ottawa had controlled since the late nineteenth century. Although Oliver conceded that the legality of the transfer was "not open to question," Ottawa was morally bound to return what it no longer needed.[23] Oliver hoped to sell those lands to support *his* railway construction projects. King had replied politely, and promised nothing.

Now, on May 28, 1926, Oliver was back at King's door. On the brink of his seventieth birthday, the premier sent a sixteen-page outline of his province's claims.[24] King likely sighed: here was the familiar recitation of legal history, virtually unchanged since McBride had pestered the harried Sir Robert Borden before the war. King did not appreciate Oliver's garrulous and insistent ways. As he once confided to his diary: "I went to the office quite tired, found John Oliver, Premier of B.C. there, another old man who uses up one's vitality in conversation."[25] Oliver's claims could wait.

King had a minority to nurture and voters to woo. His spring budget had raised the amount of income that was exempt from taxation, and reduced the tariff on automobiles—which pleased farmers on the endless prairies. Even the small-town *Nanton News* in Alberta now sported advertisements for the *"Improved* Chevrolet Coach," a "low-priced closed car that is worthy of your ownership."[26] King's budget had been balanced and the economy was moving forward. He had delivered "four years of careful administration."[27] Despite his talk of wartime debt, which remained high, his revenues were now comfortable, partly because he had mostly avoided federal involvement in such social issues as unemployment.

But King was still dodging his fellow first ministers. When the premiers finally assembled in Ottawa in early June of 1926, they snubbed King, excluding him from their deliberations. The prime minister, in turn, barely acknowledged them. On the first evening, he invited the *Liberal* premiers and their delegations to an informal dinner where he almost certainly explained his views on subsidies. Then he went back to work. He could rely on the powerful Taschereau to squelch any demands for a general increase. Predictably, the nine provinces concluded their conclave on that issue with a wishy-washy expression of "sympathy" for any faltering province: Ottawa should "favourably consider" relief to those underdogs.[28] They did not demand a general increase. King could now afford to be magnanimous when individual provinces, particularly the Prairie provinces, came calling. He had smartly preserved his spare cash.

But the prime minister was also too distracted to deal with those petitioners. He was sitting on a scandal that would shock the idealistic Progressive MPs whom he depended upon for survival: Canadian Customs officers had not only turned a blind eye to smugglers who were exploiting the failure of American prohibition—they were in league with them. In 1925 King had retired his customs minister, Jacques Bureau, who had benignly neglected the outrage, partly because he could not control his own drinking. (A Montreal customs official had even sent whiskey to Bureau, "and the minister's chauffeur was driving a smuggled car he had been able to buy cheaply.")[29] Bureau went to the Senate, and Bureau's more conscientious replacement, Georges-Henri Boivin, had appointed a Special Committee to investigate customs operations. But Boivin had also imprudently delayed the imprisonment of a New Brunswick smuggler whom he eventually sent to jail—and that *apparently* kind-hearted deed would create more scandal.[30]

The Special Committee report, which was tabled on June 18, was a blockbuster with tales of smuggled cars and barges loaded with confiscated liquor that somehow disappeared from customs officers' custody. It was strongly critical of the government, but it did not directly attack Boivin because committee members were divided. The Progressives were deeply troubled, but the Conservatives were elated, and they pushed for a motion to censure the government *and* Boivin. In late June, after more than four days and nights of scathing debate, during which one Liberal MP even left his wife's corpse to return to Ottawa, King managed to adjourn the House. He had dodged the vote on that motion of censure, although "the customary veneer of parliamentary decorum had disappeared."[31] What to do now? The motion would still be hanging over King's head when the House returned. Even if he survived the vote, the damage would be done.

The prime minister reluctantly opted for a ploy. He asked Governor General Lord Julian Byng to dissolve Parliament, which would trigger an election. To King's astonishment, Byng refused. King argued that, since he had not been defeated in the House of Commons, the governor general was obliged to accept his prime minister's advice. But Byng was disturbed by King's call for dissolution to avoid a motion of censure, and he argued that Conservative leader Meighen, who had won the most seats in the October 1925 election, deserved the chance to form a government. In a shocking development, King resigned, and Meighen took power on June 29, 1926. Early on the morning of July 2, the Tories lost a crucial vote when King challenged the legal right of Meighen's ministers to administer departmental funds. (New ministers traditionally resigned and ran again in by-elections.) Meighen in turn asked Byng for a dissolution.

The rest would become astonishing history. Meighen campaigned for clean government. King campaigned against Byng, charging that he had meddled in Canadian affairs. On September 14, 1926, King won 116 seats, which was not a majority, but King was secure in power because there were also ten Liberal-Progressive MPs—who were virtually Liberals—along with two Independents, three Labour MPs and twelve Progressives. Although the Conservatives snared slightly more of the popular vote than they had held in 1925, they were reduced to ninety-one MPs. The Liberals gained three seats in the Maritimes, one extra seat in Saskatchewan and three extra seats in Manitoba, but they lost one in Alberta. British Columbia defeated two of their three MPs. There was a regional message in those results: most Liberal and Liberal-Progressive gains were from Ontario and Manitoba, where the Progressive vote had slipped away.

King was back on top. A month after the election, he headed to the Imperial Conference in London, where he reinforced Canadian autonomy and preserved Canada's presence within the Commonwealth. In effect, King staked out equality of status for the Dominions: the parallels with the West's complaints about constitutional inequality were striking. He also ensured that communications between Ottawa and the United Kingdom would no longer go through the governor general, who would in future only represent the monarch.[32] That was a swipe at Viscount Byng of Vimy, who had returned to England in late September. King's interventions abroad were a triumph, quelling any lingering doubts about his leadership.

Once back in Ottawa, the prime minister rejoiced in the snow and "the great open spaces of Canada. . . . I have been smothering in England,— the environment of wealth & ease is far from a healthy one."[33] Now he had the time and the money to soothe Canada's regional malcontents. Sir Andrew Duncan had produced his report on the Maritimes in September 1926, although it was not tabled in the House of Commons until December 10, months later. Predictably, in submissions and at commission hearings, the Maritime provinces had once again staked out their claims to the West's lands and resources. They had also complained that although the Prairie provinces were able to tap their school lands trust to finance education, those in the East had no similar fund. Duncan was conciliatory. The Maritimes had not revived this old issue out of a "spirit of antagonism to the western provinces," but to prove that Ottawa's financial deals with the West were "much more liberal and fair" than its deals with the East.[34] After those attempts at soothing words, the British industrialist simply ignored regional rivalries as of no concern to him.

His diagnosis was stern—the Maritime provinces were partly to blame for their woes—but his thirty-nine prescriptions were generous. He called for an *interim* lump sum increase in the federal subsidy, pending in-depth federal examination. He also called for immediate freight rate reductions of twenty per cent and renewed transportation subsidies for Maritime coal. Although King worried that Duncan had "gone too far" on subsidies, the report suited his conniving soul: "All I need to do is to stand firm on this report, and count on getting back Maritime support to keep us strong in future years."[35] Now that the prime minister "got" regional divisions, the regional would always be political.

In the spring of 1927, however, King's cabinet objected to the cost. Maritime experts would later charge that the cabinet "changed Duncan's program for Maritime rehabilitation into a plan for Maritime pacification."[36] King raised the region's annual subsidies by $1.6 million—"but presented them only as temporary grants conditional upon Maritime good behaviour"—and cut most freight rates by twenty per cent. He ignored many other proposals. But he shrewdly declared that he was adopting Duncan's measures "virtually in their entirety."[37] At the time, few Maritimers noticed the gap between what he said and what he did. It was a huge political win. The Maritime Rights Movement would slip into decline: if Maritimers wanted to keep those temporary grants, they had to play ball with a master manipulator.

King puttered through the spring, happily improving his beloved property at Kingsmere. He decided to hire a caretaker, along with his Russian-born farmhand, and he was also buying sheep, even though he was surprised at their huge size: "They were too less beautiful than I had thought sheep were."[38] (He would sell his flock in 1928 when he learned that dogs and diseases preyed upon them.[39]) He was also buying more land, and the transaction brought out his prejudices. King's neighbours at Kingsmere, along with their lawyers, had him over a barrel whenever they hinted that "undesirables" were interested in their lands. The prime minister now hurriedly purchased a large plot from a frail neighbour embroiled in a lawsuit. He was worried that the elderly woman would collapse, and the property would be sold to meet legal claims before he could finalize his deal. "The greatest danger & menace is a sale to Jews, who have a desire to get in at Kingsmere & who would ruin the whole place."[40]

King could connive and scheme; he could bully and charm and bribe; and he could usually rationalize his behaviour. He was quite capable of detecting messages from the Beyond that would put a moral gloss on his deeds. But he rarely bothered to conceal his bigotry. It would always be difficult to separate King's unpleasant personality traits from his undeniable accomplishments.

The prime minister was pecking away toward regional harmony. In early March, he appointed a Royal Commission to examine British Columbia's demand for the return of its railway lands. On the advice of Interior Minister Charles Stewart, the commissioner was former Saskatchewan premier William Martin, who was now a Saskatchewan Court of Appeal judge. It was still a small world: former Alberta premier Stewart and Martin had been colleagues in the Gang of Three. After the frustrations of the November 1918 Conference, Martin would understand the fierce emotion behind British Columbia's land claims, even though that province's legal case was shaky. As King had done with the Maritimes, he and Stewart were using a Royal Commission to rally support for settling Oliver's concerns: an authoritative report would defuse any objections from the Rest of Canada.

Reform was in the air. In late March, King's government finally approved old-age pensions of up to $20-a-month for needy British subjects who had reached seventy years of age. The Senate had defeated King's earlier pension legislation, arguing that such payments could ruin the moral fibre of the elderly. But now that King was securely back in power, they prudently relented. However, Ottawa would only issue the pension cheques when a provincial government consented to share the costs. The first province to agree was British Columbia: Oliver, now gravely ill with cancer, left his sick bed to shepherd the deal through the Legislature. As his biographer James Morton noted, the politician who had worked as a child in British lead mines "no doubt had a fellow-feeling for those of his years who were thrown on the cold mercy of a destitute old age."[41]

The resource talks with the Prairie premiers inched along. In mid-April, the Supreme Court of Canada affirmed the validity of the separate-school guarantee in the Alberta Act. Justice Minister Lapointe had delayed the transfer of natural resources —and panicked King and Brownlee—for nothing. Meanwhile, Manitoba premier John Bracken asked for more talks. The prime minister stalled, unwilling to reward Bracken with a deal before the provincial election: he was pinning his hopes on the new Liberal leader, Hugh Robson, who was a prominent Winnipeg lawyer, an occasional ditherer and a perennial optimist.

Most Manitobans were oblivious to this behind-the-scenes drama. By mid-June, Westerners were transfixed by the search for the so-called "Gorilla Killer," who had murdered two women in Winnipeg and then headed to Regina. Police had lost his trail in Moose Jaw, and now, as *The Globe* breathlessly reported on its front page, "Whole West Is Combed."[42] (American Earle Nelson was eventually caught near the United States border, returned to Winnipeg, and hanged.) Two weeks after those dramatic headlines, Bracken was on *The Globe* front page himself, when he romped to an increased majority in the June 28 election.[43] His agrarian ways had held the fancy of Manitobans.

King had no patience with Robson, but he was now happily entangled in the celebration of Canada's Sixtieth Anniversary. The festivities for the Fiftieth Anniversary had been muted during those bloodied days of the Great War, but these celebrations, which stretched over three days, were a triumph. On Dominion Day, more than fifty thousand people crowded onto Parliament Hill. The miracle of radio picked up King's speech from a microphone—"a tiny instrument scarcely larger than a child's eye, suspended in a shining ring and perched on Parliament Hill," as *The Globe* marvelled—and catapulted the live transmission to broadcasters across the nation.[44] Governor General Viscount Willingdon dedicated the new Peace Tower, which King had designated as a memorial to Canada's war dead. (The former tower had been destroyed during the terrible fire of 1916.) Willingdon also inaugurated the Carillon, and the bells chimed into the evening as electric floodlights played over Parliament Hill.

Surely anything was possible when Canadians could hear King's voice in the Yukon and Vancouver "as clear as if [he] were talking in the room."[45]

The nation partied. The thwarted hopes and dreams from the early postwar years might finally come true.

The King government slipped into the "stable routine" that only prosperity could bring. In each successive year during the last half of the 1920s, "previous records of national production and income [would be] broken."[46] Mineral production and prices rose. American capital flooded into Canada, funding the construction of branch plants behind Ottawa's tariff walls. By the end of the 1920s, nearly fifteen thousand people were working in the automobile or auto parts industry.[47] Canadian consumers were on a tear, and the more prosperous families snapped up radios, electric lights, refrigerators and stoves. Radio shipments skyrocketed from more than 48,000 in 1925 to more than 170,000 by 1930.[48] Canadians had more cars per capita than any other nation except the United States.[49] In the late 1920s, two Americans would even copyright a song, "Happy Days Are Here Again." Those days gave every promise of lasting forever. Prosperous regions could afford to be generous with each other.

In 1927, even the parsimonious King observed that he had enough cash to make serious deals with the provinces. He could already imagine the grateful voters flocking to his side. In late June, when the Liberals had swept Prince Edward Island, King was "sure" that Maritimers had rewarded *his* efforts to implement the Duncan report.[50] The prime minister had first floated the idea of a federal–provincial conference in the spring of 1927, but he had been vague about the agenda and the timing. He could not risk another blow-up like the November 1918 Conference. Now, after the Dominion Day triumph, after Canadians had bonded with each other as a national community, he set out in earnest to remedy the concerns that had plagued Confederation for generations.

He had to be cautious—if only because the political players were changing. On August 17, "Honest John" Oliver died from the cancer that the famed Mayo Clinic had diagnosed in May but could not remedy. His successor was a respected medical doctor, John Duncan MacLean, who had sat in the Legislature since 1916 and run the government during

Oliver's final months. MacLean was a man in a hurry to make his mark, even if he irritated his Liberal ally in Ottawa. And he would certainly irritate King.

Two months after Oliver's death, to King's dismay, Calgary lawyer Richard Bennett replaced Arthur Meighen as National Conservative leader in the House of Commons. Publicly gracious, King telegrammed his congratulations. Privately, ensconced at his Kingsmere retreat, however, he mused that he would have preferred any of the other five contenders. "Bennett's manner is against him, his money is an asset & he has ability. He will be a difficult opponent, apt to be very unpleasant, and give a nasty tone to public affairs. I shall just have to try to be all the pleasanter and more diligent."[51] That would be a hard resolution to keep—but he could be assured that the Calgary lawyer would not object to a settlement with the West.

First, King had to charm the cantankerous provinces. As the premiers gathered in Ottawa for this pivotal federal–provincial conference in early November, the prime minister examined his face in recent photographs for signs of moral improvement. He could see no evidence of the pieties that he espoused. The pictures belied his idealistic intentions. "There is still a heavy & weary expression," he observed. "Not the noble purpose I should like. I pray God I may yet have a countenance which will reveal something of His love and high idealism in my life. . . . Now may God guide me safely thro' the coming days."[52] Then King headed out to do very strategic battle with the provinces.

The Dominion–Provincial Conference of 1927 was a turning point in the West's seemingly endless struggle for resource control. Three days into that week-long gathering, King was delighted with the good-natured discussions of roughly forty subjects, including federal funds for highway construction and a failed plan for Senate reform. "One thing is clear, the 'Diamond Jubilee attitude' is a reality," he declared. "There is a feeling for Canada, & a pride in sharing in its government, a great belief in its future."[53]

Perhaps those Dominion Day festivities really did inspire the usually querulous first ministers. Perhaps the good economic news allowed everyone to be generous. Or, most likely, King had finally figured out how to give almost everything to almost everyone. It was an *astonishing* performance. Fifty-seven years after Louis Riel had lost his bid to control Manitoba's resources, twenty-two years after North-West Territories premier Frederick Haultain had failed to secure resource control for Alberta and Saskatchewan, nine years after the uncompromising debates of the November 1918 Conference, the provincial premiers lay down alongside each other like proverbial lambs, and benignly blessed each other's requests.

Inevitably, of course, there had been low moments. On Tuesday, November 8, Prince Edward Island premier Albert Saunders tabled a sixty-three-page memorandum that outlined his province's special case for more money. The loquacious Islander had pestered King for several years since he had become opposition leader. Now he drove him over the edge. "Saunders of P.E.I. was deplorable—a terrible piece of mendicancy, unworthy of manhood," King ranted in his diary after the premier's presentation.[54]

But the discussions about resource control and subsidies, which spilled across three days, went remarkably smoothly, perhaps because the premiers were making plans to spend someone else's money and there was enough to go around. British Columbia premier MacLean called for the return of the railway lands and for an increase in the general subsidy, which Manitoba and Nova Scotia backed. Saskatchewan premier James Gardiner, the highly partisan politician who had replaced Charles Dunning in February 1926, grandly declared he "would not care if a hundred thousand or two hundred thousand dollars were taken out of the other eight provinces to help Prince Edward Island overcome the peculiar difficulties under which it laboured by reason of its isolation."[55] Gardiner clearly had not received as many lengthy letters from Saunders as King had.

The "me-too-ism" of the stand-offs that had disrupted earlier talks evaporated. In an extraordinary reversal, New Brunswick premier John Baxter declared that the Maritimes had *no* claim on Western lands and resources: he merely asked Westerners for "a brotherly consideration for those who had shared the load so that those lands could be made valuable by railways and other developments. . . . Despite what might be heard to the contrary the Maritime Provinces still believed in Confederation."[56]

In reply, Gardiner tartly emphasized that the West was *not* "indebted to any other part of the Dominion for its lands." But then he noted that the West "owed a debt to the eastern provinces for having sent their sons and daughters to the West for settlement." Equity should prevail— along with a "strong central government with an adequate taxing power."[57] (Thirty years later, Gardiner would be the federal agriculture minister when Ottawa introduced equalization to ensure that poorer provinces could provide roughly similar services for roughly similar levels of taxation.) "Some day," Gardiner added, "the Maritimes might be the richest part of Canada, while the western provinces might have to pass through a trying period from time to time."[58]

Did Mackenzie King spike the water? By Wednesday, November 9, there was "almost complete unanimity" in favour of transferring resource control to the West. The first ministers also endorsed the Duncan report— and so those conditional subsidy payments to the Maritimes became permanent. Although British Columbia premier MacLean wanted immediate satisfaction of his claims to the railway lands, his colleagues clearly wanted to wait for the report of the Martin Royal Commission.

Perhaps most important, Ontario premier Howard Ferguson and Quebec premier Louis-Alexandre Taschereau were benevolent presences— because they also needed something from Ottawa. After years of squabbling, the two had formed a powerful coalition when Ontario had scrapped its drastic curbs on French-language schools. Now they were allies in a constitutional battle with Ottawa over the control of waterpower. Ottawa controlled navigation and shipping, and it had the right to seize control over any public works "for the general Advantage of Canada."[59] The provinces controlled the riverbeds. Ferguson wanted the right to sell surplus waterpower from such navigable rivers as the Ottawa and the St. Lawrence, and he now had Taschereau's backing. The two premiers saw dollar signs.

King stalled for time. On Wednesday, as the first ministers wrapped up their talks on subsidies and resources, Ferguson noted that Ontario taxpayers made the highest per capita contributions to federal revenues, but he did not want "to cavil about small things."[60] (Many modern-day Ontario premiers might dispute that characterization of what was small.) Perhaps, he added, Ottawa could appoint a commission to examine general subsidy levels. Quebec premier Taschereau maintained that he "was not asking for anything at all for his province."[61] He deplored the disharmony within the federation—the provinces "had continually to fight to keep the rights which they had"—and he suggested that Ottawa ask the Supreme Court to determine which level of government controlled waterpower.[62] Ferguson intervened briefly to support that suggestion.

Then, after so many decades, after the bitterness of the November 1918 Conference, the breakthrough happened, almost offhandedly. Taschereau assured the gathering that he had "no objection to the return of the natural resources to the West and to the granting of special treatment to the Maritime Provinces."[63] More remarkably, adoping a stance that completely reversed the Ontario and Quebec positions from the November 1918 Conference, those two powerful Central Canadian premiers "had no objections to the western provinces continuing to retain their subsidies in lieu of lands."[64] After decades of dogged resistance, the deal was finalized.

The details would of course require more years of negotiation. The Maritimes would not object if the Western provinces received resource control. The Western provinces would not object if those temporary subsidies for the Maritimes became permanent. No one paid much attention to British Columbia's fledgling premier MacLean, but then no one objected to the eventual return of the railway lands. Ontario and Quebec did not oppose these regional moves because King agreed to *consider* a reference to the Supreme Court on waterpower.

It was a miracle of Confederation. King viewed the gathering as "the greatest possible success & I am truly thankful for the Divine Guidance which helped to this end."[65] It was marvellous to see what federal money could buy. When the Conference ended, the master manipulator went to the Country Club for dinner with friends, and danced until nearly 2 a.m. In the morning, the prime minister did not get up until 9 a.m., and he berated himself for the foolishness of his late night. But "I enjoyed it thoroughly coming at the end of the 8 days of Conference and all having gone so well."[66]

The only malcontent was British Columbia premier MacLean, who was unwilling to wait for the Royal Commission report. Before the premier left Ottawa, he wrote to King, outlining his claims. Once again. At home, on December 5, he wrote a further letter about the return of the railway lands. On January 9, 1928, he sent a list of seven demands, including an airy request that Ottawa arrange to amend the Constitution to spell out the type of taxes that his province could levy. (The implication was that Ottawa could then *not* resort to similar taxes.) In reponse, King "spoke plainly [to MacLean] as not appreciating his attitude, which was anything but friendly."[67] But the prime minister did not allow his annoyance to scupper his plans. In late January, in the Speech from the Throne, he promised to consider the return of the railway lands. The premier was elated—but his nagging had so irritated the prime minister that King even misspelled MacLean's name in his diary.[68]

The prime minister kept working away on his regional files. In that same Speech from the Throne he promised to pursue talks with the West on resource control, a public pledge that bought private peace with the impatient Bracken. In mid-January, just two months after the triumph of the 1927 Federal–Provincial Conference, the Manitoba premier had penned another irascible letter. "It was conceded nearly a year ago," Bracken had complained, "after two formal conferences, five or six informal interviews, and almost continuous correspondence, that the attempt to reach a settlement by mutual consent had broken down."[69] Bracken could not understand why King had not arranged for arbitration. What was taking so long? King asked Bracken if he would consider "one further attempt" at talks.[70]

Alberta premier Brownlee was equally anxious. During a meeting with King in mid-January of 1928, the two men had toyed with the notion of appealing the Supreme Court's support for the legality of the separate school clause in the Alberta Act. An appeal would simply be a precaution: any decision from the Judicial Committee of the British Privy Council, which was still Canada's highest court, would end *any* quibbling. But neither side had wanted to take the lead. (The idea would eventually fizzle out on technicalities, and the Supreme Court decision would stand.) King was ambivalent about the cool Brownlee, whose public demeanour would belie his eventual involvement in one of the most sensational sex scandals in Canadian history. "Brownlee is a difficult sort of man to deal with," King mused. "One does not know what he has in mind, he is of a suspicious & of a fearful nature."[71] The prime minister resolved to postpone any talks until the end of the Parliamentary session.

King still had to deal with waterpower. After the Dominion–Provincial Conference, Ferguson and Taschereau had remained "shoulder to shoulder" in their contention that surplus waterpower was a natural resource that they controlled.[72] Ottawa had countered that the only limitation on its jurisdiction over waterpower was the provinces' control over property rights.[73] After weeks of fierce cabinet debate, King finally asked the Supreme Court of Canada to decide which level of government had jurisdiction over the waterpower on the St. Lawrence and other navigable rivers.

The sophisticated Taschereau remained detached, and wise. "It is your duty to protect Dominion rights, as it is our own duty to see that Provincial rights are also protected," he later told King. "But I am afraid that some of our friends at Ottawa sometimes take it as a personal attack against them when we do not share their views on these important matters."[74] Taschereau was likely referring particularly to Justice Minister Lapointe, but that insight could easily have applied to the natural resources talks throughout the decades.

King was fighting his own battles with perspective. He was reading *The Culture of the Abdomen*, and the exercises, he wrote, "have served to reduce my waist measurement considerably."[75] He was also fretful. He had moved into Laurier House in 1923—Lady Laurier had bequeathed it to him in 1921—and he was now grappling with servant problems. He had just fended off a request from Supreme Court chief justice

Francis Alexander Anglin for a government car. *The Globe* had delivered a "miserable editorial" on that pivotal Speech from the Throne,[76] To top it off, the new bridge in his mouth was loose, and he had to return to the dentist.

But then another of his schemes paid off. On February 6, 1928, King tabled a sixty-two-page report from the Martin Royal Commission in Parliament. The former Saskatchewan premier dismissed British Columbia's legal right to the 9.6 million acres along the railway belt and the 3.23 million acres in the Peace River tract. The province had transferred those lands to Ottawa in the nineteenth century, fair and square. But Ottawa should look upon the request "from the standpoint of fairness and justice."[77] It should return the lands to avoid "a feeling of unjust treatment."[78] Martin was clearly writing from experience. He may have been wearing the judicial robes of the Saskatchewan Court of Appeal, but he could not suppress his memories as a member of the Gang of Three at the November 1918 Conference. King now had the moral authority to resolve the province's complaints. He was sure that his departed mother was "not far away."[79]

The prime minister coasted complacently throughout the spring and into the summer of 1928. Little seemed pressing. He "was inclined to interpret national prosperity as proof that most problems had been solved."[80] He ignored a chatty nine-page letter from Prince Edward Island premier Saunders for more than three months, unperturbed. He took his time when British Columbia premier MacLean demanded meetings to rebut charges that Ottawa "was not in earnest" about the return of the railway lands.[81] When MacLean lost the July 18, 1928 election along with his own seat, the prime minister was at Kingsmere, supervising the levelling of the lawn and complaining about his plumbing bills and the slipshod care of his chickens. He did not mention MacLean's defeat in his diary, and he was still misspelling the former premier's name when he finally met his Conservative replacement, Simon Fraser Tolmie. The new B.C. premier was a former veterinarian who had served in the governments of Sir Robert Borden and Arthur Meighen. He was "loyal, affable and tactful," but he knew little about economics.[82] Although King was a loyal Liberal, he would not miss MacLean.

The only settlement that moved along smartly in 1928 was the case of Manitoba. Exasperated by King's stalling, Bracken had complained to provincial Liberal leader Robson who, in turn, had told King, "Liberals here are either despondent or bad-tempered . . . the Resources question is an overhanging pall."[83] *That* got King's attention: he wanted to foster the emerging alliance between the provincial Liberals and Bracken's United Farmers of Manitoba to ensure the re-election of his own government.

Then Bracken became ensnared in a scandal: to develop waterpower, his government had asked Ottawa for the right to exploit the Seven Sisters Falls on the Winnipeg River. Bracken had also contacted a private electrical firm to determine what it would charge for power. When those facts became public, the provincial Conservatives were furious. Bracken had not consulted the Legislature. To make matters worse, they alleged that the private firm had donated to Bracken's party in 1927. Manitobans also resented Bracken's decision to abandon his policy of *public* ownership. Feeling under threat, Bracken asked Ottawa to license the private firm quickly because further delay might create more "agitation" about Ottawa's failure to transfer resource control.[84] (An inquiry into the Seven Sisters deal would eventually clear his government—although two ministers who had bought shares during the negotiations with the private firm would resign.)

King got cracking. He wanted to avoid any taint from the Manitoba scandal, and he was also ashamed to read the evidence of Ottawa's "continuous procrastination" on the transfer of resource control.[85] But, most of all, he could foresee a political payoff: "If we go the right way about it, we may win all three Western provinces to our side."[86] In early July, Bracken and King agreed to appoint a three-person commission to examine Ottawa's use of Manitoba's resources since 1870.[87] Until the commission reported, Ottawa promised it would do whatever the province asked with the resources. Remarkably, they were still bickering over the details until the day before the deal was made public on Thursday, July 11, 1928.

That was very bad timing. Given the newsworthy competition, King barely made *The Globe*'s front page. Authorities had seized "great stocks of intoxicating liquor" at the Windsor docks, from which "fleets of rum-runners" smuggled liquor over the Detroit River into the United States.[88] Toronto authorities had arrested five men and one woman for the sensational $300,000 robbery of the Union Station postal outlet; three more American gunmen with "bad, bad reputations" had fled, and the Mounties and Pinkerton detectives were "after them hot-foot."[89] Thousands of Protestant Orangemen were marching on July 12, the anniversary of the 1690 Battle of the Boyne, and they were upsetting thousands of Roman Catholics. It was a difficult day to catch the public imagination with resource deals. But, on August 1, the three commissioners set to work.

The other two Prairie premiers soon followed Bracken's lead. In December 1928, King settled on a deal with the enigmatic Brownlee: Ottawa would transfer the lands and resources to Alberta with the exception of the national parks; Alberta would administer the school trust "in compliance with the letter and spirit of the constitution"; Alberta would still receive its subsidy in lieu of lands of $562,500.[90] King also agreed to discuss resources with Saskatchewan premier Gardiner in early 1929. He even opened negotiations on the railway lands with British Columbia premier Tolmie, and they began "where we left off" with MacLean.[91] Remarkably, King found time for an Ottawa lunch with Prince Edward Island premier Saunders. He remained unimpressed: "He has a very small mind & outlook, was very hard to talk to."[92] Twelve days after that encounter, King received a six-page letter from Saunders, which outlined his case for higher subsidies and asked for yet another appointment. King ignored him.

Nonetheless, the prime minister saw only good portents for the New Year ahead. His Bible readings had convinced him that his mother's unearthly presence was hovering ever closer, and that the Good Shepherd was guiding his actions. On Christmas Day in 1928, as he opened a warm message from Queen Mary, he marvelled at the irony: his grandfather, William Lyon Mackenzie, had been the leader of the 1837 Upper Canada Rebellion. "This [message] was pretty nice to receive & pretty nice to the grandson of 'a rebel' who would have been hanged in the name of the ancestors of the present household."[93] On New Year's Eve, he looked back on "the best year of my life. . . . Nothing has been so gratifying as the change in feeling towards myself & increased confidence in the Government which is everywhere manifest."[94]

Like most Canadians, he expected that the good times would keep on rolling. On January 2, 1929, Canada and the United States signed a treaty to preserve the scenic beauty of Niagara Falls. In the West, aircraft were charting Canada's vast northern lands. Companies were mining quartz across the northern reaches of Manitoba and Saskatchewan, and into the Northwest Territories. Oil wells were bubbling in Alberta's Turner Valley to the southwest of Calgary. Ottawa was still trying to entice homesteaders onto the Prairies, even though the best lands had long ago been taken. The federal Mines Department was enlarging its research laboratories to encourage development. Ottawa and the provinces were doing an inventory of the forests to ensure conservation.[95] Signs of progress were everywhere.

On March 25, 1929, Mackenzie King celebrated seven full years in office. He had been having terrible dreams, which he interpreted as warnings from his parents that "some evil . . . might be creeping towards me."[96] The perennial bachelor thought that the best cure for his tormented nights would be marriage, but, he confessed, "I have given up worrying about it."[97] Meanwhile, perhaps he should take a little wine or spirits before he retired at night.

His resource deals were progressing nicely. On May 30, 1929, the Royal Commission on the Transfer of the Natural Resources of Manitoba delivered a forty-six-page summary of the money that Ottawa had sent to the province, the resources that Ottawa had distributed for national purposes, what Ottawa had paid for administration, what sacrifices Manitoba would likely have made to attract settlers and what it *might* have received from land sales. The calculations were *very* complex and rather arbitrary. The conclusion was precise: Ottawa should pay $4,584,212.49 to Manitoba, transfer the resources and continue the annual subsidies in lieu of resources. Bracken and King were both satisfied. Interior Ministry officials started to negotiate the details of the agreement with their provincial counterparts, churning out lists of lands, minerals and forests.

Negotiations with the other provinces were sputtering along, however, so King sweetened the stakes. He told Saskatchewan premier Gardiner that he would likely increase his subsidies in lieu of lands *after* the transfer—as the population increased. Gardiner held out for a firm promise of more cash. King would eventually make the same promise to Alberta premier Brownlee. He had some reason for hope.

Then some not-so-funny things happened to King on the way to his deals.

Premier Gardiner had called an election in Saskatchewan for June 6, 1929, just as religious bigotry and ethnic prejudice reached a zenith and public disenchantment after twenty-four years of Liberal rule ran high. The Ku Klux Klan had stationed a Kleagle in Saskatchewan in 1927. Now the Klan was spreading its anti-foreign, anti–Roman Catholic message into a province where the population had almost doubled over the last two decades. Saskatchewan also had "the highest non-Anglo-Saxon component of any province," and Canadians of British and French descent were dwindling as a proportion of the population.[98]

Gardiner denounced the Klan, but his opponent, Conservative leader James Anderson, was devilishly subtle. Although he did not embrace the Klan, the former school inspector complained of the influence of Roman Catholics on public schools. He also vowed to work with Ottawa to select immigrants who suited Saskatchewan's economic—and racial—realities. Gardiner won the greatest number of seats, but lost his majority. When the Progressive and Independent MLAs allied themselves with Anderson, Gardiner convened a special session of the Legislature in early September. To his shock, he lost a vote of confidence and, on September 9, 1929, Anderson became the premier of a Co-operative Government. To King's dismay, Gardiner had not signed any agreement on resource control before he lost power, and the prime minister now had to work with a man that many Liberals actively disliked—and who returned their antipathy.

Despite that setback, by early December, King was ready to close his deals. The three Western premiers headed to Ottawa, and the prime minister dealt with each delegation individually. The camaraderie and connivance of the old Gang of Three were long gone; even Bracken and Brownlee did not like Anderson. King met Bracken in the Cabinet Council room on the afternoon of Monday, December 9, and there were virtually no disagreements. That same day, King was dismayed to learn that Brownlee would likely ask for a Royal Commission into Alberta's claims, but he assumed that he would get an agreement in any case. Saskatchewan, however, was another story. "Anderson is here to make trouble, not to make an agreement," King concluded. "I feel annoyed at Gardiner letting his province get into such hands."[99]

The next day, he met the new Saskatchewan premier. It was dislike at first sight. "He is a rough diamond & the men with him of a type of low cunning," King concluded.[100] Anderson handed over a memorandum, but King decided that it was designed to provoke an outright refusal, which Anderson could exploit at home. Instead, he shrewdly promised to study it, which would thwart any negative headlines.

On Wednesday, December 11, the Alberta delegation met with King, Interior Minister Charles Stewart and fledgling finance minister Charles Dunning, who had taken over the economic portfolio in late November on the death of his predecessor. Both Dunning and Stewart had been Western premiers, but their presence did not soften up Brownlee. The Alberta premier had put his deal on hold until the Manitoba Royal Commission had tabled its report in late May. After careful scrutiny, he had come to an unsettling conclusion: "Alberta was entitled to very considerably more than we had suggested when we were previously in Ottawa and had made a definite offer."[101] Although the premier had qualms about reneging on his December 1928 deal, he decided that he had little choice.

Ensconced in King's office now, Brownlee steered the discussion toward unemployment relief—because the economy was stalling following the October stock market crash, and he wanted to tackle that issue before he irritated the prime minister. When Brownlee did ask about the jobless, however, King shrugged: Ottawa could just as easily ask the provinces for help because the federal government was still paying off its wartime debts. The prime minister really did not appreciate the seriousness of the nation's plight. Brownlee then declared that Alberta was renouncing its earlier resources deal because, under the principles of the Manitoba Royal Commission report, his province was entitled to more money. "I'm afraid Mr. King was rather sadly disappointed when we took that position," Brownlee would recall in an interview more than three decades later. "He was very quiet while I was setting out our present position and finally abruptly he got to his feet and told us we'd have to put our recommendations in writing."[102]

Later in that lengthy interview in 1961, Brownlee would expand upon their dramatic confrontation. "He [King] was flushed and he was angry and he slammed his book. And very abrupt [sic] said, 'Put what you have to say in writing.' I looked up and said, 'Does that mean the conference is over?' 'Oh yes,' he said taking us entirely by surprise. He raised a big issue about putting it in writing and sending it to him and he would consider it. With that he marched out the door. By the same token I slammed my books shut and didn't wait for my associates and headed out the other door through the secretary's office."[103]

King's account of that argument essentially matched Brownlee's. As he saw it, after years of talks, the premier now wanted more money because Ottawa had given away far more of Alberta's lands for railroad construction than it had handed out in Manitoba. The exasperated prime minister realized he had spent much of his morning on the issue, and it was now one o'clock. "I got a little hot & told Brownlee to put his proposition into writing & we would give an answer after Christmas. He was anxious for a further interview today. I told him we would not give it today to write me & we would have to answer in writing."[104] The prime minister then stalked off to hear Saskatchewan premier Anderson's speech to the Canadian Club, which confirmed his prejudices. It was "a very poor not even mediocre address . . . .a kindergarten affair delivered as to an audience in a field, a very ordinary man."[105]

During his brief absence from Parliament Hill, there was more uproar. Interior Minister Stewart had raced after Brownlee and asked him to keep his temper. He invited the fuming premier to visit him in the afternoon. When Brownlee returned to his hotel after lunch with the British Trade Commissioner, he found a note from Finance Minister Dunning that was terse but placatory: "The Prime Minister is very anxious to see you. Swallow your pride and call him up."[106]

Brownlee was about to telephone when Dunning called to repeat his message. Brownlee telephoned King, who was looking for *him*. The prime minister was "quite a different man," asked him to come over, apologized "quite handsomely," and invited Brownlee to dinner at Laurier House.[107] Then King went to cabinet: "I put the argument as to settling with both Manitoba & Alberta & thereby putting Sask in position of being in a cleft stick."[108] This was King at the peak of his political powers: he wanted to isolate Anderson through deals with Alberta and Manitoba so that, when the Saskatchewan premier complained about his failure to get a deal, Canadians would conclude that *he* was actually the odd man out. It was a masterstroke. That same afternoon, as King discussed Brownlee's requests with his cabinet, Bracken called: he wanted to sign *his* agreement. King suggested that Friday would work, but Bracken pointed out that it would be Friday the Thirteenth. Ever superstitious, King agreed that Saturday would be fine.

That evening, at King's residence in Laurier House, Brownlee and King, along with Dunning, tackled their differences *after* their meal, and still settled everything before 11 o'clock. They agreed to send the contentious issue of an extra cash settlement to a Royal Commission. Subsidy payments would continue, and increase as the population rose; when there were 1.2 million Albertans, the subsidy would peak at $1.125 million. On Saturday, December 14, 1929, Bracken and Brownlee signed their deals, and posed for historic photographs. "This completes the real autonomy of these two Western provinces and gives them a fresh start, with additional assured financial assistance," King wrote.[109] The consummate politician, he anticipated that the two premiers' progressive supporters might now vote for the federal Liberals. When Brownlee returned home, more than two thousand people "met him at the railway station in Edmonton in freezing weather. Organizers lit a large bonfire, a band played, and fireworks exploded in the night."[110]

King finished the year with yet another coup. On New Year's Eve, to disarm Anderson, he offered to return Saskatchewan's resources, and to ask a Royal Commission to examine his claims for extra cash. He also offered to ask the courts if Anderson was correct when he demanded compensation for resources that Ottawa had used before Saskatchewan's creation in 1905. On that same day, the impossible Saunders wrote again about Prince Edward Island's claims: Ottawa had settled the West's demands so "some equivalent increase in subsidy in lieu of public lands should be granted to us."[111] It never stopped. (One might surmise that King appointed him to the provincial Supreme Court in the spring of 1930 to keep him away from the capital.)

Meanwhile, King and British Columbia premier Tolmie finalized a deal to transfer those long-disputed railway lands. The Martin Royal Commission had calculated that Ottawa had used only 1.37 million acres of the 10.97 million acres along the railway belt, and only 270,000 acres of the 3.5 million acres in the Peace River tract. King offered to return that land, and to continue the province's subsidies in lieu of those resources. It was a win-win proposition because Ottawa's administrative costs now far exceeded its revenues. The deal sailed through cabinet, and a grateful Premier Tolmie proudly announced the pact in mid-February of 1930. King even arranged a meeting with Ontario and Quebec on waterpower after an inconclusive Supreme Court of Canada decision on the issue. (Unfortunately, that meeting would dissolve acrimoniously when King refused to concede the provinces' ownership of the power, although he offered "to treat with the provinces as if they had right to the powers.")[112]

As King had reckoned, Anderson had to come round. In a grudging, graceless five-page letter in mid-January, the premier conceded, "We are prepared to accept the terms of your offer." Then he rejected two of the commissioners that King had suggested.[113] King countered with one major caveat that would have serious consequences for the West: the Royal Commissioners would not start their hearings until the courts decided on the time frame. Should they scrutinize Ottawa's use of the resources from 1870, or from Saskatchewan's creation in 1905? In early March, King and Anderson made the agreement. "By being most conciliatory & keeping my own ministers on an even keel & feeling good, we were able to avoid any unpleasant clash," King wrote happily.[114]

He rejoiced too soon. The thrifty prime minister, who tucked his personal savings into Government of Canada bonds, was oblivious to the implications of the economic downturn. He viewed unemployment as a social injustice, but he also regarded it as a provincial responsibility. In those first months of 1930, he was frozen in a better time. "Mackenzie King behaved like a conservative," his biographer Blair Neatby explained. "The stock market crash seemed irrelevant. The economic difficulties on the prairies and elsewhere were temporary. . . . The policies of the government were sound by all the tests of experience and orthodox theory."[115]

King was wrong, of course. Canadians were toppling into the Dirty Thirties, when drought would devastate the prairies and the unemployed would ride the rails. Interregional harmony scarcely mattered when so many Canadians were struggling to feed their households, and to preserve civility and kindness. Those early days of 1926, when King had decided that regional popularity was the key to electoral victory, seemed far away. The politicians who had taken the place of the premiers at the November 1918 Conference were now far more interested in finding jobs and food for their residents than itemizing forests.

By late spring, the constitutional legislation for all four Western deals slipped successfully through Parliament and the provincial legislatures. Each transfer required the approval of the British Parliament, because each amended the terms under which the province had entered Confederation. There was a brief panic when the British government lingered over the granting of Royal Assent, but finally, on July 10, 1930, in a display of distracted brinksmanship, Britain approved the Constitution Act 1930.[116] Five days later, on July 15, 1930, the sixtieth anniversary of the creation of the province that Riel had once so ardently demanded, King was in Winnipeg. To mark the handover of resource control, he presented a cheque for $4,822,842.73 to Bracken. "This has been a memorable day," he wrote in remarkable understatement.[117]

King was subdued because the celebration was now merely a whistle stop during his whirlwind campaign tour across the West. Two weeks later, when the voters went to the polls on July 28, 1930, fearful for their jobs and families, King lost. Conservative leader Bennett won forty-seven additional seats and a clear majority. Still largely blind to the terrible impact of the Depression, King was stunned: "The result is a great surprise."[118] He reviewed the regional tallies in his diary, leaving unsaid his obvious feeling of betrayal. On July 29, he resigned. Canadians barely noticed when Saskatchewan and British Columbia took control of their lands and resources on August 1. (Premier Anderson had campaigned against King in the election.) Because Alberta premier Brownlee had pleaded for a two-month delay, that province's official handover did not take place until October 1. King did not mark the occasions.

He had done what had once seemed impossible. Sir Wilfrid Laurier had not wanted to surrender the resources. Sir Robert Borden had never cared enough to do the tough bargaining. The aloof Arthur Meighen had treated the Western premiers as unreasonable underlings. King had discovered the solution, and he had done the hard work. As Norman Rogers had told King, the prime minister had completed the creation of the Western provinces that Laurier's immigration and transportation policies had fostered. But the celebrations would have to wait.[119]

Canada had moved on—into a decade of trouble and sorrow. In mid-October 1930, the Bennett government restricted immigration to arrivals from Great Britain or its Dominions and from the United States. As the *Calgary Daily Herald* reported, the stream of new arrivals slowed to a "dribble.... The regulation is being drastically enforced in spite of many protests."[120] In that same edition, ironically, the newspaper devoted a special section to the transfer of resource control. The publicity branch of the Alberta government even trumpeted: "Alberta's Hopes to be Realized by Transfer of Natural Resources!"[121]

It was not to be. As Christmas approached, the newspaper told its readers about a family of nine living in a one-room shanty with "scraps of wood . . . nailed in place of glass" in the windows; that family would be among the recipients of its charitable Sunshine funds.[122] Two days before Christmas, however, the newspaper was forced to appeal to Good Samaritans: the Sunshine fund held only $13,302.79; on that same day the previous year, it had reached $19,340.08.

It scarcely mattered which level of government controlled the resources: the desperation affected everyone. Polish immigrant John Dul had arrived in Canada in February 1930, before the immigration ban but after the job market had collapsed. "I tried to find jobs everywhere," he would recall. "I even went to a government official and asked if he could send me back to Poland, but he refused." In the winter of 1931, Dul and five other Polish youths moved into a sod dugout on an abandoned farm in Waskatenau, sixty miles northeast of Edmonton. "We had no money, and no hope for a job. Our only food supply was the prairie chicken. Finally, after long discussions, we decided to make moonshine." Luckily, nearby Ukrainian and Polish homesteaders were willing to trade fresh food for illegal liquor. Dul survived, and eventually managed to stake out a homestead. He would remember those days with nostalgia: "People suffered much during the depression. We had neighbours who had only one pair of shoes for an entire family of five. However, those were also times when people felt closer to each other."[123]

Mackenzie King's fabled deals with the Western premiers were already almost forgotten. As Sir Robert Borden had feared before the war, Saskatchewan and Alberta both changed Ottawa's policy of virtually free homesteads. In late 1930, Saskatchewan set out a price schedule for its lands, and required four months of continuous residence per year for three years before it would grant title. A few months later, Alberta restricted free homesteading to a few centralized areas. Only British subjects or people who had declared their intention to become subjects could settle on the land—*after* they had met a residency requirement of five years.[124]

Those regulations had little effect. Homesteaders had long since claimed most arable land, and the Depression had crippled any would-be settler's zest for challenge. Resource control was "a hollow victory."[125] There were oil wells in Alberta in 1930, but most firms lacked the money to drill, and the Royal Commissions into Ottawa's use of those resources did not get underway for years. In late 1931, an exasperated Brownlee asked for a single commission for Saskatchewan and Alberta.[126] Prime Minister R.B. Bennett dismissed that request: he did not want a commission composed of one representative from Alberta, one from Saskatchewan and only one from Ottawa. Anyway, he would not act until the courts had decided the time frame for Ottawa's obligations.

Western governments were reeling, frantic for cash. Resource control had brought no flood of instant riches—partly because the administrative costs were indeed high. Ironically, Alberta's net gain would "not exceed $200,000 annually during the first five years of the 1930s."[127] Alberta and Saskatchewan pinned their hopes on the claim that Ottawa should pay for the resources that it had tapped before 1905. They lost, at the Supreme Court of Canada in 1931 and at the Judicial Committee of the British Privy Council in 1932.

Prairie governments were beset with drought and despairing farmers. In late 1933, the four Western provinces failed to convince Ottawa to assume complete responsibility for aid to homeless transients. In 1935, Ottawa established the Prairie Farm Rehabilitation Administration to assist farmers with small irrigation projects. But it would not create a single registry for water-use plans, "for fear of being drawn into expensive, open-ended projects."[128] The times were so tough that, in 1937, Alberta asked Ottawa to cover the bills for firefighting in both Alberta and Saskatchewan, arguing that healthy forests would foster the flow of streams into drought-stricken farmland. If Ottawa would take over fire protection, Alberta would actually "hand over administrative control of the watershed lands and grant the federal government the surface rights to the lands"—while keeping the subsurface rights to the oil and minerals for itself.[129] This was stunning: the province was so desperate that it was offering to surrender rights that it had fought for decades to secure. Luckily for Alberta, the proposal died.[130]

National exports had collapsed. Farmers had watched their grain crops shrivel in the drought while the price of wheat plummeted. "By the time the bottom was reached, in 1933, more than one in four Canadians was out of work, many municipalities and even some provinces hovered on the edge of bankruptcy, and thousands of individuals had been forced to shut down their businesses or farms."[131]

Improbably, amid the bad news, it was the austere Brownlee who provided distraction with a sordid sex scandal. In the fall of 1933, Allan MacMillan, a prominent railway executive who was also the mayor of the town of Edson, and his daughter Vivian MacMillan filed an explosive statement of claim in the Supreme Court of Alberta. They alleged that Brownlee, who had promised to act as a guardian for Vivian at his Edmonton home, had seduced the eighteen-year-old woman, and maintained sexual relations for three years. Vivian asked for $10,000 in damages; her father demanded $5,000.

The trial commenced in June 1934, and the sensational testimony attracted journalists from Britain and the United States. Vivian MacMillan graphically described her sexual encounters with the premier, although she occasionally mixed up her dates. She shocked Albertans to their puritanical cores. When Brownlee took the stand, his lawyer asked if he had engaged in sexual intercourse with MacMillan. "Brownlee's answer was, 'I did not'."[132] The jury found in favour of the two MacMillans, and awarded damages. Remarkably, the judge rejected that verdict, declaring that there was no proof of injury to the plaintiffs—but the harm was done. Brownlee resigned in mid-July of 1934, and his fiscally conservative treasurer Richard Reid took his place.

Was he guilty? Vivian MacMillan would appeal the judge's casual dismissal of the jury's verdict, and she would lose at the Alberta Court of Appeal. In 1937, the Supreme Court of Canada would restore the jury's original judgment against Brownlee, but the federal cabinet would then allow Brownlee to appeal to the Judicial Committee of the British Privy Council. In 1940, the Lords would decide that the law had been properly interpreted, dismissing Brownlee's last chance for exoneration. Two decades after Brownlee's death in 1961, his biographer Franklin Foster would examine the case in depth. Many close associates regarded "the idea of [the] passionless, remote, cautious and calculating John Brownlee behaving like a love torn teenager discovering sex [as] ludicrous." But, even among some of Brownlee's closest friends, the doubts would not go away. "There was his capacity for indirection, his calculating, not to say manipulative, manner of influencing people behind the scenes, a sense that he was never completely open."[133] Foster simply could not decide.

In June 1934, as the MacMillan–Brownlee case mesmerized Westerners, Saskatchewan voters threw out their grim premier Anderson, putting Liberal Jimmy Gardiner back in power. Politicking was Gardiner's greatest skill, and he was wondrous on the campaign trail. More important, most voters "were worse off financially in 1934" than they had been when Anderson took office in 1929.[134] The premier came third in his own riding. His supporters were wiped out.

In Ottawa, Prime Minister Bennett was slow to implement King's resource deals. In late December of 1933, the Saskatchewan Royal Commission finally commenced; the Alberta Royal Commission got underway in mid-July 1934. Both three-man commissions held copious hearings, and produced hundreds of pages of testimony. Both commissions, which were composed of the same two federal appointees along with a provincial appointee, reported on March 12, 1935. (Saskatchewan and Alberta had selected different representatives.) Finally, both commissions awarded a lump sum of $5 million to each province.

There was one lone dissenter in those decisions, and he would cost both provinces dearly. Saskatchewan lawyer H. V. Bigelow, "on the basis of very dubious assumptions," decided that $58 million was the proper compensation for Saskatchewan.[135] Encouraged by that notion, Gardiner's government refused its $5 million federal cheque in the futile hope of prying more money out of Ottawa. The Alberta government, which had been ready to take its $5 million to the bank, now turned down Ottawa's offer, "because she was naturally not going to jeopardize her chances of getting more in the event that Saskatchewan did."[136]

Those were expensive mistakes, and the financial situation in both provinces deteriorated. On August 22, 1935, Alberta voters rejected Brownlee's replacement, Reid, with his message of self-help and "blame

Ottawa," in favour of Fundamentalist preacher William "Bible Bill" Aberhart and his Social Credit Party. Almost two months later, in another revolution, Canadian voters threw out Bennett's government, and put King back in power. Few politicians could withstand the powerful public urge for change as the Depression persisted into a sixth year, scarring a generation with its deprivations. In November 1935, Gardiner joined King in Ottawa as minister of agriculture; his natural resources minister William Patterson took his place.

The Prairie provinces were desperate, and in early 1936, Alberta treasurer Charles Cockroft suggested that Ottawa count the natural resources settlement of $5 million—plus interest, of course—against a new loan. A few weeks later, Premier Aberhart renewed that request. In late March of 1936, Charles Dunning, who was once again the finance minister in a King government, delivered a brusque refusal: "As in the case of Saskatchewan the natural resources award would have to be offset against debts already owing by the provinces to the Dominion"— and Alberta owed $24.7 million.[137]

A few days later, the Alberta government defaulted on its bonds. That default would mark a generation of Depression-era Alberta politicians *and* their successors. The government of Prairie farmers who had prided themselves on paying their debts was now effectively insolvent. Westerners may have *legally* controlled their resources, but whatever sparse revenues they received could not come close to meeting their debts.

The Prairie provinces would muddle through the 1930s and the Second World War mired in a sea of debt. During wartime, there were oil shortages, and gasoline rationing was an unhappy fact of life. The resources were undeniably there, but the Western provinces had difficulty exploiting them. Quebec and Ontario investors were reluctant to put money into exploration, and the British were hesitant as well. After Aberhart died in May 1943, his successor, Ernest Manning, finally turned to American oil companies, "which were not only willing to invest in Alberta but also promised to bring their expertise, . . . something the central Canadian and British investors could not do."[138]

That sensible solution paid off. On February 13, 1947, after 133 dry holes, Imperial Oil hit black gold at Leduc, twenty miles south of Edmonton. It was a triumph. Until then, Canada had relied on imports for roughly ninety per cent of its crude oil supplies. Now the Western premiers, including Alberta's astonished Manning, could catch their first real glimpse of the promise of resource control.

King was still alive when Leduc came through in 1947, and remarkably— still in power. But the prime minister was turning seventy-three later that year, and he was exhausted from the stress of war and the demands of the postwar peace. Although he saw himself as another Father of Confederation, few Canadians, most Westerners included, remembered his role in their newfound prosperity. He died of pneumonia at Kingsmere on July 22, 1950.

In 1967, Alberta premier Manning would proudly note that oil and gas revenues had put $2.25 billion into the provincial treasury over the previous twenty years.[139] It was a huge windfall that few participants at the November 1918 Conference could have even imagined. At that conference table, so many decades ago, Quebec premier Lomer Gouin had complained that the first ministers did not even know if the resources were "an asset or a liability." But, if they were valuable, the Maritime provinces should receive "adequate compensation *for their interest in them.*"[140] Such language would always upset Westerners.

More than forty years after the November 1918 Conference, one of the successors of the Gang of Three would allow himself a last laugh. In an interview in 1961, shortly before his death, former Alberta premier John Brownlee cherished the delicious memory of his 1929 deal with King. "I do not need to recall the fact that the settlement took place at a very opportune time before the real beginning of the oil development in the province," he crowed, "and it has been a source of tremendously large income for the province ever since."[141]

Afterword

—

STRUGGLES OVER RESOURCE CONTROL did not end with those low-key transfers during the Depression. The Constitution was so vague that the federal government could still find ways to meddle—as it continued to do in the following years. By the early 1970s, as oil prices escalated amid Middle East turmoil, Ottawa and the provinces could see dollar signs on the drilling rigs. The old notions about resource control that had plagued the November 1918 Conference took on new immediacy. Many Easterners still maintained that they had a stake in the West's resources. Many Westerners still resented *any* federal intrusions on their turf. Those competing assertions fuelled *very* public clashes that eerily echoed the long-ago confrontations of Louis Riel and Frederick Haultain and the Gang of Three with the Rest of Canada.

Though it would take decades before the New West of the twenty-first century would trump—at least temporarily—the battle-hardened Old West in a stunning election turnaround, the mistrust between the Rest of Canada and the resource-rich provinces (whose ranks now include Atlantic Provinces with offshore energy) has not gone away. If anything, New Democratic Party leader Thomas Mulcair has deepened the suspicion with his call for stronger federal environmental regulation of the Oil Sands.

The memory of those recent energy wars remains raw. In 1972, Alberta premier Peter Lougheed, the grandson of Senator James Lougheed who had served in the cabinets of Sir Robert Borden and Arthur Meighen, raised royalties to cope with a revenue shortfall. The Petroleum Club promptly revoked his membership. A year later, as OPEC pushed up international oil prices, Canadian consumers squawked about rising fuel costs. To ease those costs, in September 1973, Energy Minister Donald Macdonald "peremptorily summoned his Alberta counterpart, Bill Dickie, to Ottawa to hear him" freeze the domestic price of oil and clamp an export tax on Western crude earmarked for American markets.[1] The abrupt delivery of the message – without any consultation – was as aggravating as the message itself. In response, in December 1973, Lougheed's government "moved unilaterally, without consulting the industry, and announced that future royalties would rise with oil prices."[2] Ottawa lashed back: oil companies could not deduct those increased royalties when they calculated their federal income tax. For the energy firms, it was a double whammy.

As the energy war escalated, the rhetoric had an oddly familiar ring. When Pierre Trudeau's minority government froze oil prices in September 1973, Lougheed denounced Ottawa's "political aggression," which was subsidizing consumers in the Rest of Canada at the expense of the West.[3] The prime minister, for his part, viewed cheap oil as a competitive advantage for the nation's industries. He reasoned that, after decades of federal subsidies, Ottawa should get its "fair share" of oil and gas revenues.[4] Energy Minister Macdonald even asserted that Ottawa was "exercising *the right of all Canadians* to a share in what is provincial but also Canadian wealth."[5] Those politicians could have easily joined the delegations at the November 1918 Conference. Meanwhile, Westerners fumed. Cars sported bumper stickers: "Let the Eastern Bastards freeze in the dark." That slogan has become part of Western folklore.

Alberta gained allies. In 1978, the Supreme Court of Canada concluded that Saskatchewan's tax on potash reserves was unconstitutional. It also ordered the province to repay $500 million that it had collected under an oil tax that the court had already declared unconstitutional because it was an indirect tax. Saskatchewan Premier Allan Blakeney promptly ran for re-election with the promise of constitutional change to ensure greater provincial control over resources. He won handily.[6]

The worst confrontation was still to come. Eight months after Pierre Trudeau won the February 1980 election with strong majorities in Central and Eastern Canada, a high-level group of energy and finance department bureaucrats would devise the National Energy Program—which

was drafted in such secrecy that it surprised many federal cabinet ministers. Unveiled on October 28, 1980, as the centrepiece of the budget, it
created a blended price for old, cheaper oil and new, more expensive oil.
It also forced "the pace of development outside Alberta by encouraging
exploration" on the Territorial lands that Ottawa still controlled.[7] It
awarded an automatic twenty-five per cent share in every new oil development to a Crown corporation, Petro-Canada, and it slapped new taxes
on gas and oil at the wellhead, along with gas from the refinery and at
the pump.

How could Ottawa do this? Didn't the Prairie provinces control their
resources? Did the federal government have the right to clamp those
taxes on oil and gas at the wellhead? The Constitution, as always, was
ambiguous. Each level of government had "substantially enough power
to thwart the other."[8] Although Alberta controlled the resources, the
federal government could flex its power to regulate interprovincial and
international trade—or invoke its right "to make laws for the Peace,
Order, and good Government of Canada."[9] The Gang of Three would
not have been surprised to read the National Energy Program's declaration that energy was a "national patrimony."[10]

The double standard outraged the West. Ottawa "would never have
dared to set prices, control exports, or impose taxes on hydroelectric
power, an important source of revenue in Ontario and Quebec."[11] The
reaction from Alberta was swift. Lougheed announced a cutback in oil
production, withheld approval for Oil Sands mega-projects, and launched
a constitutional challenge. There were more bumper stickers: "I'd rather
push this thing a mile than buy gas from PetroCan."[12] Resentment percolated across the West. The famous bumper sticker from the 1970s became
Western shorthand in the early 1980s. "Let the Eastern Bastards freeze in
the dark"—the raw phrase somehow captured those very long decades
of frustration with the Rest of Canada.

Although Lougheed and Trudeau eventually concluded a compromise
deal on the NEP in August 1981, the resource-rich Provincial premiers now
knew they needed to spell out their rights. When nine of the ten premiers
finally reached a deal on patriation of the Constitution along with an

amending formula and a Charter of Rights and Freedoms, there was a hefty addition to the list of provincial powers. Provinces could now extract revenues "by any mode or system of taxation" from their non-renewable resources, their forestry resources, and their hydroelectric facilities. They could also "exclusively make laws" for the exploration, development, conservation and management of those resources.[13] It was a triumph for Blakeney and Lougheed—the kind their predecessors had only imagined.

More federal concessions were to come. In 1985 Conservative Prime Minister Brian Mulroney allowed the Government of Newfoundland and Labrador to tax offshore oil production as if it owned those resources— although Ottawa retained official constitutional possession. In 1986, Ottawa and Nova Scotia concluded a similar pact. Mulroney also deregulated the price of oil, allowing domestic prices to track the world price. Ironically, international prices fell because of a glut of oil, and Western Canadian economies took years to recover.

Today, scuffles over resources persist—and do so because Canada is a federation. With provincial and federal governments sharing jurisdiction over the environment, the provinces and Ottawa still jostle with each other over resource management. Theoretically, the federal government could impose a carbon tax to curb greenhouse-gas emissions, but that it is unlikely: the memory of federal Liberal leader Stéphane Dion's fate remains strong. Still, the battles over resource control that Louis Riel launched can resurface with astonishing rapidity. When Métis forces long ago stopped the Canadian surveyors, the focus was on the land that the newcomers craved and the long-time inhabitants claimed. Today, those anxieties are concentrated on the Oil Sands of northeastern Alberta, and NDP leader Thomas Mulcair has deepened that alarm with the charge that Ottawa is not enforcing existing federal environmental laws. Would he impose a carbon tax if he won power? The West is also rich with oil and gas, coal, potash, minerals, uranium and forests.

Canadians remain conflicted about this wealth, and the ambiguities run deep. Those resources offer security in an age of escalating energy prices and uncertain access. They brand Canada as the proud custodian of great good fortune. They also evoke unease about pollution and carbon emissions and blighted landscapes. Those deposits of oil and gas within the Western Canadian Sedimentary Basin and along the three coasts have shifted the power and wealth of the nation from the centre to the periphery. They are changing Canadian society. The federal architects of the NEP could foresee this future when they asserted that energy "has always been a special case" within the federation.[14] The sadness is that they could make this declaration of power and control without irony or much historical insight.

Today, the edgy world of Riel and Haultain and the Gang of Three has not vanished. Their legacy lives in the residual memory. At its best, it can evoke a sense of pride and fierce stewardship: those politicians fought to assert their rights against Ottawa's colonizing power. At its most truculent, it can create a firewall around a province, fostering notions of entitlement and resentment. When Newfoundland and Labrador premier Danny Williams lowered the Canadian flag in December 2004 to protest the inclusion of resource royalties in the calculation of his province's equalization payments, many Newfoundlanders cheered his extreme antics. It was a reflexive reaction to protect their newfound wealth, bred in the bone after centuries of status as a colonial afterthought. When Alberta's Wildrose leader Danielle Smith argued during the April 2012 provincial election that the science around climate change had not been proven, she appealed to voters who did not want to create openings for federal meddling. In effect, she evoked memories of the days of two-priced oil and the National Energy Program.

There is fortunately, however, a New West. It honours the legacy of those homesteaders who broke the soil and those ranchers who battled the droughts; those miners who trudged under mountains and those displaced Aboriginals who fought for better lives; those railroad workers who spanned the nation, those doughty suffragettes who faced down Manitoba premier Sir Rodmond Roblin, and those "foreign" immigrants who were crammed into urban slums or sod huts. It also honours the legacy of those largely Anglo-British and mostly male politicians who chafed at their treatment from the Rest of Canada—and who took pride in the West's development.

Ottawa almost certainly needed to retain control of the West's resources in the nineteenth century to attract homesteaders and to subsidize railroads (although its treatment of the Métis was shameful). But the continuation of that control into the twentieth century deserves a far more skeptical judgment. In those early decades, successive prime ministers and their strong-willed interior ministers largely disregarded the West's claims, partly because the resources were sources of power and patronage. Resource control ensured that Ottawa could foster settlement, and perhaps most important, the Rest of Canada's claims to a share of the West's wealth were too difficult and too expensive to solve. When the Group of Three united in 1913 after a difficult meeting with Prime Minister Borden, they sparked the West's first Quiet Revolution against the Rest of Canada's condescension.

Today, so much has changed. The New West has power and confidence. It also has economic clout. But it still took courage for Alberta premier Alison Redford to run as a bridge-builder with the Rest of Canada in the April 2012 provincial election. Redford promised to pursue a Canadian energy strategy. She vowed to engage "environmental stakeholders in what economic development looks like in Alberta . . . to grow an energy economy across this country." Perhaps most important, after her decisive win, she made a pivotal observation: "Albertans are proud to be Canadians. . . . We are from everywhere in this country, and are very proud, I think, to have those ties back to other provinces, and are very confident that as we move ahead and continue to succeed, Canada will as well."[15] It could be the beginning of a new era.

Notes

—

PREFACE

1. I was at that meeting, and I vividly remember Dion's words and my consternation. Dion meant that the West's economies would have to diversify—and their universities would devise new methods to curb emissions. But he did not express himself well. Or perhaps he did.

2. "Dion's carbon tax would 'screw everybody,' PM says." *Globe and Mail*, Sat., June 21, 2008. p. 6.

3. Macdonald even contributed to PEI's compulsory purchase of land from large landholders.

4. "Redford's energy vision clashes with McGuinty's view of oil-sands benefits," *Globe and Mail*, Tuesday, Feb. 28, 2012. The quotation was from Alberta premier Alison Redford.

5. The Canadian Press, Tues. Feb. 28, 2012. As cited on the CBC site: http://www.cbc.ca/news/canada/saskatchewan/story/2012/02/28/sk-brad-wall-oilsands-120228.html

CHAPTER ONE

1. "Prov. Ministers Are Welcomed By Sir Thomas White," *Ottawa Citizen*, Tues. Nov. 19, 1918, p. 1.

2. Henry Borden, ed., *Robert Laird Borden: His Memoirs* (Toronto: Macmillan Company of Canada Limited, at St. Martin's House, 1938), p. 769.

3. Ibid., p. 860.

4. U.S. president Woodrow Wilson. "Fourteen Points" speech to a joint session of the U.S. Congress on January 8, 1918. This pivotal speech outlined the U.S. war aims.

5. James Trow, *Manitoba and North West Territories: Letters*. First published in Ottawa in 1878 by the Department of Agriculture (Toronto: Canadiana House, 1970), p. 5.

6. Ibid., p. 22.

7. Ibid., p. 36.

8. Donald B. Smith, *Calgary's Grand Story* (Calgary: University of Calgary Press, 2005), p. 84.

9. Lt. Col. G.B. "Buck" Buchanan, *The March of the Prairie Men: The History of the South Saskatchewan Regiment*, chap. 1, p. 2. http://cap.estevan.sk.ca/SSR/mpm .html. 1956.

10. Richard J.A. Prechtl, *Take the Soil in Your Hands* (Saskatoon: Herrem Publishing Company, 1984), p. 71.

11. Ibid.

12. Memorandum of August 9, 1917, from H.H. Rowatt, Controller of the Mining Lands & Yukon Branch of the Ministry of the Interior. LAC. Prime Ministers' Fonds, C4405, pp. 125975–78. Vol. 225, MG26 H, p. 125977.

13. Alberta premier Arthur L. Sifton, Saskatchewan premier Walter Scott and Manitoba premier R.P. Roblin to Prime Minister Sir Robert Borden. Correspondence of Dec. 22, 1913. LAC. Prime Ministers' Fonds. C4207, pp. 4882–83. Vol. 18, MG26 H, p. 4883.

14. "Premier Borden Crosses Safely." *Toronto Mail and Empire*. This was a Reuters Dispatch paraphrase of Borden's sentiments, Monday, Nov. 18, 1918, p. 1.

15. Letter from Manitoba premier T.C. Norris to Prime Minister Arthur Meighen. Correspondence of Dec. 10, 1920. LAC. Prime Ministers' Fonds. C3428, pp. 23328–48. Vol. 40, MG26 I, p. 23329.

16. "Want Lands And Money, Too," *The Globe*. Wednesday, Nov. 20, 1918, p. 2.

17. Memorandum presented by British Columbia Premier John Oliver in "Minutes of the Proceedings in Conference" for November 1918 from *Dominion Provincial and Interprovincial Conferences from 1887 to 1926* (Ottawa: Reprinted by Edmond Cloutier, King's Printer and Controller of Stationery, 1951), p. 99.

18. Memorandum Re: Per Capita Subsidy. n.d. Archives of Ontario. Canada-Dominion-Provincial Conference. MS1652. RG 3-3-0-13, p. 5.

19. Ontario premier William H. Hearst to Prime Minister Sir Robert Borden. Correspondence of February 22, 1918. LAC. Prime Ministers' Fonds. C4207, pp. 4973–75. Vol. 18, MG26 H, pp. 4973–74.

20. Université Laval historian Richard Jones, "Gouin, Sir Lomer," *Dictionary of Canadian Biography Online*. http://www.biographi.ca/009004-110.01-e.php?PHPSESSID=a4t2aqq tg4a4f34lum8sohn462&q2=lomer+gouin&q3=&q10=&q7=&q5=&q1=&interval=20.

21. Ibid.

22. Former Prince Edward Island premier A.E. Arsenault, *Memoirs of The Hon. A.E. Arsenault*. n.d. Publisher not stated, pp. 68–69.

23. Alberta premier Charles Stewart to Prime Minister Sir Robert Borden. Telegram of Sept. 27, 1918. LAC. Prime Ministers' Fonds. C4207, p. 4977. Vol. 18, MG26 H. (My italics.)

24. Ibid.

25. "Canadian Soldiers in Favor of Farming," *Toronto Star*, Thurs., Nov. 21, 1918, p. 5.

CHAPTER TWO

1. Charles Mair in a letter to his brother Holmes Mair, in W.L. Morton, ed., *Alexander Begg's Red River Journal and Other Papers Relative to the Red River Resistance of 1869–1870* (Toronto: The Champlain Society, 1956), p. 395.

2. Ibid., p. 398.

3. Ibid., p. 396.

4. W.L. Morton, ed., "L.R." in a Feb. 25, 1869 letter to *Le Nouveau Monde*, in *Alexander Begg's Red River Journal*, pp. 399–402.

5. Louis Riel as quoted in Lewis H. Thomas, *The Dictionary of Canadian Biography* http://www.biographi.ca/009004-119.01-e.php?BioId=39918, p. 2.

6. Report from the Select Committee of the British House of Commons on the Hudson's Bay Company in August, 1857, iii–iv.

7. Address to Her Majesty the Queen from the Senate and House of Commons of the Dominion of Canada. Journals of the House of Commons, 1867–1868, p. 108, as reprinted in Douglas Owram and R.C. Macleod, eds., *The Formation of Alberta: A Documentary History* (Calgary: Historical Society of Alberta, 1979), p. 21.

8. Under-Secretary of State for the Colonies Sir Frederic Rogers to Hudson's Bay Company governor Sir Stafford Northcote. Letter from Downing Street, Feb. 22, 1869. Copy or extracts of correspondence between the Colonial Office, the Government of the Canadian Dominion, and the Hudson's Bay Company, relating to the Surrender of Rupert's Land by the Hudson's Bay Company, and for the Admission thereof into the Dominion of Canada. No. 19. P. 37: http://www.Canadiana.org/view/30656/002

9. Donald Creighton, "Macdonald, Confederation and the West," in Donald Swainson, ed., *Historical Essays on the Prairie Provinces* (Toronto: McClelland & Stewart, 1970), p. 64.

10. W.L. Morton, *Manitoba: A History* (Toronto: University of Toronto Press, 1967), p. 105.

11. Ibid.

12. Memoir by Louis Riel on the Course and Purpose of the Red River Resistance on January 22, 1874, in Morton, ed., *Alexander Begg's Red River Journal*, p. 528.

13. Abbé Georges Dugas to Bishop Alexandre-Antonin Taché in Morton, ed., *Alexander Begg's Red River Journal*, p. 409.

14. Ibid., p. 410.

15. "A Great Want," *Nor'-Wester*, Winnipeg, Sept. 21, 1869, p. 2.

16. "Lieut.-Governor McDougall," *Nor'-Wester*, Winnipeg, Oct. 26, 1869, p. 1.

17. Sir Donald A. Smith, Confidential letter to Secretary of State for the Provinces Joseph Howe. April 12, 1870. LAC. Prime Ministers' Fonds. C1523. Vol. 103. MG26 A. 41842.

18. W.F. Butler, *The Great Lone Land* (Middlesex, England: The Echo Library, 2006), 73–74. (My italics.)

19. Richard Gwyn, *Nation Maker: Sir John A. Macdonald: His Life, Our Times* (Toronto: Random House Canada, 2011), 100–01.

20. Declaration of the Inhabitants of Rupert's Land and the North-West, proclaimed on Dec. 8, 1869, in W.L. Morton, ed. *Alexander Begg's Red River Journal*, p. 448.

21. Ibid., p. 449.

22. Sir John A. Macdonald to John Rose, Dec. 5, 1869. LAC. Prime Ministers' Fonds. C28. Vol. 516, part 3, MG26 A. 649. Donald Creighton has inaccurately transcribed this quotation in *John A. Macdonald: The Old Chieftain*.

23. "A Memorandum by Sir John A. Macdonald on the Necessity of Delaying the Transfer," Dec. 16, 1869. Report of a Committee of the Honorable the Privy Council, in Canada in Sessional Papers, 1870, V, 12, in Morton, ed., *Alexander Begg's Red River Journal*, pp. 450–54.

24. Ibid., p. 451.

25. Ibid., pp. 141–44.

26. Ibid., p. 454.

27. Ibid., p. 454.

28. Sir Donald A. Smith to Secretary of State for the Provinces Joseph Howe on April 12, 1870. LAC. Prime Ministers' Fonds. C1523. Vol. 103, MG26 A. 41835.

29. "Convention at Fort Garry: Very Important Debates: The Bill of Rights" in The *New Nation*, Feb. 11, 1870, p. 2.

30. Alexander Begg in a journal entry, Sat., Feb. 5, 1870, in Morton, ed., *Alexander Begg's Red River Journal*, pp. 295–96.

31. Ibid., p. 296.

32. Métis Bill of Rights, adopted at Fort Garry on Dec. 1, 1869. http://www.history -canada.com/sections/documents/thewest/metisbillrights.htm

33. Attachment to the confidential letter of Sir Donald A. Smith to Secretary of State for the Provinces Joseph Howe. LAC. Prime Ministers' Fonds. C1523. Vol. 103, MG26 A, 41840–44.

34. George Young, *Manitoba Memories* (Toronto: William Briggs, 1897), pp. 63–64.

35. Ibid., p. 64.

36. Ibid., p. 75.

37. Old Mennonite minister John F. Funk in Lawrence Klippenstein and Julius G. Toews, eds., *Mennonite Memories: Settling in Western Canada* (Winnipeg: Centennial Publications, 1977), p. 14.

38. James Trow, *Manitoba and North West Territories: Letters*, first published in Ottawa by the Department of Agriculture (Toronto: Canadiana House, 1970), p. 22.

39. J.W. Taylor to Charles John Brydges on January 5, 1870, LAC, Prime Ministers' Fonds. C1522. Vol. 102, MG26 A. 41071–74.

40. "A Memorandum by Sir John A. Macdonald on the Necessity of Delaying the Transfer," Dec. 16, 1869, in Morton, ed., *Alexander Begg's Red River Journal*, pp. 450–54.

41. *Dictionary of Canadian Biography*: http://www.biographi.ca/009004-119.01-e.php ?BioId=38817

42. W.F. Butler, *The Great Lone Land* (Middlesex, England: Echo Library, 2006), p. 75.

43. Testimony of Charles Mair, "Report of the Select Committee of the Senate on the Subject of Rupert's Land, Red River, and the North-West Territory Together with the Minutes of Evidence" (Ottawa: The Senate, April 25, 1870), p. 36.

44. Report of the Select Committee of the Senate on the Subject of Rupert's Land, Red River, and the North-West Territory Together with the Minutes of Evidence (Ottawa: The Senate, April 25, 1870), p. 2.

45. Lord Granville to Sir John Young (Lord Lisgar) on April 23, 1870. LAC. Prime Ministers' Fonds. C1522. Vol. 101, part. 2. MG26 A. 40593–95.

46. Alexander Begg in a journal entry. Mon., April 18, 1870. *Alexander Begg's Red River Journal and Other Papers Relative to the Red River Resistance of 1869–1870*. ed. W.L. Morton (Toronto: The Champlain Society, 1956), p. 359.

47. Ibid., p. 364.

48. Hansard, Monday, May 2, 1870. 1298.

49. Manitoba Act, 1870. Clause 30.

50. Hansard, Monday, May 2, 1870. 1319.

51. Special Envoy of the Crown Sir Clifford Murdoch to Under-Secretary of State for the Colonies Sir Frederic Rogers, April 28, 1870. LAC. Prime Ministers' Fonds. C1523. Vol. 103, MG26 A. 41640.

52. Special Envoy of the Crown Sir Clifford Murdoch to Under-Secretary of State for the Colonies Sir Frederic Rogers. April 28, 1870. LAC. Prime Ministers' Fonds. C1523. Vol. 103, MG26 A. 41641.

53. "Legislative Assembly of Assiniboia: Third Session" report from June 24, 1870, *New Nation*, July 1, 1870, p. 3.

54. Manitoba Act, 1870. Clause 32, subsection 4.

55. Manitoba Act, 1870. Clause 31.

56. Albert Braz, *The False Traitor: Louis Riel in Canadian Culture* (Toronto: University of Toronto Press, 2003), p. 197.

CHAPTER THREE

1. Instructions issued to Lieutenant-Governor Adams Archibald by Under-Secretary of State for the Provinces E.A. Meredith, Aug. 4, 1870, in E.H. Oliver, *The Canadian North-West* (Ottawa: Government Printing Bureau, 1915), p. 974.

2. Ibid., p. 975.

3. Ibid., p. 974. (My italics.)

4. Ordinance passed by the Lieutenant-Governor and Council of Rupert's Land and the North-Western Territories for the prevention of smallpox, Oct. 22, 1870. Oliver, ed. *The Canadian North-West*, p. 977.

5. Ruth Swan, "Robert A. Davis: 1874–1878" in Barry Ferguson and Robert Wardhaugh, eds., *Manitoba Premiers of the 19th and 20th Centuries* (Regina: Canadian Plains Research Center, University of Regina, 2010), p. 42.

6. Gerald Friesen, *The Canadian Prairies: A History* (Toronto: University of Toronto Press, 1987), p. 182.

7. Preamble to the Constitution Act, 1871.

8. Memorandum cited in Alan F.J. Artibise, *Winnipeg: A Social History of Urban Growth, 1875–1914* (Montreal and London: McGill-Queen's University Press, 1975), p. 62.

9. Account of Dr. Lachlan Taylor, cited in George Young, *Manitoba Memories: Leaves From My Life in the Prairie Province, 1868–1884* (Toronto: William Briggs, 1897), p. 255.

10. Ibid., p. 253.

11. Peter T. Barkman, "Building the Reserves: Steinbach Mills," in Lawrence Klippenstein and Julius G. Toews, eds., *Mennonite Memories: Settling in Western Canada.* (Winnipeg: Centennial Publications, 1977), p. 59.

12. Jacob Y. Schantz, Report to the Department of Agriculture in Ottawa, cited by Lawrence Klippenstein in "Native Neighbors" in *Mennonite Memories: Settling in Western Canada*, pp. 42–43.

13. W.F. Butler, *The Great Lone Land* (Teddington: Echo Library, 2006), p. 120.

14. Peter Erasmus, *Buffalo Days and Nights* (Calgary: Fifth House Ltd., 1999), p. 226.

15. Ibid., p. 244.

16. Ibid., p. 249.

17. Minutes of the North-West Territories Council, Dec. 4, 1874. Transcript in Oliver, ed., *The Canadian North-West*, pp. 1031–33.

18. James Trow, *Manitoba and North West Territories: Letters*, first published in Ottawa by the Department of Agriculture (Toronto: Canadiana House, 1970), p. 77.

19. Grant MacEwan, *Fifty Mighty Men* (Saskatoon: Western Producer Prairie Books, 1958), pp. 266, 272.

20. G.A. Friesen, "John Norquay: 1878–1887," in Barry Ferguson and Robert Wardhaugh, eds., *Manitoba Premiers of the 19th and 20th Centuries* (Regina: Canadian Plains Research Center, University of Regina, 2010), p. 51.

21. Dominion Lands Regulations. LAC. Prime Ministers' Fonds. C1525. Vol, III, MG26A. 45097.

22. Friesen, "John Norquay: 1878–1887," p. 58.

23. Canada Sessional Papers, 1882. Vol. 15, No. 10, paper 82A, p. 1.

24. James Clinkskill, *A Prairie Memoir: The Life and Times of James Clinkskill 1853–1936.* (Regina: Canadian Plains Research Center, 2003), p. 3.

25. Ibid., pp. 11, 15.

26. Clinkskill, *A Prairie Memoir*, p. 25.

27. Canada Sessional Papers, 1883. Vol. 16, no. 12, paper 108, p. 10.

28. Ibid., p. 11.

29. Canada Sessional Papers, 1885. Volume 18, Number 12, paper 61, p. 3.

30. Ibid., p. 3.

31. "The Pow-Wow," *Winnipeg Daily Sun*, Friday, Dec. 19, 1884. p. 1.

32. Ibid.

33. Ibid.

34. Ibid.

35. "Messrs. Norquay and Murray Still Hanging On," *Winnipeg Daily Sun*. Sat., Jan. 10, 1885, p. 1.

36. "Better or Worse?" *Winnipeg Daily Sun*. Feb. 24, 1885, p. 1.

37. "The Better Terms," *Winnipeg Daily Sun*, March 2, 1885, p. 1.

38. Telegram from George Purvis to the Marquess of Lansdowne. LAC. Prime Ministers' Fonds. C1523. Vol. 106. MG26 A. 42401.

39. "Premier Norquay" in the *Daily Manitoban*, Mon. Evening, Aug. 17, 1885, p. 1.

40. Ibid.

41. "Canada's Immigration," *Winnipeg Daily Sun*, March 25, 1885, p. 1.

42. Ron Graham, *The Last Act: Pierre Trudeau, the Gang of Eight, and the Fight for Canada* (Toronto: Penguin Group (Canada), 2011) p. 124.

43. As quoted in David Breen, *The Canadian Prairie West and the Ranching Frontier 1874–1924* (Toronto: University of Toronto Press, 1983), p. 13.

44. Riel's Revolutionary Bill of Rights (1885) in Maggie Siggins, *Riel: A Life of Revolution*: www.mcgill.ca/files/maritimelaw/U.doc

45. Prime Minister Sir John A. Macdonald to Manitoba Lieutenant-Governor James Cox Aikins, June 25, 1887, as quoted in Donald Creighton, *John A. Macdonald: The Old Chieftain* (Toronto: Macmillan of Canada, 1965), p. 478.

46. Mercier headed a breakaway, largely Liberal group of Quebec nationalists after 1885, the Parti National.

47. W.L. Morton, *Manitoba: A History* (Toronto: University of Toronto Press, 1957), p. 219.

48. Frederick Haultain, *Assembly Journals, 1889*, as quoted in Lewis Herbert Thomas, *The Struggle for Responsible Government in the North-West Territories 1870–1897*. (Toronto: University of Toronto Press, 1956), p. 177.

49. The North-West Territories Legislature to Interior Minister Edgar Dewdney, "Memorandum concerning the form of the Government and the finances of the Territories." Nov. 21, 1889, in Oliver, ed. *The Canadian North-West*, p. 1117.

50. "Sir John's Death In Calgary," *Calgary Daily Herald*, Mon., June 8, 1891, p. 1. http://www.ourfutureourpast.ca/newspapr/np_page2.asp?code=n8up0289.jpg

51. An Act to Amend the Acts respecting the North-West Territories. 30 Sept. 1891. Clause 6, subsection 12. (My italics.)

52. A.I. Silver in the *Dictionary of Canadian Biography* On-Line. http://biographi.ca/009004-119.01-e.php?id_nbr=7041

53.   Charles Herbert Mackintosh to Prime Minister Sir John Thompson, Jan. 4, 1894, as quoted in Thomas, *The Struggle for Responsible Government*, p. 235.

54.   Clinkskill, *A Prairie Memoir*, p. 102.

55.   Interior Minister Thomas Mayne Daly to North-West Territories Executive Council Chairman Frederick Haultain, July 29, 1894, in Douglas R. Owram, ed., *The Formation of Alberta: A Documentary History* (Calgary: Historical Society of Alberta, 1979), p. 67.

56.   North-West Territories Executive Council Chairman Frederick Haultain to Interior Minister Thomas Mayne Daly. July 18, 1894 in Owram, ed., *The Formation of Alberta: A Documentary History*, p. 67. (My italics.)

57.   Interior Minister Thomas Mayne Daly to North-West Territories Executive Council Chairman Frederick Haultain, July 28, 1894, ibid., p. 68.

58.   North-West Territories Executive Council Chairman Haultain to Interior Minister Thomas Mayne Daly. August 5, 1894, ibid., p. 68.

59.   Author Unnamed, "North and West: Homesteading at Rosthern," in Klippenstein and Toews, eds., *Mennonite Memories*, p. 183.

60.   "The Territories Asserting Its Claim—More Power and More Money Necessary in Its Government," *Daily Nor'Wester*, Oct. 10, 1896, p. 7.

61.   Jim Shilliday, *Canada's Wheat King: The Life and Times of Seager Wheeler* (Regina: Canadian Plains Research Center, University of Regina, 2007), p. 59.

CHAPTER FOUR

1.   Haultain is the only premier to be portrayed with a cigarette, so the guides have dubbed the space in front of his portrait "the smoking section."

2.   Grant MacEwan, *Frederick Haultain: Frontier Statesman of the Canadian Northwest* (Saskatoon: Western Producer Prairie Books, 1985), p. 12.

3.   Ninette Kelley and Michael Trebilcock, *The Making of the Mosaic: A History of Canadian Immigration Policy* (Toronto: University of Toronto Press, 1998), pp. 111–12.

4.   Father Nestor Dmytriw, "Canadian Ruthenia," in Joanna Matejko, ed., *Land of Pain, Land of Promise: First Person Accounts by Ukrainian Pioneers 1891–1914*, researched and translated by Harry Piniuta (Saskatoon: Western Producer Prairie Books, 1978), p. 41.

5.   Ibid., p. 47.

6.   Maria Adamowska, "Beginnings in Canada," in Matejko, ed., *Land of Pain, Land of Promise*, pp. 60–61.

7.   Johanna Allan, "Marian Plachner of Skaro," in *Polish Settlers in Alberta: Reminiscences and Biographies* (Toronto: Polish Alliance Press Ltd., 1979), p. 146.

8.   Patrick Gammie Laurie, *Dictionary of Canadian Biography* http://www.biographi.ca /009004-119.01-e.php?&id_nbr=6844&interval=25&&PHPSESSID=r4ea7j7pio708a celtte4rhpk1

9.   "More Galicians: Over Seven Hundred Arrived last Night from the East—Bound for Saskatoon and Yorkton," *Daily Nor'Wester*, May 17, 1898, p. 1.

10. "Galician Riff-Raff: The Comment of an Alberta Paper on Mr. Sifton's Pets," *Daily Nor'Wester*, May 17, 1898, p. 4.

11. James Clinkskill, *A Prairie Memoir: The Life and Times of James Clinkskill, 1853–1936* (Regina: Canadian Plains Research Center, University of Regina, 2003), p. 128.

12. North-West Territories Premier Sir Frederick Haultain to Sir Wilfrid Laurier. March 13, 1897. LAC, Prime Ministers' Fonds. C748. Vol. 40, MG26 G. 13029.

13. North-West Territories Premier Sir Frederick Haultain to Sir Wilfrid Laurier. March 13, 1897. LAC, Prime Ministers' Fonds. C748. Vol. 40, MG26 G. 13042.

14. Dmytro Romanchych, "The Dauphin District," in Matejko, ed., *Land of Pain, Land of Promise*, p, 101.

15. Father Nestor Dmytriw, "Canadian Ruthenia," in Matejko, ed., *Land of Pain, Land of Promise*, p. 37.

16. D.J. Hall, Clifford Sifton: *The Lonely Eminence, 1901–1929*, vol. 2 (Vancouver: University of British Columbia Press, 1985), p. 26.

17. *Canadian Annual Review of Public Affairs*, 1902 as quoted in Kelley and Trebilcock, *The Making of the Mosaic*, p. 118.

18. Sir Frederick Haultain to Sir Clifford Sifton, Jan. 14, 1899. Sessional Paper No. 23 in Doug Owram and R.C. MacLeod, *The Formation of Alberta: A Documentary History*. (Calgary: The Historical Society of Alberta, 1979), p. 114.

19. Sir Frederick Haultain to Sir Clifford Sifton, Jan. 14, 1899. Sessional Paper No. 23 in Owram and MacLeod, eds., *The Formation of Alberta*, p. 114.

20. "Estimates Go Through," *The Globe*, Sat., July 15, 1899, p. 23.

21. Sir Frederick Haultain to Sir Clifford Sifton, July 16, 1899. Unpublished Sessional Paper No. 1 of the North West Territories, 1900 in Owram and MacLeod, eds., *The Formation of Alberta*, p. 115.

22. Sir Clifford Sifton to Sir Frederick Haultain. August 1, 1899. Ibid.

23. Manitoba Premier Thomas Greenway to Prime Minister Sir Wilfrid Laurier, Feb. 8, 1898. LAC, Prime Ministers' Fonds. C754. Vol. 64, MG26 G. 20458.

24. Prime Minister Sir Wilfrid Laurier to Manitoba premier Thomas Greenway, Feb. 11, 1898. LAC, Prime Ministers' Fonds. C754. Vol. 64, MG26 G. 20460.

25. "Haultain comes out in favour of provincial status," *Calgary Weekly Herald*, Nov. 2, 1899, in Owram and MacLeod, eds., *The Formation of Alberta*, pp. 121–23.

26. Dmytro Romanchych, "The Dauphin District," in Matejko, ed., *Land of Pain, Land of Promise*, p. 103.

27. Father Anthony Sylla, "Krakow—Father Olszewski's Settlement (1899–1910)" in Joanna Matejko, ed., *Polish Settlers in Alberta: Reminiscences and Biographies* (Toronto: Polish Alliance Press Ltd., 1979) 270–71.

28. Giovanni Veltri, *The Memoirs of Giovanni Veltri*. John Potestio, ed. (Toronto: Multicultural History Society, 1987), p. 68.

29. Ibid., pp. 72–73.

30. Alan F. J. Artibise, *Winnipeg: A Social History of Urban Growth 1874–1914* (Montreal and London: McGill-Queen's University Press, 1975), p. 177.

31. D.J. Hall, *Clifford Sifton: The Young Napoleon, 1861–1900*, vol. 1 (Vancouver: University of British Columbia Press, 1981), p. 296.

32. Sir Frederick Haultain to Sir Clifford Sifton, January 30, 1901. Canada Sessional Paper 116a, 1904, in Owram and MacLeod, eds., *The Formation of Alberta*, p. 135.

33. Sir Clifford Sifton to Sir Frederick Haultain, March 21, 1901. Canada Sessional Paper 116a, 1904, in Owram and MacLeod, eds., *The Formation of Alberta*, p. 136.

34. D.J. Hall, *Clifford Sifton: The Young Napoleon*, p. 7.

35. "Northwest Territories: The Proposal to Give Them Provincial Status," *The Globe*, Thurs., Oct. 10, 1901, p. 8.

36. "Canadians in London," *The Globe*, Wed., Oct.9, 1901, p. 1.

37. "Some Problems of the West," *The Globe*, Sat., Oct. 19, 1901, p. 28.

38. "Manufacturers' Annual Banquet," *The Globe*, Thurs., Nov. 7, 1901, p. 1.

39. Sir Frederick Haultain to Sir Clifford Sifton, Memorandum of December 7, 1901. Canada Sessional Paper 116, 1903, in Owram and MacLeod, eds., *The Formation of Alberta*, p. 143.

40. Ibid., p. 145. (My italics.)

41. Sir Clifford Sifton to Sir Frederick Haultain, March 27, 1902. Canada Sessional Paper 116a, 1904, in Owram and MacLeod, eds., *The Formation of Alberta*, p. 188.

42. Babijak Janos, "The Hungarian Colony of Esterhaz, Assiniboia, North-west Territories, Canada: Letters from the Settlers and Illustrations from Photographs taken on the Spot in the month of July, 1902," in Martin Louis Kovacs, *Esterhazy and Early Hungarian Immigration to Canada* (Regina: Canadian Plains Research Center, 1974), p. 103.

43. Ibid.

44. Robert Harvey, "Chapman of Reston," in *Pioneers of Manitoba* (Winnipeg: Prairie Publishing Company, 1970), p. 16.

45. Joseph Prechtl, "My Homesteader Experience," in Richard J.A. Prechtl, *Take The Soil In Your Hands* (Saskatoon: Herrem Publishing Company, 1984), pp. 7 and 9.

46. Sir Frederick Haultain to Sir Clifford Sifton, April 2, 1902. Canada Sessional Paper 116a, 1904, in Owram and MacLeod, eds., *The Formation of Alberta*, p. 189.

47. Speech in Lethbridge on September 18, 1902 as reported in the *Lethbridge News*, Sept. 24, 1902, in Owram and MacLeod, eds., *The Formation of Alberta*, p. 212.

48. Clark University economist J.A. Maxwell gave a perfect definition of the debt allowance in the *Journal of Political Economy*, vol. 41, no. 6 (Dec. 1933), p. 782, note 9: "The Dominion assumed the debt of a province when it was admitted to the Confederation. But, in order to get uniform treatment, a "debt allowance" was declared for each; and where the debt allowance was larger than the actual debt assumed, the Dominion paid interest at 5 per cent on the surplus."

49. Sir Frederick Haultain to Sir Clifford Sifton, Jan. 31, 1903. Canada Sessional Paper 116b, 1903, in Owram and MacLeod, eds., *The Formation of Alberta*, pp. 216–23.

50. Sir Clifford Sifton to Sir Frederick Haultain, telegram, July 23, 1903, in E.H. Oliver, ed., *The Canadian North-West* (Ottawa: Government Printing Bureau, 1915), p. 1226.

51. Louis Rosenberg, "Canada's Jews," as quoted in Simon Belkin *Through Narrow Gates: A Review of Jewish Immigration, Colonization and Immigrant Aid Work in Canada (1840–1940)* (Montreal: Eagle Publishing Company Limited, 1966), p. 77.

52. D.J. Hall, *Clifford Sifton: The Lonely Eminence, 1901–1929*, vol. 2 (Vancouver: University of British Columbia Press, 1985), p. 57.

53. "The Old Old Story," *Calgary Herald*, March 21, 1904, in Owram and MacLeod, eds., *The Formation of Alberta*, p. 250.

54. Kenneth Norrie, Douglas Owram and J.C. Herbert Emery, *A History of the Canadian Economy*, fourth edition (Toronto: Thomson Nelson, 2008), p. 250.

55. Interior Minister Sir Clifford Sifton to Finance Minister William Fielding, Aug. 20, 1903, in Owram and MacLeod, eds., *The Formation of Alberta* (Calgary: Historical Society of Alberta, 1979), p. 241.

56. Sir Wilfrid Laurier to Sir Frederick Haultain, Sept. 30, 1904, Canada Sessional Paper 53, 1905, in Owram and MacLeod, eds., *The Formation of Alberta*, p. 257.

57. *Canadian Annual Review*, 1904, p. 202.

58. Sir Clifford Sifton to Sir Wilfrid Laurier. Jan. 22, 1905. LAC. Prime Ministers' Fonds. C819. Vol. 352, MG26 G. pg. 93970.

59. Sir Clifford Sifton to Sir Wilfrid Laurier. Jan. 22, 1905. LAC. Prime Ministers' Fonds. C819. Vol. 352, MG26 G. 93970.

60. Sir Wilfrid Laurier to Sir Clifford Sifton. Jan. 26, 1905. LAC. Prime Ministers' Fonds. C819. Vol. 352, MG26 G. 93975.

61. Sir Clifford Sifton to Sir Wilfrid Laurier. Feb. 1, 1905. LAC. Prime Ministers' Fonds. C819. 94354 to 94361. Vol. 354, MG26 G. 94355.

62. Ibid.

63. Sir Clifford Sifton to Sir Wilfrid Laurier. Feb. 1, 1905. LAC. Prime Ministers' Fonds. C819. Vol. 354, MG26 G. 94356.

64. Sir Clifford Sifton to Sir Wilfrid Laurier. Feb. 1, 1905. LAC. Prime Ministers' Fonds. C819. Vol. 354, MG26 G. 94357.

65. Sir Robert Borden, Feb. 21, 1905. House of Commons debates in Hansard. 1462–1463.

66. Sir Frederick Haultain to Sir Wilfrid Laurier, March 11, 1905. Laurier papers 95679 -95691, in Owram and MacLeod, eds., *The Formation of Alberta*, p. 308.

67. C.W. Cross to Liberal MP Peter Talbot, July 3, 1905. LAC, Prime Ministers' Fonds. C824. Vol. 373, MG26 G. 99255.

68. Prime Minister Sir Wilfrid Laurier to NWT Public Works Minister George Bulyea, July 25, 1905. Laurier papers 100389-91. in Owram and MacLeod, eds., *The Formation of Alberta*, p. 366.

CHAPTER FIVE

1. D.J. Hall, *Clifford Sifton, vol. 2: The Lonely Eminence, 1901–1929* (Vancouver: University of British Columbia Press, 1985), p. 191.

2. Frank Oliver, House of Commons, April 12, 1901, Hansard. 2938.

3. Frank Oliver, House of Commons, April 12, 1901, Hansard. 2939.

4. Haultain would be knighted in 1916.

5. Frank Oliver on the second reading of the Autonomy Bill, March 24, 1905. Hansard, 3152–69 in Doug Owram and R.C. MacLeod, *The Formation of Alberta: A Documentary History* (Calgary: Historical Society of Alberta, 1979), p. 325.

6. Cartoon in the Calgary *Eye Opener*, Sept. 2, 1905. As viewed on the site of HighBeam Research: http://www.highbeam.com/doc/1g1-132046306.html

7. *Canadian Annual Review*, 1907, p. 289, as quoted in Ninette Kelley and Michael Trebilcock, *The Making of the Mosaic: A History of Canadian Immigration Policy* (Toronto: University of Toronto Press, 1998), p. 136.

8. Alberta Public Works Minister William Henry Cushing to Interior Minister Frank Oliver, Sept. 13, 1907. LAC. Prime Ministers' Fonds. C852, Vol. 478. 129238–40.

9. Gordon L. Barnhart, "Walter Scott: 1905–1916," in Gordon L. Barnhart, ed., *Saskatchewan Premiers of the Twentieth Century* (Regina: Canadian Plains Research Center, University of Regina, 2004), p. 34.

10. Howard Palmer with Tamara Palmer, *Alberta: A New History* (Edmonton: Hurtig Publishers, 1990), p. 135.

11. Alberta premier A.C. Rutherford to Prime Minister Sir Wilfrid Laurier, Jan. 10, 1906. LAC, Prime Ministers' Fonds. C830, Vol. 397, MG26 G, 105855–56.

12. Alberta premier A.C. Rutherford to Prime Minister Sir Wilfrid Laurier, April 10, 1906. LAC, Prime Ministers' Fonds. C834, Vol. 410, MG26 G, 109447–48.

13. Patricia Roome, "Alexander C. Rutherford: 1905–1910," in Bradford J. Rennie, ed., *Alberta Premiers of the Twentieth Century*, pp. 15–16.

14. J.M.S. Careless, "Aspects of Urban Life in the West, 1870–1914," in A.W. Rasporich and H.C. Klassen, eds., *Prairie Perspectives 2: Selected Papers of the Western Canadian Studies Conferences, 1970, 1971* (Toronto and Montreal: Holt, Rinehart and Winston of Canada, 1973), p,. 25.

15. Ibid., p. 38.

16. Ibid., p. 36.

17. J.F.C. Wright, *Saskatchewan: The History of a Province* (Toronto: McClelland & Stewart, 1955), p. 128.

18. Paul Boothe & Heather Edwards, *Eric J. Hanson's Financial History of Alberta, 1905–1950* (Calgary: University of Calgary Press, 2003), p. 33.

19. W.L. Morton, *Manitoba: A History* (Toronto: University of Toronto Press: 1957) 296–97.

20. Canadian History Portal, Chinook Multimedia Inc., sponsored by Industry Canada. http://www.canadianhistory.ca/iv/1867-1914/laurier_boom/takeoff3.html

21. Kelley and Trebilcock, *The Making of the Mosaic*, p. 140.

22. Sarah Ellen Roberts, *Alberta Homestead: Chronicle of a Pioneer Family*, Lathrop E. Roberts ed. (Austin: University of Texas Press, 1968), p, 18.

23. Ibid., p. 22.

24. Ibid.

25. Ibid., p. 35.

26. Prince Edward Island premier Arthur Peters to Prime Minister Sir Wilfrid Laurier, Jan. 23, 1906. LAC, Prime Ministers' Fonds. C831. Vol. 399, MG26 G. 106378–79.

27. "M. Peters est pour le moins aussi impatient que nous." (My translation.) Quebec Premier Lomer Gouin to Prime Minister Sir Wilfrid Laurier, Jan. 27, 1906. LAC, Prime Ministers' Fonds. C831. Vol. 400, MG26 G. 106532–33.

28. André Pratte, *Wilfrid Laurier* (Toronto: Penguin Group, 2011), p. 5.

29. Sir James Whitney, "Memorandum on Behalf of the Province of Ontario," in Interprovincial Conference minutes, 1906, Sessional Paper 29a, House of Commons, 1907, pp. 6–7.

30. British Columbia premier Richard McBride, "Memorandum re: British Columbia's Claims for Special Consideration," Wed. Oct. 10, 1906. Interprovincial Conference minutes, 1906, Sessional Paper 29a, House of Commons, 1907, pp. 7–11.

31. William Rayner, *British Columbia's Premiers in Profile: The good, the bad and the transient* (Surrey, B.C.: Heritage House Publishing, 2000), p. 100.

32. Sir Wilfrid Laurier Address to Conference on Friday, Oct. 12, 1906. Interprovincial Conference minutes, 1906, Sessional Paper 29a, House of Commons, 1907, p. 12.

33. Amendment proposed by Saskatchewan premier Walter Scott, seconded by Alberta premier Alexander Rutherford to the motion of Ontario premier James Whitney, Sat., Oct. 13, 1906. *Dominion Provincial and Interprovincial Conferences From 1887 to 1926.* (Ottawa: Edmond Cloutier, King's Printer and Controller of Stationery, 1951), p. 60.

34. British Columbia premier Richard McBride to Quebec premier Lomer Gouin, Oct. 13, 1906. *Dominion Provincial and Interprovincial Conferences From 1887 to 1926.* (Ottawa: Edmond Cloutier, King's Printer and Controller of Stationery, 1951), p. 62.

35. Conference chairman Lomer Gouin to British Columbia premier Richard McBride, Sat., Oct. 13, 1906. *Dominion Provincial and Interprovincial Conferences From 1887 to 1926.* (Ottawa: Edmond Cloutier, King's Printer and Controller of Stationery, 1951), p. 63. (My italics.)

36. British North America Act, 1907. Schedule.

37. Patricia K. Wood, *Nationalism from the Margins: Italians in Alberta and British Columbia* (Montreal and Kingston: McGill-Queen's University Press, 2002), p. 26.

38. Father Anthony Sylla, "Christmas among the Miners in Canmore," in Joanna Matejko, ed., *Polish Settlers in Alberta* (Toronto: Drukiem Polish Alliance Press, 1979), p. 292.

39. A. Ross McCormack, "Networks Among British Immigrants and Accommodation to Canadian Society: Winnipeg, 1900–1914," in Gerald Tulchinsky, ed., *Immigration in Canada: Historical Perspectives* (Mississauga: Copp Clark Longman Ltd., 1994), p. 213.

40. Ibid., p. 217.

41. Ibid., p. 218.

42. W.L. Morton, *Manitoba: A History* (Toronto: University of Toronto Press: 1957), p. 281.

43. Resolution of the Legislative Assembly of Manitoba, Feb. 24, 1909. LAC, Prime Ministers' Fonds. C874. Vol. 563. MG26 G. 152759–66.

44. Manitoba Premier Rodmond Roblin to Prime Minister Sir Wilfrid Laurier, Nov. 19, 1909. LAC, Prime Ministers' Fonds. C882. Vol. 599. MG26 G. 162394–401.

45. Prime Minister Sir Wilfrid Laurier to Manitoba premier Rodmond Roblin, Nov. 30, 1909. LAC, Prime Ministers' Fonds. C882. Vol. 599. MG26 G. 162394–401.

46. Manitoba premier Rodmond Roblin to Prime Minister Sir Wilfrid Laurier, Oct. 17, 1910. LAC, Prime Ministers' Fonds. C895. Vol. 647. MG26 G. 175842.

47. Manitoba premier Rodmond Roblin to Prime Minister Sir Wilfrid Laurier, November 5, 1906. LAC, Prime Ministers' Fonds. C839. Vol. 432. MG26 G. 175842. 115338.

48. British North America Act of 1867. Clause 91, Section 24.

49. Richard Spaulding, "Executive Summary," in Peggy Martin-McGuire, *First Nation Land Surrenders on the Prairies, 1896–1911*, prepared for the Indian Claims Commission (Ottawa, 1998), p. xlvi. This recent federal study looked at twenty-five land surrenders around the turn of the last century. http://www.indianclaims.ca /pdf/FNLS_summary_eng.pdf

50. Ibid.

51. D.J. Hall, "Clifford Sifton and Canadian Indian Administration, 1896–1905," in Gregory P. Marchildon, *Immigration & Settlement, 1870–1939* (Regina: Canadian Plains Research Center, University of Regina, 2009), p. 199.

52. J. William Brennan, "The 'Autonomy Question' and The Creation of Alberta and Saskatchewan, 1905," in Howard Palmer and Donald Smith, eds., *The New Provinces: Alberta and Saskatchewan 1905–1980* (Vancouver: Tantalus Research, 1980), p. 55.

53. Alberta Public Works minister William Henry Cushing to Alberta premier Alexander Rutherford, Feb. 14, 1910: http://www.abheritage.ca/telephone/era /cushing.html

54. Patricia Roome, "Alexander C. Rutherford: 1905–1910" in Rennie, ed., *Alberta Premiers of the Twentieth Century*, p. 12.

55. Simon Belkin, *Through Narrow Gates: A Review of Jewish Immigration, Colonization and Immigrant Aid Work in Canada (1840–1940)* (Montreal: Eagle Publishing, 1966), p. 81.

56. Ernest R. Forbes, *Maritime Rights: The Maritime Rights Movement, 1919–1927: A Study in Canadian Regionalism* (Montreal: McGill–Queen's University Press, 1979), p. 15.

57. Christopher Armstrong, "Ceremonial Politics: Federal–Provincial Meetings Before the Second World War," in R. Kenneth Carty and W. Peter Ward, eds. *National Politics and Community in Canada* (Vancouver: University of British Columbia Press, 1986), p. 118.

58. Minutes of the Proceedings in Conference of the Representatives of the Provinces of the Dominion of Canada, December 1910, in *Dominion Provincial and Interprovincial Conferences from 1887 to 1926* (Ottawa: Edmond Cloutier, King's Printer and Controller of Stationery, 1951), p. 67.

59. Saint John MP J.W. Daniels, House of Commons Debates, in January 1911, 2732–34, as quoted in Forbes, *Maritime Rights*, p. 16.

60. Resolution of Alwyn Bramley-Moore, as quoted in David Hall, "Arthur Sifton: 1910–1917," in Rennie, ed., *Alberta Premiers of the Twentieth Century*, p. 29.

61. *Alberta premier Arthur L. Sifton to Prime Minister Sir Wilfrid Laurier, Correspondence of March 20,1911. LAC. Prime Ministers' Fonds. C906. Vol. 690, MG26 G, 189016–28.

62. "Call Of West For Unity Of Provinces: Premier A.L. Sifton of Alberta at Canadian Club," *The Globe*, Tues., Jan. 24, 1911, p.7.

63. Alberta premier Arthur L. Sifton to Prime Minister Sir Wilfrid Laurier, March 20, 1911. LAC. Prime Ministers' Fonds. C906. Vol. 690, MG26 G. 189017.

64. Alberta premier Arthur L. Sifton to Prime Minister Sir Wilfrid Laurier, March 20, 1911. LAC. Prime Ministers' Fonds. C906. Vol. 690, MG26 G. 189020.

65. Alberta premier Arthur L. Sifton to Prime Minister Sir Wilfrid Laurier, March 20, 1911. LAC. Prime Ministers' Fonds. C906. Vol. 690, MG26 G. 189022.

66. Alberta premier Arthur L. Sifton to Prime Minister Sir Wilfrid Laurier, March 20, 1911. LAC. Prime Ministers' Fonds. C906. Vol. 690, MG26 G. 189022.

67. Alberta premier Arthur L. Sifton to Prime Minister Sir Wilfrid Laurier, March 20, 1911. LAC. Prime Ministers' Fonds. C906. Vol. 690, MG26 G. 189023.

68. Alberta premier Arthur L. Sifton to Prime Minister Sir Wilfrid Laurier, March 20, 1911. LAC. Prime Ministers' Fonds. C906. Vol. 690, MG26 G. 189025.

69. Prime Minister Sir Wilfrid Laurier to Alberta premier Arthur L. Sifton, Aug. 7, 1911. LAC. Prime Ministers' Fonds. C906. Vol. 690, MG26 G. 189030–33. (Draft of Aug. 5, 1911, is at LAC. Prime Ministers' Fonds. C906. pages 188571–79. Vol. 689, MG26 G. Two other drafts dated Aug. 7, 1911, are at LAC. Prime Ministers' Fonds. C906. pages 189007–15. Vol. 690, MG26 G.)

70. Prime Minister Sir Wilfrid Laurier to Alberta premier Arthur L. Sifton, Aug. 7, 1911. LAC. Prime Ministers' Fonds. C906. Vol. 690, MG26 G. 189032.

71. Prime Minister Sir Wilfrid Laurier to Alberta premier Arthur L. Sifton, Aug. 7, 1911. LAC. Prime Ministers' Fonds. C906. Vol. 690, MG26 G. 189032.

72. Alberta premier Arthur L. Sifton to Prime Minister Sir Wilfrid Laurier, Aug. 18, 1911. LAC. Prime Ministers' Fonds. C906. Vol. 690, MG26 G. 189006.

73. Prime Minister Sir Wilfrid Laurier to Alberta premier Arthur L. Sifton, Aug. 11, 1911. LAC. Prime Ministers' Fonds. C906. Vol. 689, MG26 G. 188579.

74. Alberta premier Arthur L. Sifton to Prime Minister Sir Wilfrid Laurier, Aug. 18, 1911. LAC. Prime Ministers' Fonds. C906. Vol. 690, MG26 G. 189005.

75. Alberta premier Arthur L. Sifton to Prime Minister Sir Wilfrid Laurier, Aug. 18, 1911. LAC. Prime Ministers' Fonds. C906. Vol. 690, MG26 G. 189006.

76. Alberta premier Arthur L. Sifton to Prime Minister Sir Wilfrid Laurier, Aug. 18, 1911. LAC. Prime Ministers' Fonds. C906. Vol. 690, MG26 G. 189005.

77. Kelley and Trebilcock, *The Making of the Mosaic*, p. 113.

CHAPTER SIX

1. Sir Robert Borden in the House of Commons, Feb. 21, 1905. Hansard, 1463.

2. Memorandum of Debates and Correspondence relating to the Administration and Control of the Public Domain in the Provinces of Manitoba, Saskatchewan and Alberta. Prepared for Prime Minister Arthur Meighen. Prime Ministers' Fonds. LAC. C4306. Vol. 57, MG26 H. 023518.

3. "Conservative Party Policy Announced By Mr. Borden," *The Globe*, Wed., Aug. 21, 1907, p. 1.

4. Jim Blanchard, *Winnipeg 1912* (Winnipeg: University of Manitoba Press, 2005), p. 16.

5. Ibid., p. 119.

6. "Winnipeg: The Gateway of the Canadian West," *Canadian Annual Review*, 1912, pp. 89–90.

7. Jim Shilliday, *Canada's Wheat King: The Life and Times of Seager Wheeler* (Regina: Canadian Plains Research Center, University of Regina, 2007), p. 89.

8. "Alberta: Free Home for Settlers," in Arthur Lewis Sifton Fonds, LAC, MG 27, Series IID19, Vol. 12. Alberta Publicity.

9. Manitoba premier Rodmond Roblin to Prime Minister Sir Robert Borden along with a pamphlet or published item, which the Executive Government of Manitoba approved on Oct. 31, 1911. LAC. C4207, Vol. 18, MG26-H, 5233.

10. British Columbia premier Richard McBride to Prime Minister Sir Robert Borden, Nov. 6, 1911. Prime Ministers' Fonds. LAC. C4207, Vol. 19, MG26-H, 5251.

11. "Conservative Party Policy Announced By Mr. Borden," *The Globe*, Wed., Aug. 21, 1907, p. 1.

12. Henry Borden, ed. *Robert Laird Borden: His Memoirs* (Toronto: The Macmillan Company of Canada Limited, at St. Martin's House, 1938), p. 360.

13. Ibid., p. 365.

14. Ibid., p. 376.

15. Robert Laird Borden, excerpt from his diary for June 25, 1913, as quoted in Henry Borden, ed., *Robert Laird Borden: His Memoirs* (Toronto: Macmillan of Canada Limited, at St. Martin's House, 1938), p. 378.

16. Unsigned memo noting that Alberta premier Arthur Sifton forwarded a copy of his March 20 1911 letter to Prime Minister Sir Wilfrid Laurier to Prime Minister Sir Robert Borden on Nov. 8, 1911. LAC. Prime Ministers' Fonds. C4207. Vol. 18, MG26 H, 5017.

17. Saskatchewan premier Walter Scott to Prime Minister Sir Robert Borden, Nov. 8, 1911. LAC. Prime Ministers' Fonds. C4207, Vol. 18, MG26-H, 4846–47.

18. Prime Minister Sir Robert Borden to Acting Saskatchewan premier James Alexander Calder, Jan. 9, 1912. LAC. Prime Ministers' Fonds. C4209, Vol. 22, MG26-H, 7267.

19. Militia Minister Sam Hughes to Prime Minister Sir Robert Borden, Dec. 29, 1911. LAC. Prime Ministers' Fonds C4408, Vol. 231, MG26-H, 129019.

20. British Columbia premier Richard McBride to Prime Minister Sir Robert Borden, Feb. 5, 1912. LAC. Prime Ministers' Fonds. C4408, Vol. 231, MG26-H, 129041.

21. Acting Saskatchewan premier James Alexander Calder to Prime Minister Sir Robert Borden. Telegraph of March 6, 1912. LAC, Prime Ministers' Fonds. C4209, Vol. 22, MG26-H, 7274–76.

22. Acting Saskatchewan premier James Alexander Calder to Prime Minister Sir Robert Borden. Telegraph of March 9, 1912. LAC. Prime Ministers' Fonds. C4209, Vol. 22, MG26-H, 7281.

23. New Brunswick premier James Kidd Flemming to Prime Minister Sir Robert Borden, March 25, 1912. LAC. Prime Ministers' Fonds, C4207, Vol. 18, MG26-H, 4867.

24. Ibid., 4868.

25. Ibid., 4869.

26. "Government At War With 'Executioners,'" *The Globe*, Mon., March 25, 1912, p. 1.

27. Lewis Carroll, *Through the Looking Glass* in *The Annotated Alice* (New York: Bramhall House, 1960), p. 247. (Italics in text.)

28. Confidential memorandum from the British Colonial Office, forwarded by Prime Minister Sir Robert Borden from the Savoy Hotel in London on July 20, 1912, to Minister Without Portfolio George Perley. LAC. Prime Ministers' Fonds. C4209. Vol. 22, MG26 H. 7286.

29. *Quong Wing versus R. Supreme Court of Canada* (1913–14) 49 S.C.R. 44.

30. Saskatchewan premier Walter Scott to Ontario premier Sir James Whitney, Jan. 31, 1913. LAC. Prime Ministers' Fonds. C4209, Vol. 22, MG26 H, 7299.

31. Ibid.

32. Grant MacEwan, *Frontier Statesman of the Canadian Northwest: Frederick Haultain*. (Saskatoon: Western Producer Prairie Books, 1985) 168.

33. J.F.C. Wright, *Saskatchewan: The History of a Province* (Toronto: McClelland & Stewart, 1955), p. 160.

34. Saskatchewan premier Walter Scott to Prime Minister Sir Robert Borden, Aug. 16, 1912. LAC, Prime Ministers' Fonds. C4207, Vol. 18, MG26-H, 4870.

35. It is not clear which minister Scott meant. Parliamentary records show that Robert Rogers was Interior Minister until October 28, 1912. But they also indicate that his successor, Manitoba physician William James Roche, assumed the portfolio on October 9, 1912, and held it for five years.

36. Saskatchewan premier Walter Scott to Prime Minister Sir Robert Borden, Dec. 21, 1912. LAC, Prime Ministers' Fonds. C4209, Vol. 22, MG26-H, 7291.

37. Ibid.

38. Ibid.

39. Memorandum, "The Lands Question of Saskatchewan," n.d. and no signature. LAC. Prime Ministers' Fonds, C4207, Vol. 18, MG26-H, 5026.

40. Memorandum, "The Lands Question of Saskatchewan," n.d. and no signature. LAC. Prime Ministers' Fonds, C4207, Vol. 18, MG26-H, 5025.

41. Prime Minister Sir Robert Borden to Saskatchewan premier Walter Scott, Jan. 9, 1913. LAC. Prime Ministers' Fonds, C4209, Vol. 22, MG26 H, 7293.

42. Saskatchewan premier Walter Scott to Ontario premier Sir James Whitney, Jan. 31, 1913. LAC, Prime Ministers' Fonds. C4209, Vol. 22, MG26 H, 7299.

43. Prime Minister Sir Robert Borden in the House of Commons on Dec. 5, 1912, as quoted in Borden, ed., *Robert Laird Borden: His Memoirs*, p. 404.

44. Borden, ed., *Robert Laird Borden: His Memoirs*, p. 409.

45. Ibid., p. 413.

46. Ibid., p. 415.

47. Robert F. Harney, "The Padrone System and Sojourners" in Gerald Tulchinsky, ed., *Immigration in Canada: Historical Perspectives* (Toronto: Copp Clark Longman Limited, 1994), p. 261.

48. John Liss, "Polish Protestants in Richmond Park," in Joanna Matejko, ed., *Polish Settlers in Alberta: Reminiscences and Biographies* (Toronto: Polish Alliance Press Ltd, 1979), pp. 298–99.

49. Ibid., p. 299.

50. Herman Ganzevoort, "Introduction," in Willem de Gelder, *A Dutch Homesteader on the Prairies*, Michael Bliss, ed. (Toronto: University of Toronto Press, 1973), p. vii.

51. Willem de Gelder, *A Dutch Homesteader on the Prairies*, Michael Bliss, ed. (Toronto: University of Toronto Press, 1973), p. 71.

52. Ibid., p. 79.

53. Ibid., p. 66.

54. Ibid., p. 77.

55. Borden, ed., *Robert Laird Borden: His Memoirs*, p. 379.

56. Prime Minister Sir Robert Borden to Minister Without Portfolio George Halsey Perley, July 7, 1913. LAC. Prime Ministers' Fonds. C4409, Vol. 231, MG26-H, 129124.

57. A.B. Gillis to Prime Minister Sir Robert Borden, Aug.11, 1913. LAC, Prime Ministers' Fonds. C4207, Vol. 18, MG26 H. 4876.

58. "Suggestions for Preparation." LAC, Prime Ministers' Fonds. C4212, Vol. 28, MG26 H. 10912–14. There is no date, but an accompanying letter dated Oct. 17, 1913, from Secretary of State Louis Coderre, acknowledges receipt of Borden's memorandum.

59. "Memorandum as to preparation." n.d. and no signature. LAC, Prime Ministers' Fonds. C4212, Vol. 28, MG26 H. 10905.

60. "Memorandum Re Western Conditions." n.d. and no signature. LAC. Prime Ministers' Fonds. C4212, Vol. 28, MG26 H. 10908–10911.

61. "Memorandum Re Western Condition." n.d. and no signature. LAC, Prime Ministers' Fonds. C4212, Vol. 28, MG26 H. 10908. (My italics.)

62. Ibid., 10909.

63. Ibid., 10911.

64. Borden, ed., *Robert Laird Borden: His Memoirs*, p. 383.

65. "Demand of the Provinces for Increased Subsidies," *The Globe*, Wed. Oct. 29, 1913, p. 2.

66. Minutes of the Interprovincial Conference 1913 in Dominion Provincial and Interprovincial Conferences from 1887 to 1926 (Ottawa: Edmond Cloutier, King's Printer and Controller of Stationery, 1951), p. 76.

67. Alberta premier Arthur Sifton and Saskatchewan premier Walter Scott to Prime Minister Sir Robert Borden, Oct. 28, 1913. LAC, Prime Ministers' Fonds. C4212, Vol. 28, MG26-H. 10916.

68. Prime Minister Sir Robert Borden to Manitoba premier Sir Rodmond Roblin, Oct. 29, 1913. LAC. Prime Ministers' Fonds. C4212. Vol. 28, MG26 H. 10917. (My italics.)

69. Minutes of the Interprovincial Conference 1913 in Dominion Provincial and Interprovincial Conferences from 1887 to 1926 (Ottawa: Edmond Cloutier, King's Printer and Controller of Stationery, 1951), p. 77.

70. Memorandum from Prince Edward Island, Nova Scotia and New Brunswick. Wed., Oct. 29, 1913. Minutes of the Interprovincial Conference 1913 in Dominion Provincial and Interprovincial Conferences from 1887 to 1926 (Ottawa: Edmond Cloutier, King's Printer and Controller of Stationery, 1951), p. 78.

71. Prime Minister Sir Robert Borden to Alberta premier A.L. Sifton, Saskatchewan premier Walter Scott and Manitoba premier Sir R.P. Roblin. March 5, 1914. LAC. Prime Ministers' Fonds. C4207. Vol. 18, MG26 H. 4906.

72. Ibid.

73. Ibid.

74. Borden, ed., *Robert Laird Borden: His Memoirs*, p. 384.

75. Alberta premier Arthur L. Sifton, Saskatchewan premier Walter Scott and Manitoba premier R.P. Roblin to Prime Minister Sir Robert Borden, Dec. 22, 1913. LAC. Prime Ministers' Fonds. C4207. Vol. 18, MG26 H. 4883.

76. Ibid.

77. Nova Scotia premier G.H. Murray to Prime Minister Sir Robert Borden, Jan. 19, 1914. LAC. Prime Ministers' Fonds. C4207. Vol. 18, MG26 H. 4897.

78. Ibid.

79. Ibid., 4898.

80. Ibid.

81. Prince Edward Island premier John A. Mathieson to Prime Minister Sir Robert Borden. Feb. 2, 1914. LAC. Prime Ministers' Fonds. C4207. Vol. 18, MG26 H. 4901.

82. Ibid., 4902.

83. Ibid.

84. New Brunswick premier J.K. Flemming to Prime Minister Sir Robert Borden. Feb. 17, 1914. LAC. Prime Ministers' Fonds. C4207. Vol. 18, MG26 H. 4903.

85. Ibid., 4904. (My italics.)

86. Ibid.

87. Borden, ed., *Robert Laird Borden: His Memoirs*, pp. 438–39.

88. Ibid., p. 439.

89. Prime Minister Sir Robert Borden in the House of Commons on Feb. 24, 1914, as quoted in Borden, ed., *Robert Laird Borden, His Memoirs*, p. 439.

90. Prime Minister Sir Robert Borden to Premiers A.L. Sifton, Walter Scott and Sir R.P. Roblin. March 5, 1914. LAC. Prime Ministers' Fonds. C4207. Vol. 18, MG26 H. 4908–09.

91. William Rayner, *British Columbia's Premiers in Profile: The Good, the Bad and the Transient* (Surrey: Heritage House Publishing, 2000), p. 107.

92. British Columbia premier Sir Richard McBride to Prime Minister Sir Robert Borden, June 20, 1914. Prime Ministers' Fonds. LAC. C4207. Vol. 19, MG26-H, 5357.

93. Borden, ed., *Robert Laird Borden: His Memoirs*, p. 456.

94. British Columbia premier Sir Richard McBride to Prime Minister Sir Robert Borden, Sept. 22, 1914. LAC, Prime Ministers' Fonds. C4207, Vol. 19, MG26-H, 5375.

95. Borden, ed., *Robert Laird Borden: His Memoirs*, 446.

CHAPTER SEVEN

1. Borden was knighted in June 1914.

2. Jim Blanchard, *Winnipeg's Great War: A City Comes of Age* (Winnipeg: University of Manitoba Press, 2010), p. 39.

3. The War Measures Act, 1914: clause 6; and clause 6, subsection (b).

4. Art Grenke, "The German Community of Winnipeg and the English-Canadian Response to World War I," *Canadian Ethnic Studies*, vol. 20, no. 1, 1988, p. 25.

5. Donald B. Smith, *Calgary's Grand Story* (Calgary: University of Calgary Press, 2005), p. 150.

6. Ibid.

7. "Germans Are Poisoning Soldiers of Canada," *The Globe*, Wed., April 28, 1915, p. 1.

8. Hugh Robert Ross, *Thirty-Five Years in the Limelight: Sir Rodmond P. Roblin and His Times* (Winnipeg: Farmer's Advocate of Winnipeg Limited, 1936), p. 151.

9. http://www.gov.mb.ca/mit/legtour/legbld.html

10. W.L. Morton, *Manitoba: A History* (Toronto: University of Toronto Press, 1957), p. 343.

11. Blanchard, *Winnipeg's Great War*, pp. 145–46.

12. Ibid., p. 97.

13. "Manitoba Is For Norris and Clean Government," *The Globe*, Sat., Aug. 7, 1915, p. 1.

14. British Columbia premier Sir Richard McBride to Prime Minister Sir Robert Borden, Jan. 18, 1915. LAC. Prime Ministers' Fonds. C4207, Vol. 19, MG26-H, 5376.

15. Prime Minister Sir Robert Borden to British Columbia premier Sir Richard McBride, March 12, 1915. LAC. Prime Ministers' Fonds. C4207, Vol. 19, MG26-H, 5378.

16. Ibid.

17. Saskatchewan premier Walter Scott, Alberta premier Arthur L. Sifton and Manitoba premier T.C. Norris to Prime Minister Sir Robert Borden, Nov. 30, 1915. LAC. Prime Ministers' Fonds. C4207. Vol. 18, MG26 H. 4930.

18. Ibid.

19. Ibid..

20. "Authorized Canadian Force Raised From Quarter to Half-Million Men," *The Globe*, Sat., Jan. 1, 1916, p. 1.

21. "War Summary," *The Globe*, Sat., Jan. 1, 1916, p. 1.

22. Saskatchewan premier Walter Scott, Alberta premier Arthur L. Sifton and Manitoba premier T.C. Norris to Prime Minister Sir Robert Borden, Feb. 9, 1916. LAC. Prime Ministers' Fonds. C4207. Vol. 18, MG26 H. 4941.

23. Henry Borden, ed., *Robert Laird Borden: His Memoirs* (Toronto: Macmillan of Canada, at St. Martin's House, 1938), p. 543.

24. Government of Canada website: http://www.collineduparlement-parliamenthill.gc.ca/histoire-history/1916-eng.html

25. Prime Minister Sir Robert Borden to Saskatchewan premier Walter Scott, Alberta premier Arthur L. Sifton and Manitoba premier T.C. Norris. March 10, 1916. LAC. Prime Ministers' Fonds. C4207. Vol. 18, MG26 H. 4948.

26. Prime Minister Sir Robert Borden to Manitoba premier T.C. Norris, April 17, 1916. LAC. Prime Ministers' Fonds. C4207. Vol. 18, MG26-H, 4952–4953. Borden refers to the trio's joint letter—so this letter is clearly intended as the response to all three Western premiers.

27. Gordon L. Barnhart, "Walter Scott," in Gordon L. Barnhart, ed., *Saskatchewan Premiers of the Twentieth Century* (Regina: Canadian Plains Research Center, University of Regina, 2004) 29.

28. Ted Regehr, "William M. Martin," in Barnhart, ed., *Saskatchewan Premiers of the Twentieth Century*, p. 47.

29. Ibid., p. 26.

30. http://www.wwıcemeteries.com/wwıfrenchcemeteries/adanac.htm

31. http://www.wwıcemeteries.com/wwıfrenchcemeteries/adanac.htm

32. Howard Palmer and Tamara Palmer, "The Religious Ethic and the Spirit of Immigration: The Dutch in Alberta," in Gregory P. Marchildon, ed., *Immigration and Settlement, 1870–1939* (Regina: Canadian Plains Research Center, University of Regina, 2009), p. 326.

33. Ibid., 326.

34. Richard J.A. Prechtl. *Take The Soil In Your Hands* (Saskatoon: Herrem Publishing Company, 1984), p. 71.

35. Ibid.

36. http://archives.queensu.ca/Exhibits/archres/wwi-intro/women.html

37. Donald H. Avery, *Reluctant Host: Canada's Response to Immigrant Workers, 1896–1994.* (Toronto: McClelland & Stewart, 1995), p. 74.

38. Mark Minenko, "Without Just Cause: Canada's First National Internment Operations, " in Lubomyr Luciuk and Stella Hryniuk, eds., *Canada's Ukrainians: Negotiating an Identity* (Toronto: Ukrainian Canadian Centennial Committee in association with University of Toronto Press, 1991), p. 302.

39. Allen Seager, "Class, Ethnicity, and Politics in the Alberta Coalfields, 1905–1945," in Dirk Hoerder, ed.. *"Struggle a Hard Battle": Essays on Working-Class Immigrants* (DeKalb, Illinois: Northern Illinois University Press, 1986), p. 309.

40. Memorandum from the Dominion Parks Branch, forwarded by Interior Minister William James Roche to Prime Minister Sir Robert Borden, June 2, 1917. LAC. Prime Ministers' Fonds. C4404. Vol. 222, MG26 H. 124463.

41. British Columbia premier Harlan Carey Brewster to Prime Minister Sir Robert Borden, Aug. 1, 1917. LAC. Prime Ministers' Fonds. C4404. Vol. 222, MG26 H. 124764–65.

42. Letter from unknown solicitor for District 18 of the United Mine Workers of America. Dec. 31, 1917. Distributed to Borden and four ministers including Immigration and Colonization Minister James Calder and Minister Without Portfolio Gideon Decker Robertson, who had ties to Conservative union members. LAC. Prime Ministers' Fonds. C4400. Vol. 213, MG26 H. 120422.

43. Stella Hryniuk, "Pioneer Bishop, Pioneer Times: Nykyta Budka in Canada," in CCHA, *Historical Studies*, vol. 55 (1988), pp. 21–41

44. Father Anthony Sylla, "The Detention Camp in the Area of Castle Mountain," in Joanna Matejko, ed., *Polish Settlers in Alberta: Reminiscences and Biographies* (Toronto: Polish Alliance Press, 1979), p. 327.

45. Waclaw Fridel, "'An Alien' in Wabamun," in Matejko, ed., *Polish Settlers in Alberta,* p. 328.

46. The Alberta Government would finally accept that Polish Canadians "were neither Austrians nor enemies of the Allies" *after* the war, when Polish patriots proclaimed

their newly liberated state in December 1918. Father Anthony Sylla, "The Detention Camp in the Area of Castle Mountain" in Matejko, ed., *Polish Settlers in Alberta: Reminiscences and Biographies*, p. 327.

47. Shell Transport Company executive R.N. Benjamin through his attorney Adam T. Shillington to Interior Minister William James Roche. The letter is dated July 30, 1917. LAC. Prime Ministers' Fonds. C4405. Vol. 225, MG26 H. 125967–79.

48. Ibid.

49. United Farmers of Alberta to Prime Minister Sir Robert Borden, June 19, 1914. LAC, Prime Ministers' Fonds. C4383, Vol. 183, MG26-H, 100372-100374. The memorandum itself is dated January 19, 1914, but the Association did not forward it until June.

50. http://www.collectionscanada.gc.ca/firstworldwar/025005-2400.003.05-e.html

51. Borden, ed., *Robert Laird Borden: His Memoirs*, p. 693.

52. John English, *Borden: His Life and World* (Toronto: McGraw-Hill Ryerson Limited, 1977), p. 136.

53. Borden, ed., *Robert Laird Borden: His Memoirs*, p. 746.

54. Ibid., p. 748.

55. Ibid., p. 741.

56. Ibid., 765.

57. Ibid., pp. 766–67.

58. Ibid., p. 767.

59. Desmond Morton, http://www.thecanadianencyclopedia.com/index.cfm?PgNm= TCE&Params=A1ARTA0008716

60. Lloyd Duhaime, http://www.duhaime.org/LawMuseum/CanadianLegalHistory /LawArticle-168/1917-The-Birth-of-Income-Tax.aspx

61. Lloyd Duhaime, http://www.duhaime.org/LawMuseum/CanadianLegalHistory /LawArticle-168/1917-The-Birth-of-Income-Tax.aspx

62. Borden, ed., *Robert Laird Borden: His Memoirs*, p. 767.

63. Alberta premier Charles Stewart to Prime Minister Sir Robert Borden. Telegram of Jan. 16, 1918. LAC. Prime Ministers' Fonds. C4406. Vol. 225, MG26 H. 126264.

64. Alberta premier Charles Stewart to Prime Minister Sir Robert Borden, Feb. 7, 1918. LAC. Prime Ministers' Fonds. C4336, Vol. 105, MG26 H. 57941.

65. Ibid.

66. Ibid.

67. "Alberta Needs Immigration," *The Globe*, Thurs., Feb. 14, 1918, p. 6.

68. Saskatchewan attorney-general W.A. F. Turgeon to Prime Minister Sir Robert Borden, telegram of Feb. 10, 1918. LAC. Prime Ministers' Fonds. C4336. Vol. 105, MG26 H. 57951.

69. New Brunswick premier W.E. Foster to Prime Minister Sir Robert Borden. Telegram of Feb. 11, 1918. LAC. Prime Ministers' Fonds. C4336. Vol. 105, MG26 H. 57955. The premier did not mention Manitoba.

70. Prince Edward Island premier Aubin-Edmond Arsenault to Prime Minister Sir Robert Borden. Telegram of February 11, 1918. LAC. Prime Ministers' Fonds. C4336. Vol. 105, MG26 H. 57953.

71. "Food, Labor, Settlement," *The Globe*, Fri., Feb. 15, 1918, p. 5.

72. Canadian Government memorandum on Conference of Dominion and Provincial Governments, Feb. 1918,. LAC. Prime Ministers' Fonds. C3428. Vol. 40, MG26 I. 023564.

73. Canadian Government memorandum on Conference Dominion and Provincial Governments, Feb, 1918. LAC. Prime Ministers' Fonds. C3428. Vol. 40, MG26 I. 023566.

74. Ibid., MG26 I. 023565.

75. Ibid., MG26 I. 023566.

76. Borden, ed., *Robert Laird Borden: His Memoirs*, p. 769.

77. Alberta premier Charles Stewart, Saskatchewan premier W.M. Martin and Manitoba attorney-general Thomas H. Johnson to Prime Minister Sir Robert Borden. Feb. 19, 1918. LAC. Prime Ministers' Fonds. C4207. Vol. 18, MG26 H. 4963.

78. Ibid..

79. Ibid., 4963A.

80. Ibid.

81. Prime Minister Sir Robert Borden to Alberta premier Charles Stewart, Saskatchewan premier W.M. Martin and Manitoba attorney-general Thomas H. Johnson. Feb. 20, 1918. LAC. Prime Ministers' Fonds. C4207. Vol. 18, MG26 H. 4964A. (My italics.)

82. British Columbia premier H.C. Brewster to Prime Minister Sir Robert Borden. Feb. 20, 1918. LAC. Prime Ministers' Fonds. C4207. Vol. 18, MG26 H. 4971.

83. Ibid.

84. Ontario premier Sir William H. Hearst to Prime Minister Sir Robert Borden. Feb. 22, 1918. LAC. Prime Ministers' Fonds. C4207. Vol. 18, MG26 H. 4973–74.

85. Ibid., 4974.

86. Prince Edward Island premier A. E. Arsenault to Prime Minister Sir Robert Borden. Feb. 25, 1918. LAC. Prime Ministers' Fonds. C4207. Vol. 18, MG26 H. 4972.

87. Rayner, *British Columbia's Premiers in Profile*, pp, 120–21.

CHAPTER EIGHT

1. Henry Borden, ed., *Robert Laird Borden: His Memoirs* (Toronto: Macmillan of Canada Limited, at St. Martin's House, 1938), p. 776.

2. Robert Craig Brown and Ramsay Cook, *Canada, 1896–1921: A Nation Transformed*. (Toronto: McClelland & Stewart, 1974), p. 238.

3. Universal adult suffrage was not extended to male or female status Indians until 1960. I am indebted to York doctoral candidate Sara Howdle for parsing the legal distinctions in Aboriginal status.

4. "Current Events: Daylight Savings," *The Voice*, Winnipeg, April 12, 1918, p. 1.

5. "Jottings From Billboard," *The Voice*, Winnipeg, April 12, 1918, p. 1.

6. Sir Robert Borden to British Columbia premier John Oliver, March 26, 1918. LAC. Prime Ministers' Fonds. C4403, Vol. 220, MG26 H. 123987.

7. Prime Minister Sir Robert Borden to Prince Edward Island premier Aubin-Edmond Arsenault, March 12, 1918. LAC. Prime Ministers' Fonds. C4403, Vol. 220, MG26 H. 123618.

8. "Three Alberta Men Mentioned In Casualties," *Calgary Daily Herald*, Wed., May 1, 1918, p. 1.

9. "Tag Day Results In Sum of $1,100 Being Secured," *Calgary Daily Herald*, Mon., April 22, 1918, p. 8.

10. "Maclin Motors Limited Losing Good Salesman," *Calgary Daily Herald*, Mon., April 22, 1918, p. 9.

11. Borden, ed., *Robert Laird Borden: His Memoirs*, p, 786.

12. Morris Mott, "Tobias C. Norris: 1915–1922," in Barry Ferguson and Robert Wardhaugh, eds., *Manitoba Premiers of the 19th and 20th Centuries* (Regina: Canadian Plains Research Center, University of Regina. 2010), p. 145.

13. Ibid., p. 147.

14. Carrol Jaques, "Charles Stewart: 1917–1921" in Bradford J.Rennie, ed., *Alberta Premiers of the Twentieth Century* (Regina: Canadian Plains Research Center, University of Regina. 2004), p. 52.

15. Saskatchewan premier Tommy Douglas, as quoted in Ted Regehr, "William M. Martin," in Gordon L. Barnhart, ed., *Saskatchewan Premiers of the Twentieth Century* (Regina: Canadian Plains Research Center, University of Regina. 2004), p. 40.

16. Ted Regehr, "William M. Martin," in Barnhart, ed., *Saskatchewan Premiers of the Twentieth Century*, p. 59.

17. Sir Robert Borden to Saskatchewan premier William M. Martin, April 30, 1918. LAC. Prime Ministers' Fonds. C4420, Vol. 250, MG26 H. 140769.

18. Manitoba premier Tobias Crawford Norris to Prime Minister Sir Robert Borden, May 14, 1918. LAC. Prime Ministers' Fonds. C4420. Vol. 250, MG26 H. 140815.

19. Borden, ed., *Robert Laird Borden: His Memoirs*, p. 807.

20. Robert Craig Brown and Ramsay Cook, *Canada, 1896–1921: A Nation Transformed* (Toronto: McClelland & Stewart, 1974), p. 234.

21. Allen Seager and David Roth, "British Columbia and the Mining West: A Ghost of a Chance," in Craig Heron, ed., *The Workers' Revolt in Canada: 1917–1925* (Toronto: University of Toronto Press, 1998), p. 246.

22. Alberta premier Charles Stewart, as quoted in "R.N.W.M.P. Wanted Back In Alberta," *Calgary Daily Herald*, Sat. April 6, 1918, p. 1.

23. Craig Heron, "Introduction" to Craig Heron, ed., *The Workers Revolt in Canada: 1917–1927*, p. 7.

24. "Confidential information re the causes of Drumheller strike, received from a reliable source." LAC. Prime Ministers' Fonds. C4400,.Vol. 213, MG26 H. 120442.

25. "Increase In Production Of The Mines," *Calgary Daily Herald*, Mon. April 22, 1918, p. 1.

26. Brown and Cook, *Canada, 1896–1921: A Nation Transformed*, p. 247.

27. Borden, ed., *Robert Laird Borden: His Memoirs*, p. 792.

28. Ibid.

29. Ibid., p. 851.

30. Ibid., p. 821.

31. Immigration and Colonization Minister James Calder to Prime Minister Sir Robert Borden, April 12, 1918. LAC. Prime Ministers' Fonds. C4414. Vol. 239, MG26 H. 133558–59.

32. Borden, ed., *Robert Laird Borden: His Memoirs*, p. 844.

33. "Memorandum regarding arrangements desired by Premiers Norris, Martin and Stewart in connection with their overseas visit," June 1918. LAC. Prime Ministers' Fonds. C4328. Vol. 93, MG26 H. 49037.

34. Borden, ed., *Robert Laird Borden: His Memoirs*, p. 849. (My italics.)

35. Ibid., p. 848.

36. Cabinet committee memorandum from Immigration Minister James Calder to Prime Minister Sir Robert Borden, Sept. 11, 1918. LAC. Prime Ministers' Fonds. C4365, Vol. 155, MG26 H. 83210–14.

37. Alberta premier Charles Stewart to Prime Minister Sir Robert Borden. Telegram of Sept. 27, 1918. LAC. Prime Ministers' Fonds. C4207. Vol. 18, MG26 H. 4977.

38. Ibid.

39. Ibid.

40. Borden, ed., *Robert Laird Borden: His Memoirs*, p. 860.

41. "Minutes of the Proceedings in Conference" for November 1918 from *Dominion Provincial and Interprovincial Conferences From 1887 to 1926* (Ottawa: Reprinted by Edmond Cloutier, King's Printer and Controller of Stationery. 1951), p. 95.

42. Ibid.

43. R.M. Stewart, Acting Secretary-Treasurer of The Great War Veterans' Association of Canada, to Sir Robert Borden. Oct. 28, 1918. LAC. Prime Ministers' Fonds. C4414. Vol. 240, MG26 H. 134008.

44. Ibid., 134009.

45. Saskatchewan premier William M. Martin to Prime Minister Sir Robert Borden. Nov. 1, 1918. LAC. Prime Ministers' Fonds. C4207. Vol. 18, MG26 H. 4978.

46. Prime Minister Sir Robert Borden to the First Ministers. Memorandum of Nov. 2, 1918. LAC. Prime Ministers' Fonds. C4207. Vol. 18. MG26 H. 4985A.

47. Borden, ed., *Robert Laird Borden: His Memoirs*, p. 860.

48. Privy Council president Newton W. Rowell to Prime Minister Sir Robert Borden, Nov. 7, 1918. LAC. Prime Ministers' Fonds. C4417. Vol. 245, MG26 H. 136929–30.

49. Ibid.

50. Prime Minister Sir Robert Borden to Public Works Minister Frank Carvell, Nov. 9, 1918. LAC. Prime Ministers' Fonds. C4367. Vol. 158, MG26 H. 84273.

51. Ibid.

52. Ibid.

53. Ibid.

54. Ibid.

55. Ibid.

56. Borden, ed., *Robert Laird Borden: His Memoirs*, p. 861.

57. Ibid., p. 865.

58. Ibid., p. 860.

59. "Thanksgiving Sunday Postponed to Dec. 1," *The Globe*, Fri., Nov 15, 1918, p. 3.

60. Canada: *Conference of Dominion and Provincial Governments: Ottawa: November 1918*. LAC. Prime Ministers' Fonds. C4413. Vol. 237, MG26 H. 132554.

61. Canada: *Conference of Dominion and Provincial Governments: Ottawa: November 1918. Memorandum Relating to the Soldier Settlement and Suggestions Relating to Agricultural Training For Returned Soldiers*. LAC. Prime Ministers' Fonds. C4413. Vol. 237, MG26 H. 132591, which is page 2 of Soldier Settlement memorandum.

62. Memorandum, Feb. 25, 1916, signed Frank S. Checkley, enclosed in Canada: *Conference of Dominion and Provincial Governments: Ottawa: November 1918*. LAC. Prime Ministers' Fonds. C4413. Vol. 237, MG26 H. 132563–65.

63. Canada: *Conference of Dominion and Provincial Governments: Ottawa: November 1918*. LAC. Prime Ministers' Fonds. C4413. Vol. 237, MG26 H. 132586.

64. Canada: *Conference of Dominion and Provincial Governments: Ottawa: November 1918. Memorandum Relating to the Soldier Settlement and Suggestions Relating to Agricultural Training For Returned Soldiers*. LAC. Prime Ministers' Fonds. C4413. Vol. 237, MG26 H. 132591. Page 4 of Suggestions Relating to Agricultural Training memorandum.

65. Memorandum, Nov. 18, 1918. LAC. Prime Ministers' Fonds. C4328. Vol. 94, MG26 H. 49882–83.

66. Ibid.

67. Borden, ed., *Robert Laird Borden: His Memoirs*, p. 869.

68. Ibid.

69. Ibid.

70. Headlines from Tues. Nov. 19, 1918, *The Globe*, p. 1.

71. "Minutes of the Proceedings in Conference" for November 1918 from *Dominion Provincial and Interprovincial Conferences From 1887 to 1926* (Ottawa: Reprinted by Edmond Cloutier, King's Printer and Controller of Stationery, 1951), p. 96.

72. Manitoba premier Tobias Crawford Norris to Prime Minister Arthur Meighen, Dec. 10, 1920. LAC. Prime Ministers' Fonds. C3428. Vol. 40, MG26 I. 22328–48.

73. Former Prince Edward Island premier A.E. Arsenault. *Memoirs of The Hon. A.E. Arsenault*, n.d. Publisher not stated, p. 72.

74. Ibid.

75. Ibid.

76. "Minutes of the Proceedings in Conference" for November 1918 from *Dominion Provincial and Interprovincial Conferences From 1887 to 1926* (Ottawa: Reprinted by Edmond Cloutier, King's Printer and Controller of Stationery, 1951), p. 97.

77. Ibid.

78. Report from the Representatives of Nova Scotia, New Brunswick, Prince Edward Island, Quebec, Ontario and British Columbia in "Minutes of the Proceedings in Conference" for November 1918, from *Dominion Provincial and Interprovincial Conferences From 1887 to 1926* (Ottawa: Reprinted by Edmond Cloutier, King's Printer and Controller of Stationery, 1951) 98. (My italics.)

79. Ibid., p. 98.

80. Judith Fingard, Dictionary of Canadian Biography Online: http://www.biographi.ca /009004-119.01-e.php?BioId=41977

81. Arthur T. Doyle, *Front Benches & Back Rooms: A Story of Corruption, Muckraking, Raw Partisanship and Political Intrigue in New Brunswick* (Toronto: Green Tree Publishing, 1976), p. 181.

82. Former Prince Edward Island premier A.E. Arsenault. *Memoirs of The Hon. A. E. Arsenault*, n.d. Publisher not stated, p. 106.

83. Report from the Representatives of Nova Scotia, New Brunswick, Prince Edward Island, Quebec, Ontario and British Columbia in "Minutes of the Proceedings in Conference" for November 1918 from *Dominion Provincial and Interprovincial Conferences From 1887 to 1926*. (Ottawa: Reprinted by Edmond Cloutier, King's Printer and Controller of Stationery, 1951), p. 98. (My italics.)

84. Former Prince Edward Island premier A.E. Arsenault. *Memoirs of The Hon. A. E. Arsenault*, n.d. Publisher not stated, p. 72.

85. Jean Barman, *The West Beyond the West: A History of British Columbia* (Toronto: University of Toronto Press, third edition, 2007), p. 252.

86. Memorandum presented by British Columbia premier John Oliver in "Minutes of the Proceedings in Conference" for November 1918 from *Dominion Provincial and Interprovincial Conferences From 1887 to 1926*, p. 99.

87. Ibid., p. 100.

88. Ibid.

89. Ibid.

90. Ibid.

91. Ibid.

92. Ibid.

93. "Minutes of the Proceedings in Conference" for November 1918 from *Dominion Provincial and Interprovincial Conferences From 1887 to 1926*, p. 100.

94. Statement by Minister of the Interior Arthur Meighen in "Minutes of the Proceedings in Conference" for November 1918 from *Dominion Provincial and Interprovincial Conferences From 1887 to 1926*, p. 100.

95. Ibid. (My italics.)

96. James Morton, *Honest John Oliver: The Life Story of the Honourable John Oliver, Premier of British Columbia: 1918–1927* (London, Toronto and Vancouver: J.M. Dent and Sons Ltd., 1933), p. 212.

97. "May Levy More Direct Taxation," *The Globe*, Fri., Nov. 22, 1918, p. 3.

98. Letter from Manitoba premier T.C. Norris, Saskatchewan premier W.M. Martin and Alberta premier Charles Stewart in "Minutes of the Proceedings in Conference" for November 1918 from *Dominion Provincial and Interprovincial Conferences From 1887 to 1926*, p. 102.

99. Ibid.

100. Ibid.

101. Ibid.

102. Ibid.

103. Ibid.

104. "Progress By Getting Together," *The Globe*. Sat. Nov. 23, 1918, p. 2.

105. Ibid.

106. Sir Thomas White to Prince Edward Island premier A.E. Arsenault, Feb. 11, 1919, as quoted in Christopher Armstrong, *The Politics of Federalism: Ontario's Relations with the Federal Government, 1867–1942* (Toronto: University of Toronto Press, 1981), p. 130.

Chapter Nine

1. Andre Pratte, *Wilfrid Laurier* (Toronto: Penguin Group [Canada], 2011), p. 199. Pratte indicates that Laurier's biographers do not agree on his last words—but Pratte adopts the standard sentence.

2. http://www.collectionscanada.gc.ca/008/001/008001-119.01-e.php?&document _id_nbr=260&ts_nbr=14&brws=1&&PHPSESSID=dc4is8d4el3lo41rd6e1irl704

3. "Sir Wilfrid Laurier Is Dead," *Calgary Daily Herald*, Mon., Feb. 17, 1919, p. 1. http://www.ourfutureourpast.ca/newspapr/np_page2.asp?code=N43P0771.JPG

4. http://www.collectionscanada.gc.ca/008/001/008001-119.01-e.php?&document _id_nbr=260&ts_nbr=14&brws=1&&PHPSESSID=dc4is8d4el3lo41rd6e1irl704

5. Henry Borden, ed., *Robert Laird Borden: His Memoirs* (Toronto: Macmillan of Canada, at St. Martin's House, 1938), p. 914.

6. Memorandum dated Paris, April 2, 1919, with the notation, "To be published in *Daily Mail*, Monday Apr. 7/19." LAC. Prime Ministers' Fonds. C4329. Vol. 95. MG26 H. 50028.

7. Resolution of the Alberta Legislative Assembly. February 20, 1919. Library and Archives of Canada. Prime Ministers' Fonds. C4328. Vol. 94. MG26 H. 49643.

8. Memorandum of August 9, 1917, from H.H. Rowatt, Controller of the Mining Lands & Yukon Branch of the Ministry of the Interior. LAC. Prime Ministers' Fonds. C4405. Vol. 225, MG26 H. 125977.

9. Ibid.

10. Sir Reginald MacLeod to Prime Minister Sir Robert Borden. April 17, 1919. LAC. Prime Ministers' Fonds. C4328. Vol. 94, MG26 H. 49823.

11. Imperial Oil Ltd. president W.J. Hanna to Prime Minister Sir Robert Borden. Telegram of February 11, 1919. LAC. Prime Ministers' Fonds. C4328. Vol. 94, MG26 H. 49596.

12. "Many Mourn Big Canadian," *The Globe*, Wed., March 26, 1919, p. 8.

13. Borden, ed. *Robert Laird Borden: His Memoirs,* 'p. 942.

14. R.G. Moyles, Introduction to *Challenge of the Homestead: Peace River Letters of Clyde and Myrle Campbell, 1919–1924*, R.G. Moyles ed. (Calgary: Historical Society of Alberta, 1988), p. viii.

15. Ibid., viii.

16. R.G. Moyles, ed., *Challenge of the Homestead: Peace River Letters of Clyde and Myrle Campbell, 1919–1924* (Calgary: Historical Society of Alberta, 1988), June 21, 1919, p. 16.

17. http://www.thecanadianencyclopedia.com/index.cfm?Params=A1ARTA0008649&PgNm=TCE

18. Roger Graham, *Arthur Meighen, Vol. 1: The Door of Opportunity* (Toronto: Clarke, Irwin, 1960), p. 238.

19. http://www.thecanadianencyclopedia.com/index.cfm?Params=A1ARTA0008649&PgNm=TCE

20. Borden, ed., *Robert Laird Borden: His Memoirs*, p. 972.

21. In 1904, King Edward VII had added "Royal" to the force's name.

22. Customs Minister Arthur L. Sifton to Sir Robert Borden. June 7, 1919. LAC. Prime Ministers' Fonds. C4419. Vol. 248, MG26 H. 139626.

23. Prime Minister Sir Robert Borden to H.A. Lovett, president of North American Collieries Ltd., June 16, 1919. LAC. Prime Ministers' Fonds. C4419. Vol. 248, MG26 H. 139627.

24. Ibid., 139627–28.

25. As quoted in letter from Prince Edward Island premier John Bell to Prime Minister Mackenzie King, Dec. 12, 1922. LAC. Prime Ministers' Fonds. C2242. Vol. 69, MG26 J1. 59499.

26. Memorandum to Prime Minister Sir Robert Borden, Sept. 1919. LAC. Prime Ministers' Fonds. C4342. Vol. 114, MG26 H. 63188-63189.

27. Ibid., 63192. (My italics.)

28. Resolution of the Nova Scotia House of Assembly as forwarded to Privy Council President N.W. Rowell. Forwarding letter dated September 25, 1919. Resolution passed on April 22, 1919. LAC. Prime Ministers' Fonds. C4207. Vol. 18. MG26 H 1. 4988.

29. Outgoing Ontario premier William Hearst to Prime Minister Sir Robert Borden, Oct. 27, 1919. LAC. Prime Ministers' Fonds. C4342. Vol. 114. MG26 H. 62890.

30. "Drury Returns To Plow," *The New York Times*, Nov. 1, 1919, p. 9.

31. Borden, ed., *Robert Laird Borden: His Memoirs*, p. 1015.

32. Moyles, ed., *Challenge of the Homestead*, Jan. 5, 1920, p. 44.

33. Ibid., p. 42.

34. Ibid., Jan. 13, 1920, p. 48.

35. Ibid.

36. Ibid., Jan. 24, 1920. p. 50.

37. Ibid.

38. Alberta premier Charles Stewart to Acting Prime Minister Sir George Foster. Feb. 9, 1920. LAC. Prime Ministers' Fonds. C4207. Vol. 18. MG26 H. 4996.

39. Ibid.

40. Borden, ed., *Robert Laird Borden: His Memoirs*, p. 1026.

41. "Premier Makes Good Progress," *The Globe*, Thurs., April 15, 1920, p. 2.

42. Nova Scotia House of Assembly speech by Commissioner of Public Works and Mines James C. Tory on April 14, 1920. Later transmitted to Prime Minister Arthur Meighen. LAC. Prime Ministers' Fonds. C3428. Vol. 40. MG26 I. 023571a.

43. Ibid., 023571s.

44. Ibid., 023571b.

45. Ibid.

46. Alberta premier Charles Stewart to Acting Prime Minister Sir George Foster. April 19, 1920. LAC. Prime Ministers' Fonds. C4207. Vol. 18, MG26 H. 5000.

47. Ibid.

48. Saskatchewan Conservative (Unionist) MP W.D. Cowan to Acting Prime Minister Sir George Foster, April 30, 1920. LAC. Prime Ministers' Fonds. C4207. Vol. 18, MG26 H. 5002A.

49. Ibid.

50. Moyles, ed., *Challenge of the Homestead*, April 23, 1920, p. 56.

51. Ibid., May 2, 1920, p. 58.

52. Ibid., p. 59.

53. Ibid.

54. "Premier Back Home, But There Is Doubt If He Will Remain" *The Globe*, Thurs. May 13, 1920, p. 1.

55. Ibid.

56. *The Globe*, Wed. May 19, 1920, p. 1.

57. "New Taxes Fall on Shoulders of Those Best Able to Meet Payment," *Edmonton Journal*, Wed., May 19, 1920, p. 1.

58. "Gov't To Evade Fate At Polls For Two Years," *The Globe*, Thurs. May 20, 1920, p. 1.

59. Borden, ed., *Robert Laird Borden: His Memoirs*, p. 1031.

60. Roger Graham, *Arthur Meighen, Vol. 2: And Fortune Fled* (Toronto: Clarke, Irwin, 1963), p. 5.

61. Former Prince Edward Island premier A.E. Arsenault. *Memoirs of The Hon. A. E. Arsenault*, n.d. Publisher not stated, p. 77.

62. Ibid., p. 78.

63. Moyles, ed., *Challenge of the Homestead*, Aug. 21, 1920, p. 76.

64. Ibid., Aug. 22, 1920, p. 77.

65. Ibid., Aug. 21, 1920, p. 73.

66. Calgary businessman A.E. Cross to Prime Minister Arthur Meighen. Nov. 11, 1920. LAC. Prime Ministers' Fonds. C3428. Vol. 40, MG26 I. 23299.

67. Prime Minister Arthur Meighen to R.E. Gosnell, Dec. 2, 1920. LAC. Prime Ministers' Fonds. C3428. Vol. 40, MG26 I. 23303.

68. Ibid., 23304.

69. As quoted in letter from Prince Edward Island premier John Bell to Prime Minister Mackenzie King, Dec. 12, 1922. LAC. Prime Ministers' Fonds. C2242. Vol. 69, MG26 J1. 59499.

70. Prime Minister Arthur Meighen to R.E. Gosnell, Dec. 2, 1920. LAC. Prime Ministers' Fonds. C3428. Vol. 40, MG26 I. 23304.

71. Ontario Conservative MP Francis Keefer to Prime Minister Arthur Meighen. Memo Re Natural Resources, Dec. 10, 1920. LAC. Prime Ministers' Fonds. C3428. Vol. 40, MG26 I. 23572.

72. Memo prepared by Ontario Conservative MP Francis Keefer for then-Interior Minister Arthur Meighen, n.d. but the memo makes reference to the "Conference on the 19th instant," which indicates November 1918. LAC. Prime Ministers' Fonds. C3428. Vol. 40, MG26 I. 23273 and 23283.

73. Ibid., 23280.

74. Ontario Conservative MP Francis Keefer to Prime Minister Arthur Meighen. Memo Re Natural Resources, Dec. 10, 1920. LAC. Prime Ministers' Fonds. C3428. Vol. 40, MG26 I. 23576.

75. Moyles, ed., *Challenge of the Homestead*, Nov. 29, 1920, p. 98.

76. Ibid., Dec. 9, 1920, 101.

77. Prime Minister Arthur Meighen to Manitoba premier T.C. Norris. The letter is not dated beyond December 1920, but Meighen identifies the date as December 7 in a subsequent letter to Norris. LAC. Prime Ministers' Fonds. C3428. Vol. 40. MG26 I. 23262.

78. Ibid.

79. Ibid. (My italics.)

80. Ibid. (My italics.)

81. Manitoba premier T.C. Norris to Prime Minister Arthur Meighen, Dec. 10, 1920. LAC. Prime Ministers' Fonds. C3428. Vol. 40. : MG26 1. 23328–29.

82. Ibid., 23329.

83. Ibid., 23332.

84. Ibid., 23347.

85. Ibid., 23348.

86. Prime Minister Arthur Meighen to Manitoba premier T.C. Norris, Dec. 24, 1920. LAC. Prime Ministers' Fonds. C3428. Vol. 40, MG26 I. 23325.

87. Ibid., 23326.

88. Ibid.

89. Ibid.

90. David J. Hall on Arthur Sifton: http://www.biographi.ca/009004-119.01-e.php?&id_nbr=8365

91. Chester Martin, *"Dominion Lands" Policy* (Toronto: Macmillan of Canada, at St. Martin's House, 1938), p. 490.

92. Ibid., p. 483.

93. Brief of Manitoba, May 1921, as quoted in Martin, *"Dominion Lands" Policy*, p. 484.

94. Graham, *And Fortune Fled*, p. 134.

95. Moyles, ed., *Challenge of the Homestead*, March 9, 1921, p. 122.

96. Ibid., June 8, 1921, p. 146.

97. Ibid., March 9, 1921, p. 125.

98. Ibid., p. 124.

99. Ibid., June 8, 1921, p. 145.

100. Ibid., June 16, 1921, p. 148.

101. Graham, *And Fortune Fled*. p. 111.

102. Ibid.

103. As quoted in Graham, *And Fortune Fled*, p. 112.

104. David C. Jones, "Herbert W. Greenfield: 1921–1925" in Bradford J. Rennie, ed., *Alberta Premiers of the Twentieth Century* (Regina: Canadian Plains Research Center, University of Regina, 2004), p. 62.

105. Moyles, ed., *Challenge of the Homestead*, Oct. 31, 1921, p. 168.

106. Ibid., Nov. 27, 1921, p. 174.

107. Ibid., Nov. 27, 1921, p. 173.

108. Ibid.

109. Graham, *And Fortune Fled*, p. 113.

110. Ibid.

111. As calculated by Graham, *And Fortune Fled*, p. 166.

112. Graham, *And Fortune Fled*, p. 167.

113. Moyles, ed., *Challenge of the Homestead*, April 12, 1922, p. 194.

114. Ibid., p. 195.

115. Ibid., June 14, 1922, p. 209.

116. Ibid., Jan. 1, 1924, p. 299.

117. Ibid., p. 299.

118. Ibid., p. 301.

119. Ibid., Dec. 7, 1924, p. 323.

CHAPTER TEN

1. Diary of William Lyon Mackenzie King, July 22, 1927. http://www.collectionscanada .gc.ca/databases/king/001059-119.02-e.php?&page_id_nbr=10587&interval=20&& PHPSESSID=2v4lol8cgj384q8fq47ebkc9r2

2. Diary of William Lyon Mackenzie King, July 2, 1927. http://www.collectionscanada .gc.ca/databases/king/001059-119.02-e.php?&page_id_nbr=10552&interval=20&& PHPSESSID=pcatsrm3c51dgj9c5adafofolo

3. H. Blair Neatby, *William Lyon Mackenzie King, 1924–1932: The Lonely Heights* (Toronto: University of Toronto Press, 1963), p. 197.

4. Robert A. Wardhaugh, *Mackenzie King and the Prairie West* (Toronto: University of Toronto Press, 2000), p. 35.

5. Diary of William Lyon Mackenzie King, Thursday, December 29, 1921, p. 473.

6. Prime Minister Mackenzie King to Alberta premier Herbert Greenfield, Feb. 20, 1922. LAC. Prime Ministers' Fonds. C3428. Vol. 40, MG26 1. 23582. King wrote identical letters to the three Western Premiers.

7. Ibid.

8. Ibid., 23582–83.

9. Ibid., 23583.

10. Diary of William Lyon Mackenzie King, March 8, 1922. http://www.collections canada.gc.ca/databases/king/001059-119.02-e.php?&page_id_nbr=8131&interval= 20&&PHPSESSID=a9f846p1lma237hrkcljftgof2

11. Ibid.

12. Manitoba premier T.C. Norris to Prime Minister Mackenzie King, March 10, 1922. LAC. Prime Ministers' Fonds. C2248. Vol. 79, MG26 J1. 66914.

13. Ibid., 66918.

14. Ibid.

15. Alberta premier Herbert Greenfield to Prime Minister Mackenzie King, April 15, 1922. LAC. Prime Ministers' Fonds. C2245. Vol. 74. MG26 J1. 62718. (My italics.)

16. Ibid.

17. Saskatchewan premier Charles Dunning to Prime Minister Mackenzie King, April 10, 1922. LAC. Prime Ministers' Fonds. C2244. Vol. 72. MG26 J1. 61639.

18. Ibid.
19. David C. Jones, "Herbert W. Greenfield: 1921–1925," in Bradford J. Rennie, ed. *Alberta Premiers of the Twentieth Century* (Regina: Canadian Plains Research Center, University of Regina, 2004), p. 65.
20. Nikola Wirsta, "Endurance," in *Land of Pain, Land of Promise: First Person Accounts by Ukrainian Pioneers 1891–1914*, trans. Harry Piniuta (Saskatoon: Western Producer Prairie Books, 1978), p. 154.
21. Ibid.
22. Ibid., pp. 154–55.
23. David C. Jones, "Herbert W. Greenfield: 1921–1925," in Rennie, ed., *Alberta Premiers of the Twentieth Century*, p. 68.
24. Diary of William Lyon Mackenzie King, Thurs., April 20, 1922. http://www.collections canada.gc.ca/databases/king/001059-119.02-e.php?&page_id_nbr=8197&interval= 20&&PHPSESSID=473eung9d952nprolqfu3fcmoo
25. http://spacingottawa.ca/2011/01/13/the-grand-dames-ottawa's-historic -apartment-buildings/
26. Diary of William Lyon Mackenzie King, Friday, April 21, 1922. http://www.collections canada.gc.ca/databases/king/001059-119.02-e.php?&page_id_nbr=8197&interval= 20&&&&&&&&PHPSESSID=a9f846p1lma237hrkcljftgof2
27. Diary of William Lyon Mackenzie King, Friday, April 21, 1922. http://www.collections canada.gc.ca/databases/king/001059-119.02-e.php?&page_id_nbr=8198&interval= 20&&&&&&&&PHPSESSID=473eung9d952nprolqfu3fcmoo
28. Diary of William Lyon Mackenzie King, Sunday, April 23, 1922. http://www.collectionscanada.gc.ca/databases/king/001059-119.02-e.php?& page_id_nbr=8198&interval=20&&PHPSESSID=473eung9d952nprolqfu3fcmoo
29. Diary of William Lyon Mackenzie King, Tuesday, April 25, 1922.
30. Diary of William Lyon Mackenzie King, Friday, April 28, 1922. http://www.collectionscanada.gc.ca/databases/king/001059-119.02-e.php?& page_id_nbr=8209&interval=20&&PHPSESSID=473eung9d952nprolqfu3fcmoo
31. Diary of William Lyon Mackenzie King, Sunday, April 30, 1922. http://www.collectionscanada.gc.ca/databases/king/001059-119.02-e.php?& page_id_nbr=8213&interval=20&&PHPSESSID=473eung9d952nprolqfu3fcmoo
32. Morris Mott, "Tobias C. Norris: 1915–1922," in Barry Ferguson and Robert Wardhaugh, eds., *Manitoba Premiers of the 19th and 20th Centuries* (Regina: Canadian Plains Research Center, University of Regina, 2010), p. 154.
33. Robert Wardhaugh and Jason Thistlewaite, "John Bracken: 1922–1943," Ferguson and Wardhaugh, eds., in *Manitoba Premiers of the 19th and 20th Centuries*, p. 170.
34. W.L. Morton, *Manitoba: A History* (Toronto: University of Toronto Press, 2nd edition, 1967), p. 383.
35. Ibid.

36. Diary of William Lyon Mackenzie King, Tuesday, November 14, 1922.
http://www.collectionscanada.gc.ca/databases/king/001059-119.02-e.php?&
page_id_nbr=8318&interval=20&&PHPSESSID=a9f846p1lma237hrkcljftgof2

37. Diary of William Lyon Mackenzie King, Tuesday, Nov. 14, 1922.
http://www.collectionscanada.gc.ca/databases/king/001059-119.02-e.php?&
page_id_nbr=8319&interval=20&&PHPSESSID=473eung9d952nprolqfu3fcmoo

38. Diary of William Lyon Mackenzie King, Wednesday, November 15, 1922.
http://www.collectionscanada.gc.ca/databases/king/001059-119.02-e.php?&
page_id_nbr=8321&interval=20&&&&PHPSESSID=a9f846p1lma237hrkcljftgof2

39. Diary of William Lyon Mackenzie King, Thursday, Nov. 16, 1922. My italics.

40. Diary of William Lyon Mackenzie King, Thursday, Nov.16, 1922.
http://www.collectionscanada.gc.ca/databases/king/001059-119.02-e.php?&
page_id_nbr=8323&interval=20&&PHPSESSID=473eung9d952nprolqfu3fcmoo

41. Diary of William Lyon Mackenzie King, Friday, Nov. 17, 1922.
http://www.collectionscanada.gc.ca/databases/king/001059-119.02-e.php?&
page_id_nbr=8323&interval=20&&PHPSESSID=473eung9d952nprolqfu3fcmoo

42. Diary of William Lyon Mackenzie King, Friday, April 21, 1922.
http://www.collectionscanada.gc.ca/databases/king/001059-119.02-e.php?&
page_id_nbr=8197&interval=20&&PHPSESSID=473eung9d952nprolqfu3fcmoo

43. Diary of William Lyon Mackenzie King, Friday, November 17, 1922.
http://www.collectionscanada.gc.ca/databases/king/001059-119.02-e.php?&
page_id_nbr=8323&interval=20&&PHPSESSID=473eung9d952nprolqfu3fcmoo

44. Alberta premier Herbert Greenfield to Prime Minister Mackenzie King, Jan. 2,
1923. LAC. Prime Ministers' Fonds. C2253. Vol. 87. MG26 J1. 73600.

45. Prime Minister Mackenzie King to Alberta premier Herbert Greenfield, Jan. 12,
1923. LAC. Prime Ministers' Fonds. C2253. Vol. 87. MG26 J1. 73602.

46. Prime Minister Mackenzie King to Alberta premier Herbert Greenfield, Feb 15,
1923. LAC. Prime Ministers' Fonds. C2253. Vol. 87. MG26 J1. 73606.

47. I am indebted to York University doctoral candidate Christine Sismondo for help
in clarifying the U.S. situation.

48. Craig Heron, *Booze: A Distilled History* (Toronto: Between The Lines, 2003), p. 271.

49. "From Edge of Arctic Off Norway's Shores Comes Raging Blast," *The Globe*,
Thurs. Feb. 15, 1923, p. 1.

50. H. Blair Neatby, *William Lyon Mackenzie King, 1924–1932: The Lonely Heights*, p. 5–6.

51. Ibid.

52. Ibid.

53. "The Behavior of Today's Young Girls," *Calgary Daily Herald*, Sat., March 3, 1923, p. 18.

54. "Six County Ridings May Pass In Ontario To Fill Cities' Bill," *The Globe*, Tues.,
March 6, 1923. Page one.

55. "Wants Nova Scotia Separate Dominion," *The Globe*, Fri., April 20, 1923, p. 1.

56. E.R. Forbes, "The Origins of the Maritime Rights Movement," in E.R. Forbes, ed., *Challenging the Regional Stereotype: Essays on the 20th Century Maritimes* (Fredericton: Acadiensis Press, 1989), p. 102.

57. Ibid., p. 113.

58. Prince Edward Island premier John Bell to Prime Minister Mackenzie King, December 12, 1922. LAC. Prime Ministers' Fonds. C2242. Vol. 69, MG26 J1. 59498.

59. Ibid., 59499.

60. Nova Scotia premier Ernest Armstrong to Prime Minister Mackenzie King, April 4, 1923. LAC. Prime Ministers' Fonds. C2250. Vol. 83. MG26 J1. 70270.

61. Robert A. Wardhaugh, *Mackenzie King and the Prairie West* (Toronto: University of Toronto Press, 2000), p. 89.

62. Diary of William Lyon Mackenzie King, Wed., May 2, 1923. http://www.collections canada.gc.ca/databases/king/001059-119.02-e.php?&page_id_nbr=8556&interval=20&&PHPSESSID=iuvk1pvvfbpnmhlb8nejihqv76

63. "Public Ownership Running Gauntlet of Propagandists," *The Globe*, Mon., May 7, 1923, p. 1.

64. "Bread-winner and Housewife (Helps Folk To Make Ends Meet) Partners in Fielding Budget," *The Globe*, Sat., May 12, 1923, p. 1.

65. Diary of William Lyon Mackenzie King, Fri., May 11, 1923. http://www.collections canada.gc.ca/databases/king/001059-119.02-e.php?&page_id_nbr=8564&interval=20&&PHPSESSID=sletgvt3np6ut6q8lnlrbcpqo1

66. Manitoba premier John Bracken to Prime Minister Mackenzie King, May 10, 1923. LAC. Prime Ministers' Fonds. C2251. Vol. 84. MG26 J1. 71027.

67. Diary of William Lyon Mackenzie King, Thurs., Sept. 13, 1923. http://www.collectionscanada.gc.ca/databases/king/001059-119.02-e.php?&page_id_nbr=8595&interval=20&&PHPSESSID=sletgvt3np6ut6q8lnlrbcpqo1

68. http://www.international.gc.ca/history-histoire/world-monde/1921-1939.aspx?lang=eng&view=d

69. Manitoba premier John Bracken to Prime Minister Mackenzie King, Jan. 1, 1924. LAC. Prime Ministers' Fonds. C2262. Vol. 97. MG26 J1. 82448.

70. Diary of William Lyon Mackenzie King, Thurs., Jan. 3, 1924. LAC. http://www.collectionscanada.gc.ca/databases/king/001059-119.02-e.php?&page_id_nbr=8661&interval=20&&PHPSESSID=3n8p1dkohe1g1bmbqtt8kjj901

71. Diary of William Lyon Mackenzie King, Fri., Jan. 4, 1924. LAC. http://www.collectionscanada.gc.ca/databases/king/001059-119.02-e.php?&page_id_nbr=8662&interval=20&&PHPSESSID=3n8p1dkohe1g1bmbqtt8kjj901

72. Prime Minister Mackenzie King to Manitoba premier John Bracken, Jan. 29, 1924. LAC. Prime Ministers' Fonds. C2262. Vol. 97. MG26 J1. 82451.

73. Manitoba premier John Bracken to Prime Minister Mackenzie King, Feb. 1, 1924. LAC. Prime Ministers' Fonds. C2262. Vol. 97. MG26 J1. 82454.

74. Kenneth Norrie, Douglas Owram and J.C. Herbert Emery, *A History of the Canadian Economy* (Toronto: Nelson, a division of Thomson Canada Limited, fourth edition, 2008), p. 280.

75. Ibid.

76. Ibid.

77. Ibid., p. 299.

78. Alberta premier Herbert Greenfield to Prime Minister Mackenzie King, March 30, 1924. LAC. Prime Ministers' Fonds. C2265. Vol. 101. MG26 J1. 85334.

79. Prime Minister Mackenzie King to Alberta premier Herbert Greenfield, April 1, 1924. LAC. Prime Ministers' Fonds. C2265. Vol. 101. MG26 J1. 85335.

80. Diary of William Lyon Mackenzie King, Thurs., April 10, 1924. http://www.collectionscanada.gc.ca/databases/king/001059-119.02-e.php?& page_id_nbr=8819&interval=20&&PHPSESSID=837hhhuahpdgvt26dt5coscqd7

81. Paul Boothe and Heather Edwards, eds., *Eric J. Hanson's Financial History of Alberta: 1905–1950* (Calgary: University of Calgary Press, 2003), p. 91.

82. Alberta premier Herbert Greenfield to Prime Minister Mackenzie King, Dec. 16, 1924. LAC. Prime Ministers' Fonds. C2265. Vol. 101. MG26 J1. 85387.

83. Prime Minister Mackenzie King to Alberta premier Herbert Greenfield, June 30, 1925. LAC. Prime Ministers' Fonds. C2276. Vol. 115. MG26 J1. 97992–94.

84. Franklin L. Foster, "John E. Brownlee: 1925–1934," in Rennie, ed., *Alberta Premiers of the Twentieth Century*, p. 83.

85. Diary of William Lyon Mackenzie King, Tuesday, August 18, 1925. http://www.collectionscanada.gc.ca/databases/king/001059-119.02-e.php?& page_id_nbr=9365&interval=20&&PHPSESSID=pb19fbbpvqj6ev5593urod3i9

86. Diary of William Lyon Mackenzie King, Tues., Aug. 18, 1925. http://www.collections canada.gc.ca/databases/king/001059-119.02-e.php?&page_id_nbr=9365&interval= 20&&PHPSESSID=9ljadif6ilvoo894k3c83fpvk6

87. Neatby, *William Lyon Mackenzie King, 1924–1932: The Lonely Heights*, p. 69.

88. Saskatchewan premier Charles Dunning to Prime Minister Mackenzie King. Feb. 11, 1926. LAC. Prime Ministers' Fonds. C2288. Vol. 131. MG26 J1. 111159.

89. Diary of William Lyon Mackenzie King, Mon., Feb. 15, 1926. http://www.collections canada.gc.ca/databases/king/001059-119.02-e.php?&page_id_nbr=9709&interval= 20&&PHPSESSID=ishv5u58m94qlv7pu4agbijmo5

90. Ibid.

91. All headlines drawn from the *Calgary Daily Herald*, Sat., Feb. 13, 1926, p. 1. http://www.ourfutureourpast.ca/newspapr/np_page2.asp?code=NAXP0725.JPG

92. J.F.C. Wright, *Saskatchewan: The History of a Province* (Toronto: McClelland & Stewart, 1955), p. 203.

93. Howard Palmer with Tamara Palmer, *Alberta: A New History* (Edmonton: Hurtig, 1990), p. 222.

94. All headlines and ads drawn from the *Calgary Daily Herald*, Sat., Feb. 13, 1926, pp. 1, 10 and 12. http://www.ourfutureourpast.ca/newspapr/np_page2.asp?code=NAXP0725.JPG

95. *Calgary Daily Herald*, Saturday, February 27, 1926. page one http://www.ourfuture ourpast.ca/newspapr/np_page2.asp?code=NAXp0989.jpg

CHAPTER ELEVEN

1. Prime Minister Mackenzie King to Queen's University professor Norman McLeod Rogers, Jan. 25, 1930. LAC. Prime Ministers' Fonds. C2322. Vol. 180. MG26 J1. 153886.

2. Ibid., 153887.

3. Ibid. (My italics.)

4. Diary of William Lyon Mackenzie King, Thurs., Dec. 31, 1925. http://www.collections canada.gc.ca/databases/king/001059-119.02-e.php?&page_id_nbr=9627&interval=20&&PHPSESSID=pti4nh7erq2jf5rn240oh5dtur7

5. Diary of William Lyon Mackenzie King, Sat., Jan. 2, 1926. http://www.collections canada.gc.ca/databases/king/001059-119.02-e.php?&page_id_nbr=9639&interval=20&&PHPSESSID=ishv5u58m94qlv7pu4agbijmo5

6. Alberta Act, 1905. Clause 17 (2).

7. Diary of William Lyon Mackenzie King, Sat., Feb. 13, 1926. http://www.collections canada.gc.ca/databases/king/001059-119.02-e.php?&page_id_nbr=9707&interval=20&&PHPSESSID=ishv5u58m94qlv7pu4agbijmo5

8. Diary of William Lyon Mackenzie King, Thurs., Feb. 25, 1926. http://www.collections canada.gc.ca/databases/king/001059-119.02-e.php?&page_id_nbr=9731&interval=20&&PHPSESSID=ishv5u58m94qlv7pu4agbijmo5

9. Ibid.

10. Quebec premier Louis-Alexandre Taschereau to Prime Minister Mackenzie King. March 22, 1926. LAC. Prime Ministers' Fonds. C2294. Vol. 139. MG26 J1. 118744.

11. Prince Edward Island premier J.D. Stewart to Quebec premier Louis-Alexandre Taschereau, March 13, 1926. LAC. Prime Ministers' Fonds. C2294. Vol. 139. MG26 J1. 118747.

12. Prime Minister Mackenzie King to Quebec premier Louis-Alexandre Taschereau, March 26, 1926. LAC. Prime Ministers' Fonds. C2294. Vol. 139. MG26 J1. 118749.

13. "Willys-Overland Fine Motors Cars," advertisement in *The Globe*, Fri., April 2, p. 1

14. "Get Your Allotment of Shares In Highland Oil Company Limited, Before It Is Too Late," *Calgary Daily Herald*, Sat., April 3, 1926, p. 10.

15. "History Turns Back Its Pages And Men Lead In Easter Show," *The Globe*, Mon., April 5, 1926, p. 1.

16. Prime Minister Mackenzie King to Queen's University professor Norman McLeod Rogers, Jan. 25, 1930. LAC. Prime Ministers" Fonds. C2322. Vol. 180. MG26 J1. 153886.

17. "Sir Andrew Duncan Heads Commission of Investigation," *Halifax Herald*, Thurs., April 8, 1926. LAC. Prime Ministers' Fonds. C3457. Vol. 102. MG26 I. 58487.

18. Ibid.

19. Diary of William Lyon Mackenzie King, Thurs., April 8, 1926. http://www.collections canada.gc.ca/databases/king/001059-119.02-e.php?&page_id_nbr=9805&interval= 20&&&PHPSESSID=5dmic1fije5c1beg7ikvj8vlu7

20. Ibid.

21. Alberta premier John Brownlee to Prime Minister Mackenzie King, May 22, 1926. LAC. Prime Ministers' Fonds. C2286. Vol. 128. MG26 J1. 109377.

22. Prime Minister Mackenzie King to Winnipeg Liberal lawyer Albert Blellock Hudson. May 17, 1926. LAC. Prime Ministers' Fonds. C2289. Vol. 132. MG26 J1. 112598.

23. British Columbia premier John Oliver to Prime Minister Mackenzie King, Jan. 20, 1926. LAC. Prime Ministers' Fonds. C2292. Vol. 136. MG26 J1. 116073.

24. British Columbia premier John Oliver to Prime Minister Mackenzie King, May 28, 1926. LAC. Prime Ministers' Fonds. C3464. Vol. 116. MG26 1. 68045–46, plus enclosures.

25. Diary of William Lyon Mackenzie King, Wed., March 23, 1927. http://www.collectionscanada.gc.ca/databases/king/001059-119.02-e.php?& page_id_nbr=10391&interval=20&&PHPSESSID=bhfis4v5a5lho93trv2c7jmhf7

26. Advertisement for "The Improved Chevrolet Coach," *Nanton News*, Thurs., May 13, 1926, p. 4. http://www.ourfutureourpast.ca/newspapr/np _page2.asp?code=N9QP0155.JPG

27. H. Blair Neatby, *William Lyon Mackenzie King: 1924–1932: The Lonely Heights*. (Toronto: University of Toronto Press, 1963), p. 124.

28. Proceedings of Tues. June 8 at the 1926 Interprovincial Conference. *Dominion Provincial and Interprovincial Conferences From 1887 to 1926* (Ottawa: Reprinted by Edmond Cloutier, King's Printer and Controller of Stationery, 1951), p. 110.

29. Allan Levine, *King: William Lyon Mackenzie King: A Life Guided by the Hand Of Destiny* (Vancouver/Toronto: D&M Publishers, 2011), 142.

30. Allan Levine maintains that the smuggler was also "a talented Liberal fundraiser" who had helped the local Liberal in the 1925 federal election. Levine, *King: William Lyon Mackenzie King*, p. 154.

31. Neatby, *The Lonely Heights*, p. 143.

32. Neatby, ibid., p. 187.

33. Diary of William Lyon Mackenzie King, Wed., Dec. 8, 1926. http://www.collections canada.gc.ca/databases/king/001059-119.02-e.php?&page_id_nbr=10233&interval= 20&&&PHPSESSID=3r1j7fkr543p9dqkl7cut4n9m1

34. "Report of the Royal Commission on Maritime Claims" (Ottawa: F.A. Acland, Printer to the King's Most Excellent Majesty, 1926), p. 18.

35. Diary of William Lyon Mackenzie King, Wed., Sept. 22, 1926. http://www.collections
     canada.gc.ca/databases/king/001059-119.02-e.php?&page_id_nbr=10127&interval=
     20&&PHPSESSID=ganapt4ehvqasv8sn6m917t0n1

36. Ernest R. Forbes, *Maritime Rights: The Maritime Rights Movement, 1919–1927:
     A Study in Canadian Regionalism*. (Montreal: McGill-Queen's University Press,
     1979) 176.

37. Ernest R. Forbes, *Maritime Rights: The Maritime Rights Movement, 1919–1927:
     A Study in Canadian Regionalism* (Montreal: McGill-Queen's University Press,
     1979), p. 177.

38. Diary of William Lyon Mackenzie King, Sun., April 3, 1927. http://www.collections
     canada.gc.ca/databases/king/001059-119.02-e.php?&page_id_nbr=10408&interval=
     20&&PHPSESSID=u27d8brrfsms2manc5ua6cg0c5

39. Neatby, *The Lonely Heights*, p. 200.

40. Diary of William Lyon Mackenzie King, Mon., April 11, 1927. http://www.collections
     canada.gc.ca/databases/king/001059-119.02-e.php?&page_id_nbr=10421&interval=
     20&&PHPSESSID=u27d8brrfsms2manc5ua6cg0c5

41. James Morton, *Honest John Oliver: The Life Story of the Honourable John Oliver,
     Premier of British Columbia: 1918–1927* (London, Toronto and Vancouver: J.M. Dent
     and Sons, 1933), 238.

42. "Trail of Strangler Is Lost By Police: Whole West Is Combed," *The Globe*,
     Wed., June 15, 1927, p. 1.

43. "Bracken Government Is Sustained at Polls In Manitoba Contest," *The Globe*,
     Wed., June 29, p. 1

44. "Huge Broadcast Is Great Success," *The Globe*, Sat., July 2, 1927, p. 1.

45. "Great Broadcast Is Huge Success," *The Globe*, Sat., July 2, 1927, p. 2. The
     newspaper reversed the adjectives on the page two continuation of the story.

46. Neatby, *The Lonely Heights*, p. 210.

47. Kenneth Norrie, Douglas Owram and J.C. Herbert Emery, *A History of the
     Canadian Economy* (Toronto: Nelson, Thomson, fourth edition, 2008), p. 291.

48. Ibid., p. 288.

49. Ibid., p. 289.

50. Diary of William Lyon Mackenzie King, Sun., June 26, 1927. http://www.collections
     canada.gc.ca/databases/king/001059-119.02-e.php?&page_id_nbr=10544&interval=
     20&&&PHPSESSID=lovqqoo8v2nooil3ejkostfq53

51. Diary of William Lyon Mackenzie King, Wed., Oct. 12, 1927. http://www.collections
     canada.gc.ca/databases/king/001059-119.02-e.php?&page_id_nbr=10726&interval=
     20&&PHPSESSID=m6edq4297mqqb52b1089tlmlc3

52. Diary of William Lyon Mackenzie King, Wed., Nov. 2, 1927. http://www.collections
     canada.gc.ca/databases/king/001059-119.02-e.php?&page_id_nbr=10758&interval=
     20&&PHPSESSID=m6edq4297mqqb52b1089tlmlc3

53. Diary of William Lyon Mackenzie King, Sat., Nov. 5, 1927. http://www.collections canada.gc.ca/databases/king/001059-119.02-e.php?&page_id_nbr=10763&interval= 20&&PHPSESSID=m6edq4297mqqb52b1089tlmlc3

54. Diary of William Lyon Mackenzie King, Tues., Nov. 8, 1927. There is a question mark in brackets after the word "mendicancy" to indicate the typist's uncertainty about King's handwriting here. http://www.collectionscanada.gc.ca/databases /king/001059-119.02-e.php?&page_id_nbr=10767&interval=20&&PHPSESSID=m 6edq4297mqqb52b1089tlmlc3

55. Official Précis for Wednesday morning, Nov. 9, 1927. *Dominion–Provincial Conferences: November 3–10, 1927; December 9–13, 1935; January 14–15, 1941* (Ottawa: Reprinted by Edmond Cloutier, King's Printer, 1951) Gardiner's remarks were actually made on Tues., Nov. 8, p. 23.

56. Official Précis for Tuesday afternoon, Nov. 8, 1927. *Dominion–Provincial Conferences: November 3–10, 1927; December 9–13, 1935; January 14–15, 1941* (Ottawa: Reprinted by Edmond Cloutier, King's Printer, 1951), p. 22.

57. Ibid., p. 23.

58. Ibid.

59. Constitution Act, 1867. Clause 92, subsection 10 (c).

60. Official Précis for Wed. morning, Nov. 9, 1927. *Dominion–Provincial Conferences: November 3–10, 1927; December 9–13, 1935; January 14–15, 1941* (Ottawa: Reprinted by Edmond Cloutier, King's Printer, 1951), p. 25.

61. Ibid.

62. Ibid.

63. Ibid., p. 26.

64. Ibid.

65. Diary of William Lyon Mackenzie King, Thurs., Nov. 10, 1927. http://www.collections canada.gc.ca/databases/king/001059-119.02-e.php?&page_id_nbr=10771&interval= 20&&PHPSESSID=m6edq4297mqqb52b1089tlmlc3

66. Diary of William Lyon Mackenzie King, Fri., Nov. 11, 1927. http://www.collections canada.gc.ca/databases/king/001059-119.02-e.php?&page_id_nbr=10771&interval= 20&&PHPSESSID=m6edq4297mqqb52b1089tlmlc3

67. Diary of William Lyon Mackenzie King, Mon., Jan. 9, 1928. http://www.collections canada.gc.ca/databases/king/001059-119.02-e.php?&page_id_nbr=10869&interval= 20&&PHPSESSID=rbltfmgrl6j93bopbdh46kocs4

68. King wrote "McLean." Diary of William Lyon Mackenzie King, Mon., January 9, 1928. http://www.collectionscanada.gc.ca/databases/king /001059-119.02-e.php?& page_id_nbr=10869&interval=20&&PHPSESSID =rbltfmgrl6j93bopbdh46kocs4

69. Manitoba premier John Bracken to Prime Minister Mackenzie King, Jan. 10, 1928. LAC. Prime Ministers' Fonds. C2302. Vol. 150. MG26 J1. 128324.

70. Prime Minister Mackenzie King to Manitoba premier John Bracken, Feb. 28, 1928. LAC. Prime Ministers' Fonds. C2302. Vol. 150. MG26 J1. 128327.

71. Diary of William Lyon Mackenzie King, Fri., Jan. 13, 1928. http://www.collections canada.gc.ca/databases/king/001059-119.02-e.php?&page_id_nbr=10874&interval= 20&&PHPSESSID=rbltfmgrl6j93bopbdh46kocs4

72. That alliance would dissolve when the Ontario Government cancelled Ontario Hydro's contracts with private Quebec contractors in 1935. I am grateful to McMaster University scholar Viv Nelles, who holds the L.R. Wilson Chair in Canadian history, for drawing this to my attention.

73. "Courts Will Decide Ownership Of Power Premier Promises" in *The Globe*, Tues., January 31, 1928. page one.

74. Quebec premier Louis-Alexandre Taschereau to Prime Minister Mackenzie King, March 2, 1928. LAC. Prime Ministers' Fonds. C2307. Vol. 158. MG26 J1. 134873–74.

75. Diary of William Lyon Mackenzie King, Thurs., Jan. 19, 1928. http://www.collections canada.gc.ca/databases/king/001059-119.02-e.php?&page_id_nbr=10884&interval= 20&&PHPSESSID=rbltfmgrl6j93bopbdh46kocs4

76. Diary of William Lyon Mackenzie King, Fri., Jan. 27, 1928. http://www.collections canada.gc.ca/databases/king/001059-119.02-e.php?&page_id_nbr=10897&interval= 20&&PHPSESSID=rbltfmgrl6j93bopbdh46kocs4

77. Report of the Royal Commission Pursuant to Order-in-Council of March 8, 1927, formally submitted on Feb. 16, 1928. Ministry of the Interior RG 15. 2088, p. 49.

78. Report of the Royal Commission Pursuant to Order-in-Council of March 8, 1927, formally submitted on Feb. 16, 1928. Ministry of the Interior RG 15. 2088, p. 49.

79. Diary of William Lyon Mackenzie King, Mon., Feb. 6, 1928.

80. Neatby, *The Lonely Heights*, p. 270.

81. British Columbia premier John MacLean to Prime Minister Mackenzie King, May 15, 1928. LAC. Prime Ministers' Fonds. C2304. Vol. 154. MG26 J1. 131468.

82. William Rayner, *British Columbia's Premiers in Profile: The Good, the Bad and the Transient* (Surrey, B.C.: Heritage House Publishing, 2000), p. 141.

83. Manitoba Liberal leader Hugh Amos Robson to Prime Minister Mackenzie King, May 18, 1928. LAC. Prime Ministers' Fonds. C2306. Vol. 156. MG26 J1. 133150.

84. Manitoba premier John Bracken to Prime Minister Mackenzie King, June 15, 1928. LAC. Prime Ministers' Fonds. C2302. Vol. 150. MG26 J1. 128342.

85. Diary of William Lyon Mackenzie King, Tues., July 3, 1928.

86. Diary of William Lyon Mackenzie King, Sat., July 7, 1928.

87. Memorandum on Manitoba Federal Resources. LAC. Prime Ministers' Fonds. C2302. Vol. 150. MG26 J1. 128372.

88. "Huge Stocks of Liquor Are Seized at Border on Drayton's Order," *The Globe*, Thurs., July 12, 1918, pp. 1 and 2.

89. "Police Net Six in Mail Robbery and Dig Up Buried Booty," *The Globe*, Thurs., July 12, 1918, p. 1.

90. Prime Minister Mackenzie King to Alberta premier John Brownlee, Dec. 29, 1928. LAC. Prime Ministers' Fonds. C2302. Vol. 151. MG26 J1. 128478–79.

91. Diary of William Lyon Mackenzie King, Fri., Dec. 14, 1928. http://www.collections canada.gc.ca/databases/king/001059-119.02-e.php?&page_id_nbr=11388&interval= 20&&PHPSESSID=ei2bl16th71b5or00fqhlbers1

92. Diary of William Lyon Mackenzie King, Wed., Dec. 5, 1928. http://www.collections canada.gc.ca/databases/king/001059-119.02-e.php?&page_id_nbr=11376&interval= 20&&PHPSESSID=ei2bl16th71b5or00fqhlbers1

93. King mistakenly refers to the Queen as "Alexandra," who had been the wife of the previous king and who had died in 1925. Diary of William Lyon Mackenzie King, Tues., Dec. 25, 1928. http://www.collectionscanada.gc.ca/databases/king/001059-119.02-e.php ?&page_id_nbr=11407&interval=20&&PHPSESSID=kiggju04of8jod5vab7gugm284

94. Diary of William Lyon Mackenzie King, Mon., Dec. 31, 1928. http://www.collections canada.gc.ca/databases/king/001059-119.02-e.php?&page_id_nbr=11418&interval= 20&&PHPSESSID=jr81e22a39s024438q0aah8up4

95. Most resource data in this paragraph from Interior Minister Charles Stewart to Privy Council Secretary Norman Rogers, May 21, 1929. LAC. Prime Ministers' Fonds. C2314. Vol. 169. MG26 J1. 143998.

96. Diary of William Lyon Mackenzie King, Sun., March 24, 1929. http://www.collections canada.gc.ca/databases/king/001059-119.02-e.php?&page_id_nbr=11541&interval=2 0&&PHPSESSID=vr2k19jbuafpv6medfoaucogi1

97. Diary of William Lyon Mackenzie King, Mon., March 25, 1929. http://www.collections canada.gc.ca/databases/king/001059-119.02-e.php?&page_id_nbr=11544&interval= 20&&&PHPSESSID=vr2k19jbuafpv6medfoaucogi1

98. David Smith, "James G. Gardiner," in Gordon L. Barnhart, ed., *Saskatchewan Premiers of the Twentieth Century* (Regina: Canadian Plains Research Center, University of Regina, 2004), p. 99.

99. Diary of William Lyon Mackenzie King, Mon., Dec. 9, 1929. http://www.collections canada.gc.ca/databases/king/001059-119.02-e.php?&page_id_nbr=11982&interval= 20&&PHPSESSID=4c6eu5j245f3e2i16hmrb55jf4

100. Diary of William Lyon Mackenzie King, Tues., Dec. 10, 1929. http://www.collections canada.gc.ca/databases/king/001059-119.02-e.php?&page_id_nbr=11984&interval= 20&&PHPSESSID=4c6eu5j245f3e2i16hmrb55jf4

101. Interview with former Alberta premier John Edward Brownlee, recorded between March 28, 1961 and May 12, 1961. United Farmers of Alberta Oral History Project. Glenbow Museum Archives. Standard number M-4079, p. 45. Transcript available on-line at: http://www.glenbow.org/collections/search/findingAids/archhtm/ ufa_oral.cfm

102. Ibid., pp. 46–47. Transcript available on-line at: http://www.glenbow.org/collections/search/findingAids/archhtm/ufa_oral.cfm

103. Ibid., p. 53. Transcript available on-line at: http://www.glenbow.org/collections/search/findingAids/archhtm/ufa_oral.cfm

104. Diary of William Lyon Mackenzie King, Wed., Dec. 11, 1929. http://www.collectionscanada.gc.ca/databases/king/001059-119.02-e.php?&page_id_nbr=11986&interval=20&&PHPSESSID=4c6eu5j245f3e2i16hmrb55jf4

105. Diary of William Lyon Mackenzie King, Wed., Dec. 11, 1929. http://www.collectionscanada.gc.ca/databases/king/001059-119.02-e.php?&page_id_nbr=11987&interval=20&&PHPSESSID=4c6eu5j245f3e2i16hmrb55jf4

106. Interview with former Alberta premier John Edward Brownlee, recorded between March 28, 1961 and May 12, 1961. United Farmers of Alberta Oral History Project. Glenbow Museum Archives. Standard number M-4079, pp. 53–54. Transcript available on-line at: http://www.glenbow.org/collections/search/findingAids/archhtm/ufa_oral.cfm

107. Ibid., p. 54. Transcript available on-line at: http://www.glenbow.org/collections/search/findingAids/archhtm/ufa_oral.cfm

108. Diary of William Lyon Mackenzie King, Wed., Dec. 11, 1929. http://www.collectionscanada.gc.ca/databases/king/001059-119.02-e.php?&page_id_nbr=11987&interval=20&&PHPSESSID=4c6eu5j245f3e2i16hmrb55jf4

109. Diary of William Lyon Mackenzie King, Sat., Dec. 14, 1929. http://www.collectionscanada.gc.ca/databases/king/001059-119.02-e.php?&page_id_nbr=11991&interval=20&&PHPSESSID=tvmavjqojj1knsfdmhrjviaj55

110. Howard Palmer with Tamara Palmer, *Alberta: A New History* (Edmonton: Hurtig, 1990), p. 217.

111. Prince Edward Island premier Alberta Saunders to Prime Minister Mackenzie King, Dec. 31, 1929. LAC. Prime Ministers' Fonds. C2313. Vol. 167. MG26 J1. 142440.

112. King's words for Jan. 24, 1930 are quoted in Christopher Armstrong, *The Politics of Federalism: Ontario's Relations with the Federal Government, 1867–1942* (Toronto: University of Toronto Press, 1981), p. 174. I have slightly changed the wording to match the actual diary quotation.

113. Saskatchewan Premier James Anderson to Prime Minister Mackenzie King, Jan. 17, 1930. LAC. Prime Ministers' Fonds. C2315. Vol. 170. MG26 J1. 145234.

114. Diary of William Lyon Mackenzie King, Tues., July 15, 1930. http://www.collectionscanada.gc.ca/databases/king/001059-119.02-e.php?&page_id_nbr=12362&interval=20&&PHPSESSID=hcqr6lvj2d2op507uk94gpc800

115. Neatby, *The Lonely Heights*, p. 312.

116. I am indebted for guidance on this issue to the masterful constitutional expert Sujit Choudhry, who is now at the New York University School of Law.

117. Diary of William Lyon Mackenzie King, Thurs., March 6, 1930. http://www.collections canada.gc.ca/databases/king/001059-119.02-e.php?&page_id_nbr=12362&interval= 20&&PHPSESSID=ot2ps7plgaqauh2q42g64p3qb2

118. Diary of William Lyon Mackenzie King, 3:30 a.m., Tues., July 29, 1930. http://www.collectionscanada.gc.ca/databases/king/001059-119.02-e.php?& page_id_nbr=12384&interval=20&&PHPSESSID=qfqing9s51v8ohraqjm7slhge6

119. Queen's University professor Normal McLeod Rogers to Prime Minister Mackenzie King, March 5, 1930. LAC. Prime Ministers' Fonds. C2322. Vol. 180. MG26 J1. 153902. This was *not* taken from the pamphlet, which Rogers wrote on his portable typewriter and mailed to King. It has apparently not survived.

120. "Immigration Into Canada Is Cut To Minimum by Ban," *Calgary Daily Herald*, Wed., Oct. 15, 1930, p. 1. http://www.ourfutureourpast.ca/newspapr/np_page2.asp ?code=nhip0909.jpg

121. "Province of Alberta" advertisement in the special section of the *Calgary Daily Herald*, Wed. Oct. 15, 1930, p. 24. http://www.ourfutureourpast.ca/newspapr/np _page2.asp?code=nhip0965.jpg

122. "No Glass in Windows; Family of Nine Live In One-Room Shanty," *Calgary Daily Herald*, Sat., Dec. 20, 1930. http://www.ourfutureourpast.ca/newspapr/np_ page2.asp?code=nhjp1117.jpg

123. John Dul, "Moonshining in Waskatenau" in Joanna Matejko, ed., *Polish Settlers in Alberta: Reminiscences and Biographies* (Toronto: Polish Alliance Press, 1979), pp. 80–81.

124. *Saskatoon Star Phoenix*, Oct. 31, 1930 and *Regina Star*, Oct. 30, 1930; *Edmonton Journal*, Feb. 10, 1931.

125. James G. MacGregor, *A History of Alberta*, revised edition (Edmonton: Hurtig, 1981) 263.

126. Alberta premier John Brownlee to Prime Minister R.B. Bennett, Dec. 28, 1931. LAC. Prime Ministers' Fonds. M1111. Vol. 467. MG26 K. 294936–37.

127. Paul Boothe and Heather Edwards, eds., *Eric J. Hanson's Financial History of Alberta: 1905–1950* (Calgary: University of Calgary Press, 2003), p. 97.

128. Christopher Armstrong, Matthew Evendon, and H.V. Nelles. *The River Returns: An Environmental History of the Bow*. (Montreal-Kingston: McGill-Queen's University Press, 2009) 338.

129. Christopher Armstrong, Matthew Evendon, and H.V. Nelles. *The River Returns: An Environmental History of the Bow*. (Montreal-Kingston: McGill-Queen's University Press, 2009) 112.

130. Alberta and Ottawa would establish a joint body to protect the headwaters of the Saskatchewan River in 1947.

131. Norrie, Owram and Emery, *A History of the Canadian Economy*, p. 317.

132. Franklin L. Foster, "John E. Brownlee: 1925–1934," in Bradford J.. Rennie, ed., *Alberta Premiers of the Twentieth Century* (Regina: Canadian Plains Research Center, University of Regina, 2004), p. 101.

133. Franklin Foster, *John Brownlee: A Biography* (Lloydminster: Foster Learning, 1981), p. 260.

134. Patrick Kyba, "J.T.M. Anderson" in Barnhart, ed., *Saskatchewan Premiers of the Twentieth Century*, p. 133.

135. Boothe and Edwards, eds., *Eric J. Hanson's Financial History of Alberta: 1905–1950*, p. 96.

136. Ibid.

137. From correspondence read by Finance Minister Charles Dunning to the House of Commons on April 1, 1936, as quoted in Boothe and Edwards, eds. *Eric J. Hanson's Financial History of Alberta: 1905–1950*, p. 175.

138. Edward Bell, "Ernest Manning: 1943–1968," in Rennie, ed., *Alberta Premiers of the Twentieth Century*, p. 165.

139. Ibid.

140. Former Prince Edward Island premier A.E. Arsenault. *Memoirs of The Hon. A. E. Arsenault*, n.d. Publisher not stated, p. 66.

141. Interview with former Alberta premier John Edward Brownlee, recorded between March 28, 1961 and May 12, 1961. United Farmers of Alberta Oral History Project. Glenbow Museum Archives. Standard number M-4079, pp. 48–49. Transcript available on-line at: http://www.glenbow.org/collections/search/findingAids/archhtm/ufa_oral.cfm

AFTERWORD

1. Christina McCall and Stephen Clarkson, *Trudeau And Our Times, Volume Two: The Heroic Delusion* (Toronto: McClelland & Stewart, 1994), p. 169.

2. James H. Marsh, *Alberta's Quiet Revolution: The Early Lougheed Years*. Nov. 28, 2011 p. 7: http://www.jameshmarsh.com/2011/11/alberta's-quiet-revolution-the-early-lougheed-years/

3. Ibid.

4. As quoted in James H. Marsh, ibid.

5. As quoted in Douglas Owram, "The Perfect Storm: The National Energy Program and the Failure of Federal–Provincial Relations," in Richard Connors and John M. Law, eds., *Forging Alberta's Constitutional Framework* (Edmonton: University of Alberta Press, 2005), p. 397. (My italics.)

6. Dennis Gruending, "Allan E. Blakeney," in Gordon L. Barnhart, ed., *Saskatchewan Premiers of the Twentieth Century* (Regina: Canadian Plains Research Center, University of Regina, 2004), p. 305.

7. McCall and Clarkson, *The Heroic Delusion*, p. 176.

8.   John Helliwell and Anthony Scott, as cited in Douglas Owram "The Perfect Storm, p. 396.

9.   Constitution Act, 1982. Clause 91.

10. The National Energy Program, p. 14, as cited in Douglas Owram "The Perfect Storm," p. 397.

11. Stephen Clarkson & Christina McCall, *Trudeau and Our Times, Volume One: The Magnificent Obsession* (Toronto: McClelland & Stewart, 1990), p. 305.

12. As cited in footnote 32, chapter five. McCall and Clarkson, *The Heroic Delusion*, p. 472.

13. Constitution Act 1982. Clause 92A. Subsections 4 and 1.

14. As quoted in Douglas Owram, "The Perfect Storm," p. 397.

15. "Redford ready to take the lead," *The Globe and Mail*, Wed., April 25, 2012, p. 1.

Bibliography

—

A History of the Vote in Canada, second edition. Ottawa: Office of the Chief Electoral
    Officer of Canada, Elections Canada, 2007.

Akenson, Donald H., ed. *Canadian Papers In Rural History, Vol. 1*. Gananoque, Ont.:
    Langdale Press, 1978.

Armstrong, Christopher. *The Politics Of Federalism: Ontario's Relations with the Federal
    Government, 1867–1942*. Toronto: Ontario Historical Studies Series, 1981.

Armstrong, Christopher, Matthew Evenden and H.V. Nelles. *The River Returns:
    An Environmental History of the Bow*. Montreal and Kingston: McGill-Queen's
    University Press, 2009.

Armstrong, Elizabeth H. *The Crisis of Quebec 1914–1918*. New York, Morningside
    Heights: Columbia University Press, 1937.

Arsenault, A.E. *Memoirs of The Hon. A. E. Arsenault*. No place. No date.

Artibise, Alan F.J. *Winnipeg: A Social History of Urban Growth, 1874–1914*. Montreal and
    Kingston: McGill-Queen's University Press, 1975.

Avery, Donald H. *Reluctant Host: Canada's Response to Immigrant Workers, 1896–1994*.
    Toronto: McClelland & Stewart Inc., 1995.

Barman, Jean. *The West Beyond The West: A History of British Columbia*. Toronto:
    University of Toronto Press, 1995.

Barnhart, Gordon L., *"Peace, Progress and Prosperity": A Biography of Saskatchewan's First
    Premier, T. Walter Scott*. Regina: Canadian Plains Research Center, University of
    Regina, 2000.

Barnhart, Gordon L., ed. *Saskatchewan Premiers of the Twentieth Century*. Regina:
    Canadian Plains Research Center, University of Regina, 2004.

Belkin, Simon. *Through Narrow Gates: A Review of Jewish Immigration, Colonization and
    Immigrant Aid Work in Canada (1840–1940)*. Montreal: The Eagle Publishing Co.,
    Limited, 1966.

Bercuson, David Jay, ed. *Canada And The Burden of Unity*. Toronto: Copp Clark Pitman
    Ltd., 1986.

Blanchard, Jim. *Winnipeg 1912*. Winnipeg: University of Manitoba Press, 2005.

Blanchard, Jim. *Winnipeg's Great War: A City Comes Of Age*. Winnipeg: University of Manitoba Press, 2010.

Boothe, Paul and Heather Edwards, eds. *Eric J. Hanson's Financial History of Alberta, 1905–1950*. Calgary: University of Calgary Press, 2003.

Borden, Henry, ed. *Robert Laird Borden: His Memoirs*, two volumes. Toronto: The Macmillan Company of Canada Limited, at St. Martin's House, 1938.

Boyden, Joseph. *Louis Riel & Gabriel Dumont*. Toronto: Penguin Group (Canada), 2010.

Boyko, John. *Bennett: The Rebel Who Challenged and Changed a Nation*. Toronto: Key Porter Books Limited, 2010.

Braid, Don and Sydney Sharpe. *Breakup: Why The West Feels Left Out Of Canada*. Toronto: Key Porter Books Limited, 1990.

Braz, Albert. *The False Traitor: Louis Riel In Canadian Culture*. Toronto: University of Toronto Press, 2003.

Breen, David H. *The Canadian Prairie West and the Ranching Frontier 1874–1924*. Toronto: University of Toronto Press, 1983.

Brown, Robert Craig. *Robert Laird Borden: A Biography, Vol. 1, 1854–1914*. Toronto: Macmillan of Canada, 1975.

Brown, Robert Craig and Ramsay Cook. *Canada 1896–1921: A Nation Transformed*. Toronto: McClelland and Stewart Limited, 1974.

Bumsted, J.M. *Dictionary of Manitoba Biography*. Winnipeg: University of Manitoba Press, 1999.

Butler, W.F. *The Great Lone Land*. Teddington, Middlesex: Echo Library, 2006.

Carty, R. Kenneth and W. Peter Ward, eds. *National Politics and Community in Canada*. Vancouver: University of British Columbia Press, 1986.

Cavanaugh, Catherine and Jeremy Mouat, eds. *Making Western Canada: Essays on European Colonization and Settlement*. Toronto: Garamond Press, 1996.

Clarkson, Stephen & Christina McCall. *Trudeau and Our Times, Vol. 1: The Magnificent Obsession*. Toronto: McClelland and Stewart Inc., 1990.

Clinkskill, James. *A Prairie Memoir: The Life and Times of James Clinkskill, 1853–1936*. S.D. Hanson, ed. Regina: Canadian Plains Research Center, University of Regina, 2003.

Connors, Richard and John M. Law, eds. *Forging Alberta's Constitutional Framework*. Edmonton: The University of Alberta Press, 2005.

Constitution Acts 1867 to 1982: A Consolidation. Ottawa: Department of Justice Canada, Canadian Government Publishing Centre, 1986.

Cook, Ramsay. *The Politics of John W. Dafoe and the Free Press*. Toronto: University of Toronto Press, 1963.

Cook, Sharon Anne, Lorna R. McLean and Kate O'Rourke. *Framing Our Past: Canadian Women's History in the Twentieth Century*. Montreal & Kingston: McGill-Queen's University Press, 2001.

Creighton, Donald. *John A. Macdonald: The Old Chieftain*. Toronto: The Macmillan Company of Canada Limited, 1965.

Creighton, Donald. *John A. Macdonald: The Young Politician*. Toronto: Macmillan Company of Canada Limited, 1965.

Creighton, Donald. *The Forked Road: Canada 1939–1957*. Toronto: McClelland and Stewart Limited, 1976.

Dafoe, John. *Clifford Sifton in Relation to His Times*. Toronto: Macmillan Company of Canada Limited, at St. Martin's House, 1931.

Dawson, C.A. *Group Settlement: Ethnic Communities in Western Canada*. Toronto: The Macmillan Company of Canada, Ltd., 1936.

Dawson, R. MacGregor. *William Lyon Mackenzie King: A Political Biography, 1874–1923*. Toronto: University of Toronto Press, 1958.

De Gelder, William. *A Dutch Homesteader on the Prairies*, trans. by Herman Ganzevoort. Toronto: University of Toronto Press, 1973.

Dictionary of Canadian Biography on-line. http://www.biographi.ca.

Dominion–Provincial and Interprovincial Conferences from 1887 to 1926. Ottawa: Reprinted by Edmond Cloutier King's Printer, 1951.

Dominion–Provincial Conferences: November 3–10, 1927; December 9–13, 1935 and January 14–15, 1941. Ottawa: Reprinted by Edmond Cloutier King's Printer, 1951.

Doyle, Arthur T. *Front Benches & Back Rooms: A story of corruption, muckraking, raw partisanship and political intrigue in New Brunswick*. Toronto: Green Tree Publishing Company Ltd., 1976.

Dreisziger, N.F. *Struggle and Hope: The Hungarian-Canadian Experience* with M.L.Kovacs, Paul Body and Bennett Kovrig. Toronto: McClelland and Stewart Ltd. in association with the Multiculturalism Directorate, Department of the Secretary of State, 1982.

Dreisziger, Nandor F. "Hungarians" in R.P. Magocsi, ed., *Encyclopedia of Canada's Peoples*. http://www.multiculturalcanada.ca/Encyclopedia/A-Z/h3/1

English, John. *Borden: His Life and World*. Toronto: McGraw-Hill Ryerson Limited, 1977.

Erasmus, Peter. *Buffalo Days and Nights* as told to Henry Thompson. Calgary: Fifth House Ltd., 1999.

Ferguson, Barry and Robert Wardhaugh, eds. *Manitoba Premiers of the 19th and 20th Centuries*. Regina: Canadian Plains Research Center, University of Regina, 2010.

Forbes, E.R. *Challenging the Regional Stereotype: Essays on the 20th Century Maritimes*. Fredericton: Acadiensis Press, 1989.

Forbes, Ernest R. *Maritime Rights: The Maritime Rights Movement, 1919–1927*. Montreal and Kingston: McGill-Queen's University Press, 1979.

Foster, Franklin. *John E. Brownlee: A Biography*. Lloydminster, Alberta: Foster Learning Inc., 1996.

Friesen, Gerald. *The Canadian Prairies: A History*. Toronto and London: University of Toronto Press, 1987.

Graham, Roger. *Arthur Meighen: Volume 1: The Door of Opportunity*. Toronto: Clarke, Irwin & Company Limited, 1960.

Graham, Roger. *Arthur Meighen: Volume 2: And Fortune Fled*. Toronto: Clarke, Irwin & Company Limited, 1963.

Graham, Roger. *Arthur Meighen: Volume 3: No Surrender*. Toronto: Clarke, Irwin & Company Limited, 1965.

Graham, Ron. *The Last Act: Pierre Trudeau, the Gang of Eight, and the Fight for Canada*. Toronto: Penguin Group (Canada), 2011.

Granatstein, J.L. and J.M. Hitsman, *Broken Promises: A History of Conscription in Canada*. Toronto: Oxford University Press, 1977.

Grenke, Art. "The German Community of Winnipeg and the English-Canadian Response to World War I" in *Canadian Ethnic Studies*, vol. 20, no. 1, 1988.

Gwyn, Richard. *John A.: The Man Who Made Us, 1815–1867*. Toronto: Random House Canada, 2007.

Gwyn, Richard. *Sir John A. Macdonald: Nation Maker, 1867–1891*. Toronto: Random House Canada, 2011.

Hall, D.J. *Clifford Sifton: Volume 1: The Young Napoleon, 1861–1900*. Vancouver: University of British Columbia Press, 1981.

Hall, D.J.. *Clifford Sifton: Volume 2: The Lonely Eminence, 1901–1929*. Vancouver: University of British Columbia Press, 1985.

Harvey, Dr. Robert. *Pioneers of Manitoba*. Winnipeg: The Prairie Publishing Company, 1970.

Heron, Craig. *Booze: A Distilled History*. Toronto: Between The Lines, 2003.

Heron, Craig, ed. *The Workers' Revolt in Canada 1917–1925*. Toronto: University of Toronto Press, 1998.

Hodgins, Bruce W., Don Wright and W.H. Heick. *Federalism in Canada and Australia: The Early Years*. Waterloo: Wilfrid Laurier University Press, 1978.

Hoerder, Dirk. *"Struggle a Hard Battle": Essays on Working-Class Immigrants*. DeKalb, Illinois: Northern Illinois University Press, 1986.

Hoerder, Dirk. *Creating Societies: Immigrant Lives in Canada*. Montreal and Kingston: McGill-Queen's University Press, 1999.

Hryniuk, Stella. "Pioneer Bishop, Pioneer Times: Nykyta Budka in Canada," in CCHA, *Historical Studies*, vol. 55, 1988.

Jones, David C. *Empire Of Dust: Settling and Abandoning the Prairie Dry Belt*. Edmonton: University of Alberta Press, 1991.

Kelley, Ninette and Michael Trebilcock. *The Making of the Mosaic: A History of Canadian Immigration Policy*. Toronto: University of Toronto Press Inc.. 1998.

Kendle, John. *John Bracken: A Political Biography*. Toronto: University of Toronto Press, 1979.

Klippenstein, Lawrence and Julius G. Toews, eds. *Mennonite Memories: Settling in Western Canada*. Winnipeg: Centennial Publications, 1977.

Knowles, Valerie. *Strangers At Our Gates: Canadian Immigration and Immigration Policy, 1540–2006*. Toronto: Dundurn Press, 2007.

Kovacs, Martin L. ed. *Ethnic Canadians: Culture and Education*. Regina: Canadian Plains Research Center, University of Regina, 1978.

Kovacs, Martin L. ed. *Roots and Realities Among Eastern and Central Europeans*. Edmonton: Central and East European Studies Association of Canada, 1983.

Kovacs, Martin Louis. *Esterhazy and Early Hungarian Immigration to Canada*. Regina: Canadian Plains Research Center, University of Regina, 1974.

Levine, Allan. *King: William Lyon Mackenzie King, A Life Guided By The Hand Of Destiny*. Vancouver and Toronto: Douglas & McIntyre, 2011.

Lingard, Cecil C. *Territorial Government in Canada: The Autonomy Question in the Old North-West Territories*. Toronto: University of Toronto Press, 1946.

Loewen, Royden K. *Family, Church and Market: A Mennonite Community in the Old and the New Worlds. 1850–1930*. Toronto: University of Toronto Press, 1993.

Loewen, Royden K. *Ethnic Farm Culture in Western Canada*. Ottawa: The Canadian Historical Association, Canada's Ethnic Group booklet No. 29, 2002.

Luciuk, Lubomyr and Stella Hryniuk, eds. *Canada's Ukrainians: Negotiating an Identity*. Toronto: University of Toronto Press, 1991.

MacEwan, Grant. *Fifty Mighty Men*. Saskatoon: Western Producer Prairie Books, 1982.

MacEwan, Grant. *Frontier Statesman of the Canadian Northwest: Frederick Haultain*. Saskatoon: Western Producer Prairie Books, 1985.

MacGregor, James G. *A History of Alberta: Revised Edition*. Edmonton: Hurtig Publishers, 1981.

MacKenzie, David. *Canada and the First World War: Essays in Honour of Robert Craig Brown*. Toronto: University of Toronto Press, 2005.

Mackintosh, W.A. *Economic Problems of the Prairie Provinces*. Toronto: Macmillan Company of Canada, Limited, at St. Martin's House, 1935.

Mackintosh, W.A. *The Economic Background of Dominion–Provincial Relations*. Toronto: McClelland and Stewart Limited, 1964. Originally published in 1939 as Appendix III of the Rowell-Sirois Report.

Marchildon, Gregory P., ed. *The Early Northwest*. Regina: Canadian Plains Research Center, University of Regina, 2008.

Marchildon, Gregory P., ed. *Immigration & Settlement, 1870–1939*. Regina: Canadian Plains Research Center, University of Regina, 2009.

Marsh, James H. *Alberta's Quiet Revolution: The Early Lougheed Years*. Nov. 28, 2011. http://www.jameshmarsh.com

Martin, Chester. *"The Natural Resources Question": The Historical Basis Of Provincial Claims*. Winnipeg: Philip Purcell, King's Printer for the Province of Manitoba, 1920.

Martin, Chester. *Dominion Lands Policy*. Ottawa and Toronto: The Carleton Library No. 69 and McClelland and Stewart Limited, 1973.

Martynowych, Orest. *Ukrainians in Canada: The Formative Years 1891–1924*.

Matejko, Joanna and Walter Zientarski, eds. *Polish Settlers in Alberta: Reminiscences and Biographies*. Toronto: Drukiem Polish Alliance Press Ltd, 1979.

McCall, Christina & Stephen Clarkson. *Trudeau and Our Times, Vol. 2: The Heroic Delusion*. Toronto: McClelland & Stewart Inc., 1994.

Melnyk, George, ed.. *Riel To Reform: A History of Protest in Western Canada*. Saskatoon: Fifth House Publishers, 1992.

Morton, Arthur S. *History of Prairie Settlement*. Toronto: The Macmillan Company of Canada Limited, at St. Martin's House, 1938.

Morton, James. *Honest John Oliver: The Life Story of the Honourable John Oliver, Premier of British Columbia, 1918–1927*. London, Toronto and Vancouver: J.M. Dent and Sons Ltd., 1933.

Morton, W.L. *Manitoba: A History*, Second edition. Toronto: University of Toronto Press, 1967.

Morton, W.L., ed. *Alexander Begg's Red River Journal and Other Papers Relative to the Red River Resistance of 1869–1870*. Toronto: The Champlain Society, 1956.

Moyles, R.G. ed. *Challenge of the Homestead: Peace River Letters of Clyde and Myrle Campbell, 1919–1924*. Calgary: Historical Society of Alberta, 1988.

Neatby, H. Blair. *William Lyon Mackenzie King, Vol. 2, 1924–1932: The Lonely Heights*. Toronto: University of Toronto Press, 1963.

Neatby, H. Blair. *William Lyon Mackenzie King, Vol. 3, 1932–1939: The Prism Of Unity*. Toronto and Buffalo: University of Toronto Press, 1976.

Nelles, H.V. *The Politics of Development: Forests, Mines & Hydro-Electric Power in Ontario, 1849–1941*, second edition. Montreal & Kingston: McGill-Queen's University Press, 2005.

Norrie, Kenneth, Douglas Owram and J.C. Herbert Emery. *The History of the Canadian Economy*, fourth edition. Toronto: Nelson, Thomson, 2008.

Oliver, E.H. ed. *The Canadian North-West*. Ottawa: Government Printing Bureau, 1915. Reprinted Charleston, SC: BiblioLife, 1910.

Oliver, Peter. *Public & Private Persons: The Ontario Political Culture 1914–1934*. Toronto & Vancouver: Clarke, Irwin & Company Limited, 1975.

Ormsby, Margaret A. *British Columbia: A History*. Toronto: Macmillan Company of Canada Limited, 1958.

Owram, Doug. *Promise of Eden: The Canadian Expansionist Movement and the Idea of the West 1856–1900*. Toronto: University of Toronto Press, 1992.

Owram, Douglas R. ed. *The Formation Of Alberta: A Documentary History*. Calgary: Historical Society of Alberta, 1979.

Palmer, Howard and Donald Smith, eds. *The New Provinces: Alberta and Saskatchewan, 1905–1980*. Vancouver: Tantalus Research Limited, 1980.

Palmer, Howard and Tamara Palmer, eds. *Peoples of Alberta: Portraits of Cultural Diversity*. Saskatoon: Western Producer Prairie Books, 1985.

Palmer, Howard with Tamara Palmer. *Alberta: A New History*. Edmonton: Hurtig Publishers Ltd., 1990.

Palmer, Howard, ed. *The Settlement of The West*. Calgary: University of Calgary Comprint Publishing Company, 1977.

Perin, Roberto. "Themes in Immigration History" in R.P. Magocsi, ed., *Encyclopedia of Canada's People's*. http://www.multiculturalcanada.ca/Encyclopedia/A-Z/t3

Piniuta, Harry, researcher and translator. *Land of Pain, Land of Promise: First Person Accounts by Ukrainian Pioneers 1891–1914*. Saskatoon: Western Producer Prairie Books, 1978.

Pioneers and Early Citizens of Manitoba, compiled by the Manitoba Library Association. Winnipeg: Peguis Publishers, 1971.

Pratte, André. *Wilfrid Laurier*. Toronto: Penguin Group (Canada), 2011.

Prechtl, Joseph. "My Homesteader Experience" in *Take the Soil In Your Hands* by Richard J.A. Prechtl. Saskatoon: Herrem Publishing Company, 1984.

Quiring, Brett. *Saskatchewan Politicians: Lives Past and Present*. Regina: Canadian Plains Research Center, University of Regina, 2004.

Rasporich, Anthony W. and Henry C. Klassen, eds. *Prairie Perspectives 2*. Toronto and Montreal: Holt, Rinehart and Winston of Canada, Limited, 1973.

Rasporich, Anthony W. and Henry C. Klassen, eds. *Frontier Calgary: Town, City, and Region, 1875–1914*. Calgary: McClelland and Stewart West, 1975.

Rayner, William. *British Columbia's Premiers in Profile: the good, the bad and the transient*. Surrey, B.C.: Heritage House Publishing Company Ltd., 2000.

Rennie, Bradford J., ed. *Alberta Premiers of the Twentieth Century*. Regina: Canadian Plains Research Center, University of Regina, 2004.

Report of the Royal Commission on Maritime Claims. Ottawa: F.A. Acland, Printer To The King's Most Excellent Majesty, 1926.

Report of the Royal Commission on the British Columbia Railway Lands, 1927. LAC. RG 15. File 2088. Q5-29913.

Report of the Royal Commission on the Transfer of the Natural Resources of Manitoba. Ottawa: F.A. Acland, Printer To The King's Most Excellent Majesty, 1929.

Report of the Royal Commission on the Natural Resources Of Alberta. Ottawa: J.O. Patenaude, Printer To The King's Most Excellent Majesty, 1935.

Report of the Royal Commission on the Natural Resources Of Saskatchewan. Ottawa: J.O. Patenaude, Printer To The King's Most Excellent Majesty, 1935.

Roberts, Sarah Ellen. *Alberta Homestead: Chronicle of a Pioneer Family*, ed. Lathrop E. Roberts. Austin: University of Texas Press, 1968.

Ross, Hugh R. *Thirty-Five Years In The Limelight: Sir Rodmond Roblin And His Times*. Winnipeg: Farmer's Advocate of Winnipeg Limited, 1936.

Sheppard, Robert and Michael Valpy. *The National Deal: The Fight for a Canadian Constitution*. Toronto: Fleet Books, 1982.

Shilliday, Jim. *Canada's Wheat King: The Life and Times of Seager Wheeler*. Regina: Canadian Plains Research Center, University of Regina, 2007.

Smith, Donald B. *Calgary's Grand Story*. Calgary: University of Calgary Press, 2005.

Spanner, Don. *"The Straight Furrow": The Life of George S. Henry, Ontario's Unknown Premier*. Unpublished Thesis. London: University of Western Ontario, 1994.

Spaulding, Richard. "Executive Summary" in Peggy Martin-McGuire, *First Nation Land Surrenders on the Prairies, 1896–1911*. Ottawa: prepared for the Indian Claims Commission, 1998.

Swainson, Donald, ed. *Historical Essays on the Prairie Provinces*. Toronto and Montreal: McClelland and Stewart Limited, 1970.

Thomas, Lewis Herbert. *The Struggle for Responsible Government in the North-West Territories, 1870–1897*. Second edition. Toronto: University of Toronto Press, 1978.

Thompson, John Herd. *Forging the Prairie West: The Illustrated History of Canada*. Toronto: Oxford University Press, 1998.

Thompson, John Herd with Allen Seager. *Canada 1922–1939: Decades of Discord*. Toronto: McClelland & Stewart Limited, 1985.

Tough, Frank. *"As Their Natural Resources Fail": Native Peoples and the Economic History of Northern Manitoba, 1870–1930*. Vancouver: UBC Press, 1996.

Trow, James. *Manitoba and the Northwest Territories: Letters by James Trow, M.P. 1878*. Toronto: Canadiana House reprint, 1970.

Tulchinsky, Gerald. *Taking Root: The Origins of the Canadian Jewish Community*. Toronto: Lester Publishing Ltd., 1992.

Tulchinsky, Gerald, ed. *Immigration in Canada: Historical Perspectives*. Toronto: Copp Clark Longman Ltd., 1994.

Veltri, Giovanni. *The Memoirs of Giovanni Veltri*. ed. John Potestio. Toronto: The Multicultural History Society of Ontario, 1987.

Wardhaugh, Robert A. *Mackenzie King and the Prairie West*. Toronto: University of Toronto Press, 2000.

Willms, A.M., Ramsay Cook, J.M. Bliss and Martin Robin. *Conscription, 1917*. Toronto: University of Toronto Press, Canadian Historical Readings, Volume Eight, 2010.

Wood, Patricia K. *Nationalism From The Margins: Italians in Alberta and British Columbia*. Montreal and Kingston: McGill-Queen's University Press, 2002.

Wright, J.F.C. *Saskatchewan: The History of a Province*. Toronto: McClelland and Stewart Limited, 1955.

Young, George. *Manitoba Memories: Leaves from my life in the Prairie Province, 1868–1884*. Toronto: William Briggs, 1897.

Zaslow, Morris. *The Opening Of The Canadian North: 1870–1914*. Toronto and Montreal: McClelland and Stewart Limited, 1971.

Acknowledgements

—

THIS IS THE WORK OF A JOURNALIST who took a hugely hopeful leap to return to university as a so-called mature student. I would never have found the courage if my friend Val Ross, who was very ill, had not insisted that people should follow their dreams while they have the chance. My spouse, Tom Kierans, backed her message. History was my dream. I followed it to university—and then through the archives—to tell this tale. I was lucky—and I have met some terrific people—professors and students —along the way. (I have also become rather fond of some of the long-gone people in this book!)

The first thanks must go to those who encouraged and fostered that first leap. There were three University of Toronto political scientists – David Cameron, Stephen Clarkson and Richard Simeon. There was University of Toronto sociologist Edward Harvey, University of Toronto president Dr. David Naylor, Queen's University economist Tom Courchene and University of Western Ontario economist David Laidler. York University sociologist Lorna Marsden, who has led so many rewarding lives in politics, business and academia, talked about the joy of change, as did Donald Johnston, former Canadian cabinet minister and secretary-general of the Organisation for Economic Co-operation and Development.

For someone who fears change, the folks at the other end of the leap could not have been kinder. The then-director of the York University graduate history program, Anne Rubenstein, took a chance on me. Marcel Martel and Myra Rutherdale have helped this rookie, then and now. That esteemed historian Roberto Perin, master of religious and immigration history, even hosted his class for Italian dinners. And that class – Michael Amatiello, Michael DiLuccio, Chris Grafos, Sarah Ouannou, Domenic Servello and amazing Sunny Yi along with the professor himself – pulled me through that first terrifying year. Into the doctoral years, William Westfall tactfully suggested books that emphasized the importance of religion in other lives and times – and thus explained so many political crises. Craig Heron along with Colin Coates, Marlene Shore and William Wicken opened doors to other Canadian worlds; Heron also bravely gathered his flock for dinner at his home. Program director Carolyn Podruchny always found time for so much sage advice. Gene Allen opened up a new way of viewing the world through the public sphere. My fellow students Laurie Brady and Christine Sismondo were especially insightful.

And then there is historian Jennifer Stephen, who has always understood that yesterday's fumbles have deeply influenced today's face-offs over resource control. With her encouragement, I first headed to Library and Archives Canada in early 2009 in search of lost time. At the Archives, Jennifer Devine, Martin Lanthier, Marie-Josée Néron and Daniel Somers were especially helpful. I also had assistance from Amanda Kriaski at Lougheed House and Jim Bowman and Adria Lund at the Glenbow Museum. Former Alberta Premier Peter Lougheed kick-started the process with a tour of his grandparents' home, Lougheed House, along with insights into those remarkable ancestors, Belle Hardisty Lougheed and Senator Sir James Lougheed. His spouse Jeanne and his incredible assistant Pat Welch were so kind and so very helpful. I also owe University of Western Ontario historian Roger D. Hall and Canadian constitutional expert Sujit Choudhry, who is now at the New York University law school. Cheri Tetreault at the Alberta Assembly was helpful, as was Maurice Riou at the Saskatchewan Legislature, Rachelle Gareau at the Batoche National Historic Site and Lisa Braun at West Wind Aviation, who saved one of my exploratory missions. I also owe Kim Shaver at Brechin Imaging Services, Christopher Kotecki, Kristina Rissling and Sharon Foley at the Archives of Manitoba, Megan Becker at the Provincial Archives of Alberta and Tim Novak at the Saskatchewan Archives Board.

But the book was not an academic exercise: it was an adventure into the often forgotten history of the West versus the Rest of Canada. My amazing book agent John Pearce at Westwood Creative Artists believed in the project from the beginning—when the story of a federal-provincial dust-up in 1918 sounded a tad obscure. My first editor, Michael Schellenberg, at the Knopf Random Canada Publishing Group encouraged this wander through so much time. Senior editor Paul Taunton, whose grandfather Ernie Cunningham was a champion of the West, always insisted that I was in a good place, even though I knew that I wasn't. And then he ensured that I got there. (I hope!) He also came up with the title. Copy-editor Jane McWhinney was, happily, so very careful. In my corner, I also had executive publisher Louise Dennys, publisher Anne Collins (who had also been a journalistic inspiration), associate publisher Marion Garner, designer Leah Springate (who did that amazing cover), managing editor Deirdre Molina, production manager Carla Kean, typesetters Terra Page and Erin Cooper, and avid publicist Shona Cook along with the rest of the team. It was a solace to have them there.

Finally, no one explores the past without huge support in the present. So many friends and family members would ask about Your Book, and I would mentally amend that to My "Wretched" Book. Angela Ferrante read most of the drafts, some during her weeks in Italy, cheerily dismissing calls at 11:37 p.m. (her time) as "just fine" and always offering superb advice. She was amazingly perceptive. So many others – Helen Burstyn, Warren Clements, Denise Dalphy, Patricia Finlay, Kaye Fulton, Joy and David Garrick, Carol Goar, Mizue Handa, Liz Herron, Maria Helena Higino, Anna Janigan, Michael Janigan, Diane Jermyn, Renata Kierans, Kathy Kilburn, Maxine King, Graham Lee, Kieran Lee, Sue Lockwood, Joyce McKeough, Marianne Miller, Sandra Martin, Marci McDonald, Eric Morse, Ronalda Murphy, Morton Ritts, Red Wilson – the list is huge and I feel torn even singling out these few—held my paw. Public relations wizard Marcia McClung lent her expertise and great wisdom. The late Val Ross was always there in spirit, as was my father Daniel Janigan who was so close to getting his MA in history when my mother Anne died.

But the greatest thanks go my fabulous spouse, Tom Kierans. He has always believed in me—and in the book. As a policy polymath, his help was invaluable: he read multiple drafts of the book, patiently, adding pivotal insights into federalism and finances. He spent so many days alone while I buried my nose in research texts or scrolled through archival documents. He never complained. He wandered throughout the West with me, ending up on one trip, to his astonishment, in Dawson Creek. "Am I really in northeastern British Columbia?" has become a rather funny memory. At least for me. He has tried to remain inscrutable on the topic of his travel-crazed curious spouse. Nobody could have ever done it better.

It is a cliché that success has many parents, but failure is an orphan. Any failures in this book, however, are my fault. Any successes are truly due to so many parents. Thank you.

Index

—

MARY JANIGAN is a journalist who has written extensively about Canadian public policy, including politics and economics, for the *Toronto Star*, *Maclean's* and the *Globe and Mail*. She has won the prestigious Hy Solomon award for policy analysis, and the National Newspaper Award for her clause-by-clause scrutiny of proposed Constitution changes. She has never lost her curiosity, and she has always wanted to understand how the blunders and triumphs of the past complicate the present. She lives in Toronto.